Sustainable Development Goals Series

The **Sustainable Development Goals Series** is Springer Nature's inaugural cross-imprint book series that addresses and supports the United Nations' seventeen Sustainable Development Goals. The series fosters comprehensive research focused on these global targets and endeavours to address some of society's greatest grand challenges. The SDGs are inherently multidisciplinary, and they bring people working across different fields together and working towards a common goal. In this spirit, the Sustainable Development Goals series is the first at Springer Nature to publish books under both the Springer and Palgrave Macmillan imprints, bringing the strengths of our imprints together.

The Sustainable Development Goals Series is organized into eighteen subseries: one subseries based around each of the seventeen respective Sustainable Development Goals, and an eighteenth subseries, "Connecting the Goals", which serves as a home for volumes addressing multiple goals or studying the SDGs as a whole. Each subseries is guided by an expert Subseries Advisor with years or decades of experience studying and addressing core components of their respective Goal.

The SDG Series has a remit as broad as the SDGs themselves, and contributions are welcome from scientists, academics, policymakers, and researchers working in fields related to any of the seventeen goals. If you are interested in contributing a monograph or curated volume to the series, please contact the Publishers: Zachary Romano [Springer; zachary.romano@springer.com] and Rachael Ballard [Palgrave Macmillan; rachael.ballard@palgrave.com].

Jocelyn Boiteau • Prabhu Pingali

Wasted Potential

Tackling Food Loss and Waste Across Transforming Food Systems

Springer

Jocelyn Boiteau
Tata-Cornell Institute (TCI)
Cornell University
Ithaca, NY, USA

Prabhu Pingali
Tata-Cornell Institute (TCI)
Cornell University
Ithaca, NY, USA

ISSN 2523-3084 ISSN 2523-3092 (electronic)
Sustainable Development Goals Series
ISBN 978-3-031-97410-6 ISBN 978-3-031-97411-3 (eBook)
https://doi.org/10.1007/978-3-031-97411-3

This work was supported by Cornell University/Tata-Cornell Institute

This Springer imprint is published by the registered company Springer Nature Switzerland AG
The registered company address is: Gewerbestrasse 11, 6330 Cham, Switzerland

Preface

By 2050, the global population is expected to reach nearly ten billion people, with two-thirds residing in urban areas. Meeting sustainability goals for both human and planetary health becomes increasingly crucial as climate change, global population growth, and rapid urbanization continue to strain food systems. Therefore, responsible food production and consumption are essential to ensure the availability of and access to sufficient nutritious food. Reducing food loss and waste (FLW) has been identified as a key strategy to the promotion of sustainable and healthy diets for all. Where and how much FLW occurs vary depending on the stage of transformation within specific food system contexts.

In 2015, the United Nations set Sustainable Development Goal target 12.3, which aims to reduce per capita global food waste at the retail and consumer levels by half and to reduce food losses along production and supply chains by 2030. Despite recognition of FLW as a critical issue and the establishment of global targets, there is "wasted potential," encompassing two key aspects: first, the *wasted potential for food*, which refers to the loss of potential to nourish growing populations due to food produced for human consumption that ultimately goes uneaten, and second, the *wasted potential for action*, which highlights the gap in effectively addressing FLW. The gap arises from a lack of coordinated approaches in how FLW is defined, measured, and framed within the context of transforming food systems. Discussions often reference the 2011 report of the Food and Agriculture Organization of the United Nations (FAO), which states that one-third of food produced for human consumption is lost or wasted per year, globally. However, recent estimates use different definitions, leading to varying interpretations of "food loss and waste." The lack of a harmonized definition prevents the development of consistent measurement and estimation approaches, making it challenging to compare and aggregate data accurately. Reducing the *wasted potential for food* depends on addressing the *wasted potential for action*, which, in turn, requires a comprehensive examination of FLW pathways to identify effective strategies that minimize FLW and contribute to sustainable food systems transformations.

This book systematically addresses each of these challenges to provide a clearer understanding of the frameworks, pathways, and approaches necessary for effectively tackling FLW. We provide a synthesis of the FLW pathways that facilitate the transition to sustainable food systems, while simultaneously promoting access to healthy diets at different stages of transformation. In taking a food systems approach, we examine FLW across the entire range of actors and activities, from production to consumption. Building on a foundational framework for conceptualizing FLW within food value chains, this book assesses currently available FLW data to identify key evidence gaps and opportunities to improve data collection. Furthermore, it explores FLW pathways at food value chain stages, from the moment food is ready for harvest until it is either eaten or removed from the value chain, examining the linkages to food security and diets, as well as environmental impacts. Finally, this book examines evidence-based interventions and policies to address FLW in low- and middle-income countries, considering the roles of different stakeholders and exploring how to incentivize FLW mitigation. We conclude by outlining a policy agenda, which aims to create an enabling environment for FLW reduction to ultimately enhance food security, environmental sustainability, and socioeconomic development. This comprehensive framework equips researchers, practitioners, and policymakers with the necessary tools to effectively address the FLW challenge in transforming food systems.

This book is a major output of the Tata-Cornell Institute for Agriculture and Nutrition (TCI) at Cornell University. TCI was established with a generous gift given to Cornell University from the Tata Education and Development Trust, a philanthropic branch of the Tata Group. The endowment was made possible by the vision of Mr. Ratan Tata, the former chairman of India's Tata Group and a Cornell alumnus from the Class of 1962. TCI is a long-term research initiative that develops and assesses innovative, food systems-based approaches to reducing poverty and improving nutrition and livelihoods in the developing world, with a specific focus on India.

TCI funds and advises more than a dozen PhD and Master's student researchers from disciplines across the Cornell campus. Most of the TCI scholars conduct intensive fieldwork as an integral part of their thesis research. TCI believes that the field research that addresses the problems of the rural poor is highly compatible with analytically rigorous research expected of Cornell graduate students. This book expands upon our research on FLW, which began in 2016, when Jocelyn Boiteau joined TCI as a PhD student and carried out field research along fresh tomato value chains in South India.

We thank the TCI team for their support in bringing this book together. We are grateful, particularly, to the TCI scholars for their discussions during research group meetings in the early stages of writing this book and to Dan Verderosa, Brenda Daniels-Tibke, Terry Mingle, and Patricia Mason for their support and assistance during all the stages of writing this book.

This book aims to provide valuable insights for researchers, policymakers, and practitioners working within agrifood systems. While adopting a global perspective, we believe its key takeaways are relevant to diverse countries at varying stages of food system transformation. By exploring these insights, we hope to contribute to a deeper understanding of the factors important for supporting sustainable food systems that deliver healthy diets for all. We hope you enjoy reading this book and exploring its potential for informing your own work.

Ithaca, NY, USA Jocelyn Boiteau
Ithaca NY, USA Prabhu Pingali

Contents

List of Figures

List of Tables

List of Boxes

About the Authors

Jocelyn Boiteau recently completed her postdoctoral position with the Tata-Cornell Institute for Agriculture and Nutrition (TCI) at Cornell University. She is currently the Senior Nutrition and Innovation Director at Food Systems for the Future.
Email: jmb575@cornell.edu

Prabhu Pingali is a Professor of Applied Economics and the Founding Director of the Tata-Cornell Institute for Agriculture and Nutrition (TCI) at Cornell University.
Email: plp39@cornell.edu

Abbreviations

APHLIS	African Postharvest Losses Information System
ASF	Animal-source foods
CCAFS	Climate Change, Agriculture and Food Security (CGIAR Research Program)
CDC	Centers for Disease Control and Prevention (US)
CSR	Corporate social responsibility
EE	Eastern Europe
EPA	Environmental Protection Agency (United States)
ERS	Economic Research Service (USDA)
EU	European Union
FAO	Food and Agriculture Organization of the United Nations
FBD	Foodborne disease
FBS	Food balance sheets
FDA	Food and Drug Administration (United States)
FERG	Foodborne Disease Burden Epidemiology Reference Group (WHO)
FLAPP	FAO Food Loss App
FLI	Food Loss Index
FLW	Food loss and waste
FQL	Food quality loss
FWI	Food Waste Index
GAIN	Global Alliance for Improved Nutrition
GDP	Gross domestic product
GFLI	Global Food Loss Index
GHG	Greenhouse gas
GLEAM	Global Livestock Environmental Assessment Model
HIA	High-income Asia
HIC	High-income country
HLPE	High Level Panel of Experts on Food Security and Nutrition
ICAR–CIPHET	Indian Council on Agricultural Research—Central Institute of Post Harvest Engineering & Technology
ICTSD	International Centre for Trade and Sustainable Development
IOM	Institute of Medicine (US)
LAC	Latin America and the Caribbean
LAFA	Loss-Adjusted Food Availability (USDA ERS)

LIC	Low-income country
LMIC	Low- or middle-income country
LSF	Livestock-sourced foods
MENA	Middle East and North Africa
NAO	North America and Oceania
OASIS	Operational and Administrative System for Import Support
OECD	Organisation for Economic Co-operation and Development
pFLW	Physical food loss and waste
RASFF	Rapid Alert System for Food and Feed
ReSAKSS	Regional Strategic Analysis and Knowledge Support System
SEA	Southeast Asia
SDG	Sustainable Development Goal
SME	Small- and medium-sized enterprises
SPS	Sanitary and Phytosanitary Measures (World Trade Organization)
SSA	sub-Saharan Africa
UHT	Ultra-high temperature
UMIC	Upper-middle-income country
UN	United Nations
UNEP	United Nations Environmental Programme
UNIDO	United Nations Industrial Development Organization
UNSD	United Nations Statistical Division
USAID	United States Agency for International Development
USDA	United States Department of Agriculture
WE	Western Europe
WEF	World Economic Forum
WHO	World Health Organization
WRAP	Waste and Resources Action Programme (UK)
WRI	World Resources Institute

Sustainable and Healthy Food Systems: The Role of Food Loss and Waste Reduction

1

1.1 Motivation

Food loss and waste (FLW) has been part of the global development agenda since the food crisis of the 1970s. The United Nations (UN) recognized the importance of improved processing, storage, and distribution to reduce postharvest food losses (UN 1975). The primary focus at that time was to increase food availability to consumers in the context of addressing hunger and malnutrition. Decades later, the FAO released a global study on FLW, reporting the widely cited figure that "one-third of food produced for human consumption is lost or wasted globally" each year (FAO 2011). Following this report, a resurgence of interest in FLW emerged, with a greater focus on environmental sustainability. Since 2013, there has been an uptick in the number of publications on FLW in food value chains, primarily published in outlets focused on environment and waste management (Chauhan et al. 2021). FLW was reintroduced onto the global agenda in 2015, as part of Sustainable Development Goal (SDG) 12, which aims to ensure sustainable consumption and production patterns. Among targets focused on efficient resource use, waste reduction, and minimizing of environmental impacts, SDG Target 12.3 specifically calls for halving per capita global food waste at the retail and consumer levels and reducing food losses along production and supply chains, including postharvest losses, by 2030 (UN 2015)

By 2050, the global population is expected to reach nearly ten billion people, with two-thirds residing in urban areas. Sustainable, healthy diets aim to meet the sustainability goals for both human and planetary health (Willett et al. 2019). Climate change, population growth, rapid urbanization, increasing consumer incomes, demand for diverse diets, and elongating food value chains, collectively, are putting increasing pressure on food systems. To achieve responsible production and consumption and support safe and nutritious diets, sustainable food system transformations are necessary. Strategies to achieve this transformation focus on increasing the production and supply of nutrient-rich foods, efficiently moving food along value chains to make healthy foods accessible, ensuring affordable quality diets, and driving demand for healthy diets through informed food choices (Global Panel on Agriculture and Food Systems for Nutrition 2020). Looking ahead to 2050 and beyond, addressing FLW offers an opportunity to support sustainable food system transformations. This will require a clear understanding of the extent and causes of FLW; evidence to inform public and private sector decision-making on the most effective interventions; and clarity on the potential trade-offs and unintended consequences of reducing FLW on food security, environmental impact, and socioeconomic welfare.

J. Boiteau, P. Pingali, *Wasted Potential*, Sustainable Development Goals Series,
https://doi.org/10.1007/978-3-031-97411-3_1

1.2 Approach

Much of the prior literature on FLW has focused on how much FLW, in terms of mass quantity, is generated across food value chain stages. Accurately estimating and comparing the extent of FLW is challenging due to several factors. These factors include the lack of a harmonized FLW definition and standardized measurement frameworks. Furthermore, inconsistencies arise from variations in the inclusion of specific value chain stages and the designation of final loss destinations. Moreover, stakeholders at different levels of the food system hold diverse, and sometimes conflicting, perspectives on the problems associated with FLW, which can influence the evaluation and prevention approaches. Individual studies often focus on how FLW relates to specific food systems outcomes, such as environmental impacts, economic efficiency, or food security and nutrition. These foci can shape how FLW is estimated, its drivers identified, and solutions developed. Consequently, findings may be skewed toward achieving objectives aligned with the chosen perspective. Recent works that look at the systemic causes and solutions of FLW from multidisciplinary perspectives have also demonstrated limitations, as they often have a Western-centric bias and tend to emphasize discussions of FLW at retail and consumer-level stages. Further, contributors to these works may conceptualize FLW differently, using different definitions of FLW and estimation methods, limiting the overall coherence of the analysis and interpretation of a way forward.

To effectively address the challenge of FLW, this book adopts a food systems approach to examine the full set of activities from farm to consumer. A food systems approach takes a comprehensive view, encompassing food value chain actors and the mechanisms that shape their roles. This approach enables the exploration of the drivers and interrelationships of complex food security and environmental changes (Ingram 2011). By using this framework, we can identify points for enhancing food security and analyze the synergies and trade-offs between food security, environmental impacts, and socioeconomic welfare outcomes associated with different FLW pathways.

Our central inquiry focuses on how reducing FLW can support sustainable transformations to ensure safe and nutritious diets as food systems transition from traditional to modern systems. With a particular emphasis on low- and middle-income countries (LMICs), we investigate potential entry points for FLW investment and intervention priorities.

To take a comprehensive approach to FLW reduction, we consider FLW as an intermediate goal that contributes to the integrated achievement of enhanced food security and nutrition, environmental sustainability, and socioeconomic welfare. We synthesize the latest data and scientific evidence from global to local levels to identify diverse FLW pathways. In this book, we (1) explore the potential for harmonizing the concept of FLW while highlighting the challenges in filling evidence gaps; (2) investigate the diverse pathways through which FLW impacts food security across different food system transformation contexts; and (3) examine evidence-based FLW approaches and priorities for action with emphasis on the critical role of public and private policy and institutional interventions in sustainably addressing FLW challenges.

This book is organized as follows. We first begin by examining structural transformation and the accompanying shifts within food systems. We analyze changes across food value chains, considering factors, such as modernizing food environments, shifting consumer behaviors, and changing investment priorities. We establish a foundational framework for conceptualizing FLW. This framework guides our examination of FLW problems and prevention approaches by establishing a clear boundary and scope within which FLW pathways exist. We then examine the data requirements for addressing FLW, identifying critical evidence gaps and opportunities to develop more robust data value chains. Next, we breakdown FLW pathways to examine how they contribute to the availability and accessibility of safe and nutritious foods. First, we explore the critical role of farm-level FLW in ensuring the availability of food that enters food value chains.

We then delve into post-farmgate FLW pathways to examine the impact on the accessibility of perishable and nutritious foods. Finally, we focus on losses of food safety and nutrient attributes among foods consumed, investigating their potential nutritional and health implications. In the final chapters, we explore evidence-based interventions and policies to address FLW within the context of LMICs. We consider the roles of different food system stakeholders and explore how to effectively leverage public and private incentives for optimal FLW reduction. We also examine the design and implementation of FLW interventions, considering their objectives, success metrics, and the distribution of benefits across the value chain. This book concludes by outlining a policy agenda for FLW reduction. The agenda aims to create enabling environments that incentivize FLW mitigation and foster synergies between food security, environmental, and socioeconomic objectives. By identifying FLW pathways that facilitate the transition to sustainable food systems while simultaneously promoting access to healthy diets at different stages of transformation, this book provides researchers, practitioners, and policymakers with a comprehensive framework for addressing the FLW challenge.

1.3 Key Takeaways from the Book

1.3.1 As Urbanization, Income Growth, and Dietary Diversification Drive Food System Transformations, Food Loss and Waste Presents Significant Challenges to Meeting Growing Demand for Perishable, Nutritious Foods (Chap. 2)

The structural transformation of economies, characterized by increased agricultural productivity, urbanization, industrialization, and demographic shifts, is intrinsically linked to the modernization of food systems. Chapter 2 introduces the core elements of food systems: food value chains, food environments, and consumer behaviors. It examines how these elements change with increasing urbanization, the lengthening of food value chains, and shifts in consumer incomes and preferences. As food systems modernize, healthy diets often become more accessible. Rising incomes empower consumers to demand diversified diets, encompassing nutritious, perishable foods alongside processed, and often unhealthy, options. To meet the shifting demand for fruits, vegetables, and animal-source foods (ASF), producers must diversify their production, and value chains must adapt to longer rural–urban linkages. The perishable nature of fresh produce poses significant challenges for food systems, particularly, where value chains are long and involve numerous actors, often within the context of developing infrastructure, including transportation, electricity, and market access. International food trade offers new market opportunities for value chain actors, but it also necessitates stringent adherence to food quality and safety standards for perishable products, which can limit which actors participate and which foods are traded.

As food systems transition from traditional to modern systems, Chap. 2 explores the increasing market connectedness that leads to a pattern, in which nutritious, perishable foods often become the last food groups to be readily available in formal markets. This underscores the importance of both formal and informal food environments in ensuring consumers have access to nutritious foods that align with dietary recommendations. We emphasize the significant supply gap in fruits and vegetables, which necessitates a multifaceted approach involving increased production, reduced FLW, and improved distribution to meet adequate consumption levels.

1.3.2 Consensus on a Harmonized Definition of "Food Loss and Waste" Is Greatly Needed to Align Food Loss and Waste Boundaries and Destinations with Sustainable Food Systems Objectives, to Enable Inclusion of Stakeholder Perspectives, and to Provide Consistency Across Analytical Frameworks (Chap. 3)

A significant challenge slowing progress in addressing FLW is the lack of a universally agreed upon definition. Inconsistent approaches to defining FLW and identifying its occurrence limit our ability to interpret, compare, and aggregate evidence on the extent and causes of FLW, as well as the effectiveness of prevention interventions. Chapter 3 reviews current FLW definitions used to inform global estimates, identifies opportunities for harmonization, and presents a comprehensive, globally applicable FLW pathways framework. Among the major sources of FLW data, several align more closely with the definitional framework outlined by the Global Initiative on Food Loss and Waste Reduction in 2014, of the Food and Nutrition Organization of the United Nations (FAO), than with the FLW definitions used by the Sustainable Development Goal (SDG) 12.3 Indicators, the Food Loss Index and the Food Waste Index. This work adopts the FAO 2014 definition, which defines FLW as a reduction in the quantity or quality of the edible portion of food intended for human consumption when food is redirected to nonfood uses or when there is a decrease in the nutritional value, food safety, or other quality aspect, from the time food is ready for harvest or slaughter to consumption.

Using this definition, we introduce the FLW conceptual framework to delineate the boundaries of FLW and what falls outside its scope but still remains important for creating sustainable food value chains. Plants and animals are cultivated or raised for various purposes, including food, feed, seed, industrial uses, and others. The framework focuses on plants and animals produced for human consumption, as indicated in the definition. Although plants and animals comprise both edible and inedible parts, the FLW pathways boundary encompasses only edible portions. While inedible parts do not fall within the scope of FLW, the framework acknowledges the importance of managing inedible parts and their environmental impact for achieving sustainable food value chains. FLW pathways start at the point that food is ready for harvest or slaughter and end either when food is consumed or diverted to nonfood destinations.

We acknowledge that different entities prioritize food-based or waste-based perspective differently. By distinguishing between edible and inedible parts and between consumed and diverted food, we can integrate food-based and waste-based perspectives as complementary, not competing, facilitating assessments of food security and environmental trade-offs. Under this framework, stakeholders with varying viewpoints can clearly articulate their approach to tackling specific aspects of FLW and waste management along sustainable food value chains. Although we do not recommend waiting for convergence on a harmonized definition before acting on FLW issues, we consider harmonization a priority for enhancing ongoing efforts and resources dedicated to addressing FLW.

We distinguish between two primary types of FLW: physical FLW (pFLW) and food quality loss (FQL). Although pFLW is relatively straightforward, involving a quantifiable reduction in food mass or volume, FQL is more complex due to the multifaceted nature of food quality attributes. The relationship between FQL and pFLW is contingent on acceptable quality thresholds. Chapter 3 provides a detailed discussion of food attributes and explores conceptual frameworks for FQL.

1.3.3 To Effectively Address Food Loss and Waste, UN Agencies Must Coordinate to Develop Robust Data Value Chains at the Global Level That Establish Core Indicators As Well as Guidelines for Data Collection and Estimation Approaches to Inform and Guide Effective Policies, Programs, and Investments That Are Aligned with Sustainable Food Systems Objectives (Chap. 4)

Although FLW has gained global attention, a robust data ecosystem for collecting, sharing, and utilizing FLW data remains underdeveloped. Current data efforts primarily focus on estimating the quantity of pFLW across the value chain, aligning with SDG 12.3. The Food Loss Index and Food Waste Index, the two primary indices for monitoring progress toward SDG 12.3, cover postharvest to consumer stages but employ differing methodologies and data requirements, hindering their comparability and integration for comprehensive FLW tracking across the value chain. Perhaps, a most striking challenge is the scarcity of primary data for modeling global FLW estimates, especially for perishable food commodities.

In Chap. 4, we delve into specific data gaps and the challenges of collecting primary data, particularly, in resource-constrained environments. Applying the data value chain concept, we emphasize the need for a common core set of FLW indicators that are comparable and aggregable to track progress across value chain stages and enable meaningful comparisons. This framework aims to enhance data availability, quality, and comparability, starting with a harmonized FLW definition and standardized data collection across a series of methods that can be applied to different settings. Beyond pFLW, developing robust FQL indicators and filling evidence gaps in food quality and FQL is important comprehensively to address FLW pathways. To ensure consistent FLW measurement and adequate data for prioritized indicators, three key challenges must be resolved:

1. Systematically distinguishing food safety and cultural dimensions of edibility;
2. Standardizing measurement and reporting methods for pFLW to generate comparable primary data across value chain stages, food groups, and time periods; and
3. Identifying the key quality attributes that limit food acceptance and the data sources that use standardized methods to assess intrinsic food quality attributes.

Addressing FLW data gaps will require international cooperation and collaboration. While FAO and UNEP currently serve as custodian agencies for monitoring and reporting on SDG 12.3, the 2030 target year is fast approaching. To achieve sustainable development beyond the SDGs, broader collaboration among UN agencies is necessary to support partnerships between governments, organizations, academia, and the private sector in developing and implementing data-driven strategies for reducing FLW. This requires a comprehensive approach, including agreement on clearly defined indicators and relevant, reliable data to track complex FLW pathways. Investing in these areas will enhance our understanding of FLW, inform effective interventions, and contribute to healthier and more sustainable diets.

1.3.4 Promoting Diversified Production and Food Loss and Waste Reduction at the Farmgate Serves Dual Objectives: Enhancing the Quality and Quantity of Available Nutrient-Rich Food and Improving the Marketability of Produce, Particularly, for Fresh Perishable Items (Chap. 5)

While the increasing demand for diverse diets has driven production diversification toward high-value, nutritious foods, a significant gap persists between food availability and dietary recommendations, both globally and regionally. Prioritizing the reduction of FLW for these essential food groups is crucial, alongside increasing production. However, defining the precise boundary between the end of production and the beginning of FLW pathways remains challenging. Chapter 5 examines the current evidence on farmgate FLW, emphasizing the importance of accounting for losses that occur before, during, and after harvest. As food systems transform, the classification of farm-level FLW becomes more complex due to shifting harvest timing and the location of postharvest activities (on-farm or off-farm).

For perishable fruits and vegetables, FLW pathways begin once they reach commercial maturity. Determining the readiness involves understanding plant life cycles, physiological changes, and quality attributes. After reaching commercial maturity, natural physiological processes, along with environmental, biological, and physical stressors, can impact unharvested produce, leading to quality degradation and potential physical loss if it becomes unacceptable and is left unharvested or discarded during harvest. Moreover, cosmetic quality standards can contribute to pFLW when produce with natural variations, despite being safe and edible, fails to meet these standards and is rejected. Close attention to differences in public versus private quality standards are important to consider. Therefore, addressing farmgate FLW of fruits and vegetables requires the prevention of quality loss in unharvested produce and creating demand and market opportunities for safe and edible fruits and vegetables to enter post-farmgate value chains.

ASF are a diverse group of perishable products with varied production methods and FLW pathways, compared to fruits and vegetables. FLW of ASF can occur at various stages, from animal maturity (slaughter, milk production, egg laying, aquaculture harvest, hunting) to postharvest handling. Although production-side factors, such as breeding and rearing, are outside the FLW pathways framework, they can indirectly contribute to loss. Food safety is a critical concern for ASF. Factors. like preslaughter stress, disease, and improper handling, can lead to FLW. For aquatic animals, capture methods, handling practices, and storage conditions can accelerate quality degradation. To mitigate FLW, a comprehensive approach is necessary to improve animal welfare, production practices, infrastructure, and market linkages, with a focus on temperature control for these highly perishable foods.

1.3.5 As Economic Development Enables Improved Approaches and Effective Management of Value-Added Activities throughout Food Value Chains, Reducing Food Loss and Waste Can Enhance the Accessibility and Affordability of Nutrient-Rich Foods (Chap. 6)

To meet the growing demand for diverse diets, post-farmgate value addition activities, such as processing, packaging, and distribution, become increasingly important. Although value addition can mitigate FLW by enhancing food safety and reducing spoilage, it can also contribute to FLW when value is generated through stringent cosmetic standards, limited supply, or processing methods. Food processing and packaging are important for extending shelf life, protecting products, and facilitating distribution as value chains lengthen. Balancing FLW reduction with environmental sustainability is essential, particularly, when considering single-use plastics. Aligning packaging choices with specific market and product needs can optimize food protection while minimizing environmental impact. Cold chain infrastructure also offers significant value in preserving fresh, perishable products. However, its implementation and maintenance in transitional food systems face various constraints, including infrastructure, technology, operational, economic, and regulatory challenges.

Chapter 6 explores the extent to which value addition activities influence FLW pathways for different food commodities. Fruit and vegetable value chains focus primarily on fresh products, which are highly perishable. Inadequate storage, transportation, and handling can significantly contribute to quality deterioration. Although cold chain infrastructure can mitigate these challenges, its implementation requires substantial investment and coordination. Processing, including canning, drying, refrigeration, and freezing, can extend shelf life, but each method has specific impacts on food quality and consumer acceptance. ASF presents unique challenges due to biological hazards that can lead to contamination, spoilage, and foodborne illness. To meet the growing demand for safe ASF, value chain actors must prioritize addressing time, temperature, and handling challenges throughout post-farmgate processes. However, capacity limitations and lack of enabling environments can hinder these efforts.

Food trade can contribute to food security and dietary diversity by creating new market opportunities. However, it also introduces complexities related to food quality standards. As trade increases, exporters must adhere to increasingly stringent standards, often imposed by importing countries. While these standards enhance food safety and quality, they can exclude smaller producers and limit market access for countries with less developed food systems. Import rejections due to noncompliance with food safety and quality standards can have significant economic and reputational consequences for exporting countries. To mitigate these challenges, LMICs aiming to grow high-value food exports must invest in strengthening food safety infrastructure, training, and technical assistance to ensure their exports meet international standards.

1.3.6 Pinpointing Food Safety Loss Pathways and Balancing Mitigation Efforts with Preserving Nutritional Quality Are Crucial to Ensure the Consumption of Safe and Nutritious Foods While Minimizing Foodborne Illnesses (Chap. 7)

Safe and nutritious food is a fundamental aspect of food security. Both food safety and nutrient content are essential food quality attributes. When food quality is compromised, either through contamination or nutrient loss, it can negatively impact human health and nutrition. Chapter 7 explores how urbanization and dietary shifts have transformed food systems, leading to changes in the burden of foodborne disease (FBD). Food hazards, such as biological, chemical, and physical contaminants, can contribute to food quality loss throughout the food value chain. FBDs pose significant health risks, particularly in low-income countries. Biological hazards, especially in ASF and fresh produce, combined with poor time, temperature, and handling practices, are major contributors to FBD.

As food systems become more complex, FBD risks can increase initially, especially in early stages of development where food safety infrastructure and regulations may be inadequate. Even when food safety standards emerge, challenges in enforcing food safety standards persist in informal markets and cross-border trade. However, as food systems modernize, investments in food safety and increased consumer awareness can lead to a decline in FBD. Therefore, strengthening food safety infrastructure, incentivizing value chain actors, and building consumer trust in safe food are essential to addressing these challenges in transitional food systems. This includes proactively identifying emerging food hazards associated with innovations, including growing concerns involving food packaging and the potential health impacts of microplastic consumption.

While food safety is an immediate concern, nutrient loss also occurs as foods move along the value chain. Foods undergo different processes from harvest to consumption, each of which can impact nutrient content. Similar to loss of other attributes, factors such as time, temperature, and handling influence nutrient loss, particularly, for vitamins and water-soluble nutrients. Processing techniques, such as heating and packaging, can also affect nutrient retention. However, further research is needed to fully understand the impact of nutrient loss on diet quality. Governments must consider the specific context of food systems and balance strategies to address both relevant food safety concerns and nutrient loss to ensure safe and nutritious diets.

1.3.7 Addressing Higher Level Drivers of Food Loss and Waste Necessitates Identifying Critical Loss Points and Determining Optimal Food Loss and Waste Levels that Account for Food Security Trade-Offs with Environmental Impacts and Climate Change (Chap. 8)

To achieve overarching goals, including food security, economic prosperity, and environmental sustainability, FLW reduction strategies must align with specific food system contexts and investment priorities. Identifying critical loss points within food systems is crucial for prioritizing effective interventions. Understanding the extent and distribution of FLW across regions, value chain stages, and commodities enables a deeper analysis of underlying causes and the development of targeted solutions. However, it is essential to consider the limitations of using different units to assess FLW, such as nutritional value metrics or environmental impact indicators, as this can introduce bias and hinder comparisons. Greater collaboration and data sharing between public and private sectors are necessary to identify and address critical loss points across the value chain.

Beyond identifying critical loss points, Chap. 8 examines the drivers of FLW across value chain stages and at a broader food system level. Factors such as technology, infrastructure, regulations, and environmental changes can significantly impact FLW. Although technological advancements can reduce FLW, challenges like affordability, accessibility, and appropriate implementation can limit their effectiveness. Additionally, technological innovations can sometimes introduce new challenges, such as stricter quality standards or packaging-related issues. Infrastructure limitations, such as inadequate transportation, electricity, storage, and communication, can hinder efficient food movement and storage. Public and private regulations, while important for food safety and quality, can also contribute to FLW by setting stringent standards that may exclude lower quality products. Finally, climate change can exacerbate FLW by impacting different stages of food value chains. For example, higher temperatures and altered rainfall patterns can impact harvest, storage, and transportation conditions, increasing food spoilage and loss. However, more evidence is needed to understand the role of climate change and linkages to other FLW drivers.

Determining the optimal level of acceptable FLW is key for setting realistic and effective global reduction goals. This requires a careful consideration of the trade-offs between food security needs and the climate and environmental impacts of FLW. Although countries have made progress in setting national targets and implementing policies, more comprehensive and coordinated efforts are needed to ensure targets and policies are appropriate and evidence-based. These targets and policies must align with both global and country-specific climate and environmental objectives. FLW has significant environmental and climate impacts throughout the entire food system. These impacts encompass the resources consumed in food production and post-production, the resources expended to manage and dispose of FLW (for example, transportation to nonfood destinations), and the environmental consequences of different FLW destinations (for example, compost, incineration, landfill). Furthermore, interventions aimed at reducing FLW can also have environmental and climate implications, depending on the specific strategies used. Therefore, a comprehensive assessment of optimal FLW levels must carefully consider the trade-offs between reaching food security to meet global dietary needs with existing FLW levels and the environmental and climate benefits of FLW reduction interventions.

1.3.8 Successfully Addressing Food Loss and Waste and Fostering Sustainable, Healthy Food Systems Necessitates the Private Sector's Adoption of Inclusive, Practice-Based, and Technology-Driven Interventions Integrated with Broader Sustainability Efforts (Including Innovations in Food Packaging and Alternative Energy Sources), Supported by Public Engagement and Rigorous Evaluation to Guarantee Their Contribution to Wider Sustainability Objectives (Chap. 9)

Effective FLW interventions must be tailored to specific food system contexts and prioritized sustainable food systems. Chapter 9 explores the multifaceted nature of FLW interventions, outlining comprehensive approaches that consider food security, environment, and socioeconomic objectives. FLW interventions can be categorized into prevention and management strategies. Prevention strategies focus on maintaining food quality and reducing pFLW, while management strategies aim to minimize the negative impacts of pFLW that occurs. Distinguishing between edible and inedible food parts is important for determining appropriate interventions and assessing their impact. Prevention interventions aim to keep edible food within the food value chain, ensuring it contributes to diets. In contrast, inedible food parts, which are never intended for human consumption, always require management strategies to minimize their environmental impact. Therefore, the trade-offs associated with prevention versus management of FLW are not exactly the same as the management trade-offs of inedible food parts.

Technological interventions have gained significant attention in addressing FLW, but a balanced approach that considers both technological and practice-based solutions set within an enabling environment is needed. Such an approach should strategically integrate innovations in sustainable food packaging and energy sources into FLW prevention efforts. To evaluate the effectiveness of interventions, robust methods are required to assess both FLW reduction and broader food system impacts. Assessing successful FLW reduction interventions through a food systems lens requires government and civil society evaluators to consider not only their impact on loss reduction but also potential trade-offs with other food system outcomes. Further, evaluations will benefit from taking a value chain approach to consider the motivations and responses of different actors when designing and implementing inclusive and sustainable FLW reduction strategies. Understanding the interconnectedness of FLW with other aspects of sustainable development supports global efforts to address FLW and the creation of more resilient and equitable food systems.

1.3.9 A Comprehensive Global Policy Agenda Targeting Food Demand and Supply Is Crucial for Guiding Food Loss and Waste Reduction While Mitigating Potential Feedback Loops and Cascading Effects of Food Loss and Waste Reduction that Create Trade-Offs Between Food Security, Environmental Sustainability, and Socioeconomic Welfare (Chap. 10)

This book positions a food systems approach as central to addressing FLW, essential for sustainable and healthy food system transformations. Although strategies to address FLW may sometimes appear conflicting, such as the trade-offs between food security and environmental impacts, we advocate for a multidimensional approach to conceptualize FLW and find solutions that promote both. Chapter 10 outlines a FLW policy agenda that calls for increasing demand for nutritious foods, supporting value addition innovations, enhancing market coordination, and considering environmental trade-offs while prioritizing food security. By addressing both the demand and supply sides of food systems, collaboration between UN agencies and national governments to establish evidence-based guidelines will create enabling environments that support interventions addressing specific FLW drivers and mitigate potentially negative cascading effects and feedback loops.

For FLW interventions to succeed, government and private sector actors must see clear value in investment, even though their priorities may sometimes conflict. Stimulating demand for safe and nutritious foods is critical for incentivizing private sector FLW reduction and increasing the supply of quality foods. To do so requires policies that promote healthy diets, improve food safety standards and practices to build consumer trust, and establish clear date labeling requirements. Additionally, private sector-driven innovations in value addition, such as expanding cold chain infrastructure, promoting processing of perishable foods, and increasing the marketability of "ugly" foods, can help keep nutritious foods in the food value chain. Alongside these efforts, public sector policies and initiatives to invest in infrastructure and financial services to improve market access and to encourage collaborative partner networks that balance risk between different value chain actors will be critical for equitable food system transformations. Finally, government policies must directly address food security and environmental trade-offs, partly by following evidence-based guidelines from UN agencies on emerging topics, such as sustainable food packaging, and by distinguishing management of physical FLW for different nonfood destinations. A common framework for conceptualizing FLW and prioritizing data collection will enable both public and private food systems stakeholders to identify critical pathways for intervention. Implementing comprehensive and collaborative approaches that address these pathways will be essential for reducing FLW and promoting sustainable and healthy food systems.

References

Chauhan C, Dhir A, Akram MU, Salo J (2021) Food loss and waste in food supply chains. A systematic literature review and framework development approach. J Clean Prod 295:126438. https://doi.org/10.1016/j.jclepro.2021.126438. Accessed 09 Nov 2021

FAO (Food and Agriculture Organization of the United Nations) (2011) Global food losses and food waste—extent, causes and prevention. FAO, Rome. https://www.fao.org/3/i2697e/i2697e.pdf. Accessed 22 Sep 2021

Global Panel on Agriculture and Food Systems for Nutrition (2020) Future food systems: for people, our planet, and prosperity. https://www.glopan.org/wp-content/uploads/2020/09/Foresight-2.0_Future-Food-Systems_For-people-our-planet-and-prosperity.pdf. Accessed 07 Oct 2021

Ingram J (2011) A food systems approach to researching food security and its interactions with global environmental change. Food Sec 3:417–431. https://doi.org/10.1007/s12571-011-0149-9. Accessed 01 Feb 2024

UN (United Nations) (1975) Report of the World Food Conference, Rome, November 5–16, 1974. New York, United Nations, https://digitallibrary.un.org/record/701143?ln=en&v=pdf#files. Accessed 12 Dec 2024

UN (United Nations) (2015) A/RES/70/1 Transforming our world: The 2030 Agenda for Sustainable Development. Resolution adopted by the General Assembly on 25 September 2015. https://www.un.org/en/development/desa/population/migration/generalassembly/docs/globalcompact/A_RES_70_1_E.pdf. Accessed 04 Mar 2024

Willett W, Rockström J, Loken B, Springmann M, Lang T, Vermeulen S, Garnett T, Tilman D, DeClerck F, Wood A, Jonell M, Clark M, Gordon LJ, Fanzo J, Hawkes C, Zurayk R, Rivera JA, De Vries W, Majele Sibanda L, Afshin A, Chaudhary A, Herrero M, Agustina R, Branca F, Lartey A, Fan S, Crona B, Fox E, Bignet V, Troell M, Lindahl T, Singh S, Cornell SE, Srinath Reddy K, Narain S, Nishtar S, CJL M (2019) Food in the Anthropocene: the EAT–*Lancet* Commission on healthy diets from sustainable food systems. Lancet 393(10170):P447–P492. https://doi.org/10.1016/S0140-6736(18)31788-4. Accessed 20 Aug 2021

2 Navigating Food System Transformations for Healthy Diets

2.1 Introduction

Understanding and addressing food loss and waste (FLW) requires a comprehensive understanding of food systems. This chapter provides an overview of the key elements and activities within food systems, from production to consumption. We explore how these elements interact and are influenced by various factors, such as economic development, technological advancements, and consumer preferences.

By examining how food systems transform and how food markets become increasingly connected, we gain insights into the specific challenges and opportunities associated with making healthy and sustainable diets available and accessible to all. As food systems transition from traditional to modern systems, there are significant shifts in production practices, consumption patterns, and distribution networks, in response to shifting consumer preferences. In particular, we highlight the challenges associated with diversifying into nutritious foods, particularly, those with high perishability, which are most at risk of FLW, as value chains grow longer and more intricate. This chapter provides a foundation for subsequent discussions on the extent, causes, and potential solutions to FLW by introducing key food systems concepts. By understanding the complex dynamics of food systems, we can develop effective strategies to reduce FLW and achieve sustainable and healthy food systems transformations.

2.2 Food Systems Elements and Framework

A food system is composed of elements (food supply chains, food environments, and consumer behaviors) and activities that are involved in the production, processing, distribution, preparation, and consumption of food, as well as the output of these activities, including socioeconomic and environmental outcomes (HLPE 2014). Although the flow of food products moves from production systems to consumers, food systems drivers affect each element of the food system, as these elements and processes vary over time and space (HLPE 2017). A sustainable food system is one that ensures food security and nutrition for all people, in such a way that the socioeconomic and environmental bases to generate food security and nutrition for future generations is not compromised (HLPE 2017). Reducing FLW to optimize resource use and minimize environmental impact is a potential strategy to contribute to this transformation and achieve these outcomes (Willett et al. 2019).

A food value chain is a complex network of processes and actors that bring food products from production to consumption or disposal

J. Boiteau, P. Pingali, *Wasted Potential*, Sustainable Development Goals Series,
https://doi.org/10.1007/978-3-031-97411-3_2

(Hawkes and Ruel 2012). Value chains can vary widely in structure and complexity. Generally, they are divided into three main stages: upstream, midstream, and downstream (Reardon 2016). Upstream stages involve input provision (for example, seeds, fertilizers, finance, and knowledge), production (for example, farming), and initial aggregation (for example, farmer organizations, transporters, small traders). Midstream activities encompass processing and trading, where raw materials are transformed into finished products and moved through the value chain. Downstream stages include retailing and consumption (Reardon 2016). Value addition activities can increase a product's economic value, generate economic value captured by value chain actors, and/or enhance perceived benefits relative to price for buyers (Hawkes and Ruel 2012). Therefore, value chains can be defined by the types of value added, the stages at which value is added, and the specific activities and actors involved in these processes (Tan 2001).

A food environment is where consumers interface with the food system. The interface can be in natural or built spaces, including wild and cultivated or informal and formal markets, respectively. These spaces are embedded in sociocultural and political environments and ecosystems that influence the availability, affordability, quality, promotion and sustainability of foods (Downs et al. 2020). Food environments can also be described by their scope, including community food environments, which focus on the geographic distribution of food outlets; organizational food environments, which influence specific groups of people; consumer food environments, which relate to retail food outlets; and informational food environments, which encompass media and advertising at various levels, from national to local (Ohri-Vachaspati and Leviton 2010; Tonumaipe'a et al. 2021).

Consumer behavior encompasses the factors, choices, and decisions that individuals and households make about food, including acquisition, storage, preparation, consumption, and allocation within the household (HLPE 2017). Food environments have substantial impact on these behaviors by influencing factors, like food access, affordability, marketing, and quality (Caspi et al. 2012; Swinburn et al. 2014; Hawkes et al. 2015b). Ultimately, consumer behaviors directly influence dietary patterns, affecting the quantity, quality, diversity, and safety of food consumed (HLPE 2017).

2.3 Structural Transformation and Changing Food Systems

Structural transformation, a process of economic development, involves the shift from agriculture-dominated economies to industrialized ones (Timmer 2009). This transition is marked by increased agricultural productivity, urbanization, the growth of industrial and service sectors, and demographic changes (Timmer and Akkus 2008). As agricultural productivity rises, rural workers migrate to urban areas for employment opportunities, leading to a decline in the agricultural sector's share of gross domestic product (GDP) and employment.

The broader economic transformation is interconnected with the structural transformation of food systems, mutually influencing how food systems are organized and how activities within them are conducted (Reardon and Timmer 2014; Reardon 2016). Food systems are influenced by and, in turn, influence external factors such as environmental conditions, technological innovations, political and economic policies, sociocultural norms, and demographic trends (HLPE 2017). The transformation process involves downstream shifts in consumer demand, such as urbanization and dietary changes; upstream changes in agricultural production, including farm technology, commercialization, and product composition; and intermediating developments that connect rural and urban areas, facilitating the movement of food from farms to consumers (Reardon and Timmer 2014).

Food systems are complex, transforming along a continuum and varying across global and local scales. Typologies can help to categorize and understand these diverse systems, identifying common patterns across countries and potential intervention strategies suitable for a given context (Ericksen et al. 2010; Marshall et al. 2021). Food systems transform from traditional to modern systems, with the number of categories varying depending on the specific categorization approach (IFPRI 2015; HLPE 2017; Marshall et al. 2021). Transformations typically start in urban areas, beginning with large cities, and diffuse to small and medium cities, followed by rural areas.

Although some typologies rely on conceptual frameworks, other approaches aim to use empirical data to classify different types of systems, using indicators including agricultural productivity, diversity of food supply, economic accessibility of food or level of development of the modern retail sector, and urbanization demographics (IFPRI 2015; Marshall et al. 2021). Although national-level typologies may not capture local food system variations, this approach that considers multiple variables, beyond income group alone, can be helpful for identifying priority actions (Marshall et al. 2021).

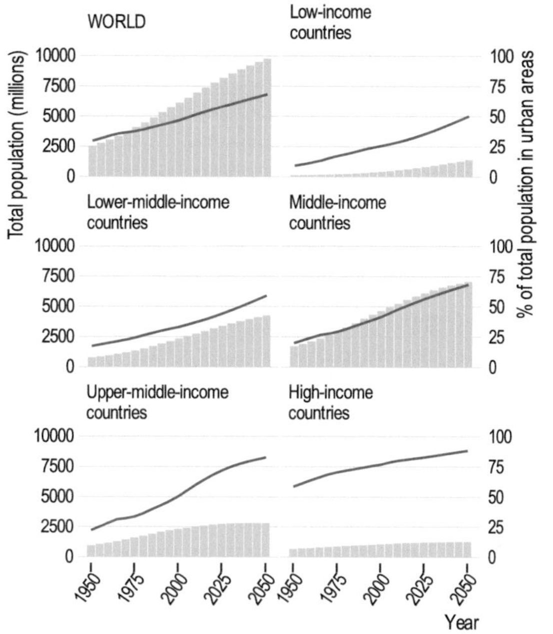

Fig. 2.1 Urban populations as a percentage of total population

Note: Bar chart represents the total population according to the left y-axis. Line graph shows the percent of total population residing in urban areas according to the right y-axis

Source: UN (2018)

2.3.1 Urbanization Drives Increasing Food Demand and Diversification

As the world becomes increasingly urbanized, the demand for food in cities is rapidly growing, coinciding with rising per capita incomes. Over half of the global population now resides in urban areas, a figure projected to reach 68% by 2050 (UN 2019, Fig. 2.1 shows population growth and urbanization trends across regional income groups in comparison to global trends. Urbanization encompasses the movement of people from rural areas to both large cities and smaller urban centers, including towns and small to medium-sized cities. While urbanization in high-income countries (HICs) and upper-middle-income countries (UMICs) is expected to slow, it will accelerate in lower-income countries (LICs) and lower-middle-income countries (LMICs), leading to significant population shifts toward urban areas in these regions (UN 2019).

As urbanization occurs and incomes rise, individuals tend to spend a smaller proportion of their income on food, while diversifying their diets to include more nutritious foods, including fresh fruits, vegetables, and animal-source foods (ASF), as well as less nutritious processed foods (Pingali 2007; Godfray 2011; Imamura et al. 2015). Factors, such as changes in women's labor force participation, improved infrastructure, and increased access to markets and food storage technologies contribute to these dietary shifts (Global Panel 2017; Pandey et al. 2020). For example, access to household-level refrigeration can influence consumer purchasing decisions for nutritious, perishable foods. With refrigeration, consumers can store more of these foods at home for extended periods (Box 2.1). Therefore, changes in storage capacity and duration, which impacts both the demand for and the handling of perishable foods, may also influence the location and magnitude of FLW. While smaller urban areas may retain strong rural connections, they often face challenges related to infrastructure and service provision, which are often concentrated in larger cities and capital regions (Reardon 2016; de Bruin et al. 2021). Therefore, careful attention to the relative increases in demand for diverse food commodities, particularly, nutritious, perishable foods in both urban and rural areas, is important for assessing supply and demand gaps and ensuring that food systems can provide healthy diets for all (Zhou and Staatz 2016; Pingali et al. 2019).

As women become more involved in work outside of the home, their opportunity cost of time increases, leading to a shift in consumer preferences toward more convenient, easy-to-prepare, and easy-to-consume foods (Senauer et al. 1986; Kennedy and Reardon 1994; Reardon et al. 2021b). Food processing can be beneficial, extending the shelf life of fresh products, eliminating harmful pathogens, and preserving nutrients (Miller and Welch 2013). However, the rise of processed foods, high in unhealthy fats, sugars, and salt, is a growing concern in both rural and urban areas (Gómez and Ricketts 2013; Global Panel 2017; Monteiro et al. 2018).

Box 2.1 Household Cold Food Storage

Refrigeration is a critical component of modern food systems, particularly, for preserving fresh, perishable foods. Both access to refrigeration technology and reliable electricity are essential for proper cold storage at the household level. However, access to refrigeration varies widely across global income regions. Although nearly all households in the United States and Europe have refrigerators, access rates in India and sub-Saharan Africa are notably lower, at around 30% and 17%, respectively (Efficiency for Access Coalition 2019). Lack of reliable household refrigeration can limit consumer choices, particularly, the frequency and amount for fresh, perishable foods purchased (Global Panel 2017).

Further, rapid urbanization can outpace improvements in electricity infrastructure, leading to unreliable power supply, especially, in densely populated areas (Efficiency for Access Coalition 2019). Therefore, it is important to consider both rural–urban disparities in electricity infrastructure, as seen in regions like sub-Saharan Africa and South Asia (Efficiency for Access Coalition 2019), and intra-urban disparities within cities, as these factors can impact access to household refrigeration and, consequently, dietary choices.

2.3.2 Changes in Food Production, Rural–Urban Linkages, and Food Environments

As economies develop and undergo structural transformation, continued investment in agriculture contributes to increased agricultural productivity and higher incomes for farmers, enabling them to participate in new market opportunities (Pingali and Rosegrant 1995; Carletto et al. 2017). Higher agricultural productivity is important for generating food surpluses for markets and freeing labor for off-farm employment. As food production systems commercialize, producers diversify their production into higher value foods to meet growing consumer demand for diversified diets (Pingali and Rosegrant 1995). Further, agriculture-producing households increasingly purchase food from the market and become net consumers.

Low-productive agricultural systems are characterized by low yields, labor-intensive practices, and limited crop diversity, often relying heavily on staple crops, while the availability of perishable, nonstaple crops is seasonally limited (Pingali et al. 2015). With investment and modernization of agricultural systems, there is increased productivity in staple grains. Although improvements in rural infrastructure and demand for diverse diets drive changes in agricultural production, the relative price of micronutrient-rich foods, fruits, vegetables, and ASF remains high compared to staples (Pingali 2010). In fully commercialized systems, high productivity and specialization in various agricultural products, including staples, horticulture, and livestock, increase the diversity of food products entering food value chains (Pingali and Rosegrant 1995).

In addition to the quantity of food produced, shifts in food quality also emerge. Farmers do not receive premium prices for variety and quality differentiation in contexts with limited market options; where production costs, rather than market demand, drive crop selection; or where quality premiums are insufficient to incentivize investment (Minten et al. 2013; Hernandez et al. 2015; Anissa et al. 2021; Reardon et al. 2021a). Farmers may capture premium prices for quality when there is strong market competition or when they implement value addition approaches (Liverpool-Tasie et al. 2017; Reardon et al. 2021a). As food systems transform, so do the quality standards and expectations, including the desired attributes, the responsible parties for setting standards, and those accountable for adhering to them. Therefore, understanding the market context is necessary for evaluating how food quality loss influences producer decisions and incentivizes them to minimize such losses.

To increase market connectedness between growing cities and urban markets to rural food-producing areas, food value chains elongate, and the arrangements of value chain stages, the interactions between, and the types of products made available to consumers all change along complex, nonlinear pathways (Gómez and Ricketts 2013; Reardon et al. 2021a). A critical component of rural-to-urban linkages is the development of midstream segments, including wholesalers, logistics operations, and processors, which facilitate movement of food from rural producers to urban consumers, influencing the quantity, quality, and diversity of food accessible to consumers (Reardon et al. 2021a).

As urbanization rises, wholesale markets emerge to link rural production areas with growing urban demand, facilitated by investments in connective road infrastructure (Reardon et al. 2021a). Although expanded road networks can facilitate the lengthening of food value chains, the quality of the roads, particularly, in rural areas, is important for ensuring efficient transportation of perishable foods to minimize FLW (HLPE 2014). In traditional food systems, local markets and rural brokers are important for connecting smallholder farmers with rural consumers across relatively short distances (Gómez and Ricketts 2013; Reardon et al. 2021a). As rural towns and small cities grow and governments invest in rural road development, the periodic, local markets begin to evolve into permanent rural markets, and rural–urban brokers become key intermediaries in connecting rural producers with urban wholesalers (Reardon 2015; Reardon et al. 2021a).

Eventually, small and medium-sized enterprises (SMEs) emerge in growing cities and supply food to central markets in major cities (Reardon et al. 2021a). As food volumes increase and value chains become more complex, consolidation occurs, leading to specialization in logistics and other value addition services (see Box 2.2) (Reardon et al. 2021a). As consolidation continues, retailers increasingly source directly from farms and food processors, or through medium and large, contracted wholesalers (Reardon 2015; Reardon et al. 2021a).

Box 2.2 A Closer Look at Midstream Value Addition

Improved market connectivity and longer food value chains requires investments in infrastructure and value-added activities to ensure the delivery of safe and nutritious food to consumers. Value-added activities in the midstream segment, such as processing, packaging, and cold chain services, facilitate the movement and accessibility of food products.

Food processing transforms foods to extend shelf life, enhance consumer acceptability, improve nutrient content and food safety, and reduce time and energy for preparation (Miller and Welch 2013; HLPE

2014). Along with food processing technologies, food packaging offers opportunities to extend the shelf life and protect food, as well as to provide information to consumers on food packaging labels (HLPE 2014). Therefore, processing and packaging stages offer opportunities to reduce FLW. As food systems transition from traditional to modern, the role of food processing also evolves. In traditional systems, most food is processed at the household level (Reardon 2015; Reardon et al. 2021a). However, with increasing urbanization and changing consumer preferences, SMEs emerge to produce minimally processed foods and processed culinary ingredients (Monteiro et al. 2018; Reardon et al. 2021a). These SMEs may initially sell unpackaged products, but later transition to branded, packaged foods. In contrast, large-scale food processors benefit from economies of scale, offering a wide range of processed and packaged foods at competitive prices to both urban and rural consumers (Gómez and Ricketts 2013; Reardon 2015; Reardon et al. 2021a). Over time, these large-scale processors often dominate the market in modern food systems.

Effective cold chain management preserves the quality and safety of perishable food products and requires maintaining lower temperatures throughout the value chain (Kitinoja 2013). Although high-income regions like the United States and the European Union have well-established cold chain infrastructure, it is more limited elsewhere (Kitinoja 2013; HLPE 2017). For comparison, the refrigerator storage capacity in developed countries was found to be 200 cubic meters per 1000 inhabitants, compared to 19 cubic meters in developing countries (Kitinoja 2013). When cold chain infrastructure is limited, higher value products are often prioritized but still lack sufficient coverage. In India, for example, less than 25% of perishable fruits and vegetables (mostly for potato storage) and 34% of meat are covered by cold chain logistics (Minten et al. 2016; IIFIIR 2019). In contrast, nearly 95% of perishable food value chains in the United States and Europe are supported by cold chain infrastructure (IIFIIR 2019).

As countries progress through structural transformation, the food environments that consumers have access to transition from natural settings, where food is sourced directly from wild or cultivated environments, to built environments, which rely on informal and formal markets (Downs et al. 2020). Informal markets, such as street vendors and wet markets, are characterized by low capital investment and often operate outside of formal regulations (FAO 2003; Reardon et al. 2003; Downs et al. 2020). In contrast, formal markets, including supermarkets, grocery stores, and online retailers, are subject to private and/or public regulations, and typically, offer a wider range of products at standardized prices (Downs et al. 2020). Downs et al. (2020) described a final food environment pattern, one that integrates cultivated environments and formal markets, to support sustainable diets and planetary health, which has yet to be demonstrated at a national scale.

The transition from informal to formal markets is driven by changes in consumer demand and value chain dynamics. As incomes rise and urbanization accelerates, consumers increasingly demand diverse, safe, and convenient food options. Increased trade, foreign direct investment, and advancements in logistics and procurement have enabled formal value chains to meet this growing demand (Reardon et al. 2003). The rise of formal, private retailers has led to the development of private food standards. These standards can serve various purposes: ensuring compliance with minimum public standards, filling gaps in weak or nonexistent public regulations, or differentiating products through higher quality standards (Reardon et al. 2003; Henson and Reardon 2005). The impact of food quality standards on avoidable FLW varies, depending on the specific quality attributes prioritized. Food safety standards, by their nature, aim to remove unsafe food from the value chain, which is important for public health. In contrast, cosmetic standards, such as those that reject produce based on "ugly" appearances, can lead to unnecessary FLW of food that is safe and edible.

As formal markets mature and expand, they offer a wider range of food products to a wider range of consumers. Initially, formal markets often offer shelf-stable, processed, and packaged foods at lower prices, compared to informal retailers, benefitting from economies of scale (Reardon et al. 2003). When expanding into fresh food, formal markets start, typically, with less perishable items (for example, potatoes), and gradually, introduce more perishable products, including fresh fruits, vegetables, and ASF (Reardon et al. 2003). The geographic expansion of formal markets typically follows a pattern, starting in larger cities before extending to medium-sized and small cities. Similarly, the socioeconomic reach of these markets often begins with higher income groups, and then extends to middle- and lower-income communities (Reardon et al. 2003). Over time, the formal retail sector shifts from domestic toward multinational chains (Reardon et al. 2012).

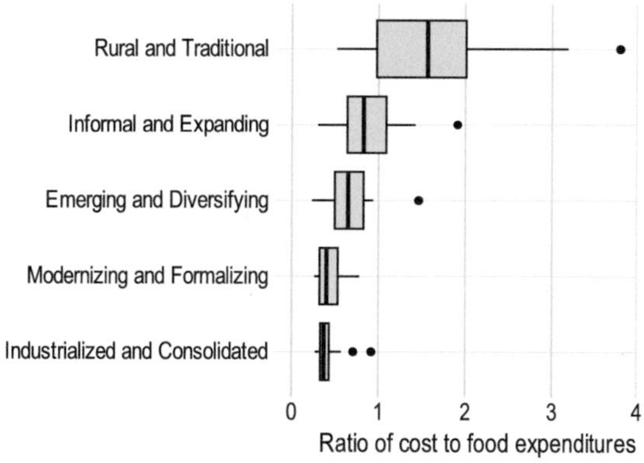

Fig. 2.2 Comparison of affordability of a healthy diet across food system types

Note: The ratio compares the cost of the least expensive diet that meets dietary guidelines to the total amount spent on food per person per day. Ratios closer to 0 indicate a more affordable healthy diet. Ratios farther from 0 suggest a less affordable healthy diet

Source: The Food Systems Dashboard (GAIN et al. 2023)

2.3.3 Diet Transformations: Supply, Demand, and Dietary Recommendations

Although urbanization, rising incomes, and changing food value chains have contributed to dietary transitions, these changes have led to increased consumption of nutritious perishable foods and to unhealthy processed foods. For example, unhealthy processed foods, formulated to be highly palatable, may compete with healthier minimally processed and freshly prepared foods, in terms of taste, convenience, and price, among other attributes (Monteiro et al. 2018). Therefore, the extent to which these dietary shifts align with dietary recommendations depends on the availability, accessibility, and consumption across different food groups and the specified dietary guidelines. The affordability of healthy diets often improves as food systems transform (Fig. 2.2). However, within a given food system, inequalities contribute to disparities in access to healthy diets.

Food-based dietary guidelines provide consumers with recommendations for healthy diet patterns and may inform national food, nutrition, and health policies and programs (Herforth et al. 2019). Dietary recommendations encourage the consumption of fruits and vegetables (excluding starchy roots and tubers), specifying minimum intake amounts and frequencies (for example, grams or portions per day). In contrast, recommended intakes for ASF are typically based on adequacy and depend on the health and nutrition linkages to the consumption of specific types of ASF. Therefore, consumers are encouraged to limit or moderate their intake of certain ASF, such as fish high in mercury, red meat, and processed meat, with diet messages tailored to local food cultures (Herforth et al. 2019). The availability, accessibility, and consumption of nutritious, perishable foods depend on factors such as production, efficient food value chains, and consumer preferences. These factors ultimately shape dietary patterns, which may or may not align with dietary recommendations.

A major challenge in achieving adequate fruit and vegetable consumption is the insufficient supply of fruits and vegetables, globally and in most countries (Siegel et al. 2014; Mason-D'Croz et al. 2019; Kalmpourtzidou et al. 2020). Among the countries with sufficient supply, intakes may still not meet diet recommendations. For example, a review by Kalmpourtzidou et al. (2020) found that very few countries meet recommended vegetable intake levels. While Asia has the highest vegetable intake, with 29% of countries meeting recommendations, despite only 61% having sufficient supply, Africa lags behind last with only 13% of African countries having adequate supply and 7% meeting intake recommendations. Therefore, increasing the supply of perishable fruits and vegetables is a crucial first step to improving consumption. However, additional strategies are needed to bridge the gap between supply and consumption, even when sufficient supply is available.

As countries urbanize and incomes rise, ASF consumption often increases. ASF represents a diverse food group that can provide important sources of quality protein and bioavailable nutrients. Therefore, meeting diet adequacy recommendations can improve nutrient intakes and reduce undernutrition. However, excessive consumption of certain types of ASF, particularly, processed meats and red meat, can have negative health consequences (Godfray et al. 2018; Troell et al. 2019; Beal et al. 2023).

At later stages of transformation, consumption of ASFs plateaus. However, unlike fruits and vegetables, where consumption often falls short of minimum recommendations, ASF consumption—particularly, meat—can exceed dietary adequacy in HICs (Godfray et al. 2018; Gouel and Guimbard 2019; Naylor et al. 2021b), prompting some dietary guidelines to recommend substituting fish or poultry in place of red meat (Herforth et al. 2019). Naylor et al. (2021b) pointed out that poultry seems to have already become the major substitute for beef in global diets.

The relationship between aquatic animal food consumption, income growth, and urbanization is less clear, compared to livestock source foods, and the lack of clarity is likely due to the greater diversity of aquatic species and data limitations (Naylor et al. 2021b; Pounds et al. 2022). Over the last two decades, growth in aquatic animal food production has been driven by aquaculture production in South and Southeast Asia, dominated by China, for both domestic and export markets (Naylor et al. 2021a). Although countries at earlier stages of structural transformation experience growth in aquatic animal food production and consumption, the starting production baseline is small (Naylor et al. 2021a). Meanwhile, countries at later stages of structural transformation produce the majority of aquatic animal foods from capture production, which has declined over the last several decades (FAO 2023). Similar to livestock source foods, per capita consumption of aquatic animal foods peaks at later stages of development (Naylor et al. 2021b; OECD and FAO 2023).

To meet the growing demand for high-value foods and enable consumers to meet diet recommendations, food systems must adapt through changes in production and post-production processes. While economic growth is crucial for increasing fruit and vegetable production in lower income regions, including sub-Saharan Africa and South Asia, investments in research and development are also essential for improving productivity and post-production technologies, organization, and policy to bridge the gap between supply and demand (Alston and Pardey 2008; Mason-D'Croz et al. 2019). Although increased meat production has been driven by both increased animal numbers and improved productivity, the expansion of livestock production may lead to land-use changes, particularly, in areas closer to urban centers. (Thornton 2010; Latino et al. 2020). Finally, in the case of aquatic animal food production, the vast majority has shifted to aquaculture, which has a range production systems, and are predominantly managed by households or SMEs (Tidwell 2012; Hernandez et al. 2018; Naylor et al. 2021a).

2.3.4 Trade Contributes to Globalized Food Systems

Food systems interact at local, regional, national, and global levels. Rising incomes, urbanization, and technological innovations act as drivers of trade, as food is moved from surplus to deficit regions, based on regional comparative advantages of production and consumer demand (FAO 2024). Although trade can, therefore, expand food security of diverse foods where there are domestic shortfalls, in reality, there have been mixed results with sometimes rising food costs and lower food quality (Ambikapathi et al. 2022). Therefore, the impacts of trade can vary depending on the value chain actors involved, the types of foods traded, the countries participating in trade, and the consequences for both consumers and local food systems in exporting and importing regions.

Liberalization of trade policies, foreign direct investment, and technology advancements in value chain logistics have contributed to increasingly open markets and globalized food systems (FAO 2024). Trade liberalization policies include those that aim to reduce financial and regulatory barriers; harmonize or increase transparency of food regulations; invest in trade infrastructure and capacity; and ensure fair competition between domestic and foreign food businesses in public food procurement (Hawkes et al. 2015a). Large companies have leveraged trade liberalization to consolidate their position through vertical and horizontal integration of value chains procurement (Hawkes et al. 2015a).

Regional variations in climate and resource endowments create comparative advantages for the production of specific crops. For example, although temperate climates are well-suited for grain production, warmer climates are more conducive to a diverse range of fruits and vegetables (FAO 2024). From 1995 to 2018, LICs experienced increases in imports for meat and fish, fruits and vegetables, and grain that outpaced increases in exports, whereas increases in exports of dairy and eggs and processed foods outpaced increases in imports (FAO 2020). During this same period, LMICs and UMICs experienced increases in imports of meat and fish, fruits and vegetables, and processed foods that outpaced increases in exports, whereas increases in the export of dairy and eggs and grain outpaced the increase of imports (FAO 2020). In contrast, HICs, generally, had smaller increases in export and imports across all foods groups between 1995 and 2018, relative to other country income groups (FAO 2020). Looking forward, the comparative advantage of regions in producing certain crops and participating in trade may shift due to the impacts of climate change on agrifood production (Cui et al. 2018).

Depending on food product composition and properties, some foods are more or less easy to trade, contributing to food value chains that geographically elongate. Although trading perishable foods requires more infrastructure investment—such as cold chain, to protect food quality—shelf-stable and processed foods are generally easier to transport, store, and trade (Kitinoja 2013; Bradford et al. 2018). Thus, processed foods, shelf-stable foods, and durable raw ingredients, like grains and oils, are more accessible to regions with limited cold chain infrastructure. Regional trade agreements and phytosanitary standards can create technical barriers to trade certain types of food, resulting in the facilitation of imports of unhealthy, processed foods (FAO 2024). Therefore, while trade increases the diversity of available foods, it also leads to a significant increase in the import of processed foods, compared to domestic production (FAO 2024).

Another feature of internationally traded food is that exporting regions must often adhere to the food quality and safety standards of importing countries, which can be challenging for regions with weaker regulatory frameworks (Jaffee et al. 2019). Therefore, the ability of value chain actors to meet stringent importing food quality standards can influence market access, the types of food products exported from a region, and the extent of FLW due to rejections. Once products are exported, they may be rejected by the importing region for failing to meet standards and are typically destroyed, diverted to other food markets with less strict requirements, or diverted to other nonfood markets and uses (for example, animal feed) (Rico-Sole 2012; Fonseca and Njie 2014; Fonseca and Vergara 2014).

2.4 Increasing Market Connectedness and Sustainable Food Systems Transformations

Food system transformations involve changes across multiple dimensions, including value chains, food environments, and consumer behavior. Increasing market connectedness at different levels, from regional to global, can have positive impacts on food system outcomes, as well as unintended negative impacts on food security and diet quality, the environment, and livelihoods (HLPE 2020). Therefore, modern food systems, which are highly connected, are not necessarily the end goal of food systems transformations; each food system type and associated knowledge systems are valuable to the delivery of healthy diets (HLPE 2017).

Although individual countries may exhibit several types of food systems (Marshall et al. 2021), it is useful to consider the broader geographic distribution of food system types and illustrate stylized descriptions of each type: traditional, transitional, and modern systems. Traditional food systems, though less prevalent, persist in remote and challenging agroclimatic regions (Reardon et al. 2021a). Transitional food systems dominate in sub-Saharan Africa and South Asia, while modern food systems are more prominent in Southeast Asia, East Asia, Latin and North America, and Europe. Emerging modern food systems account for approximately 20–25% of the food systems in Africa and South Asia (Reardon et al. 2021a). As food systems transform from traditional to modern, the organizational, financial, technological, and policy changes impact food value chains, food environments, and consumer behaviors (Hawkes 2009).

Food systems have a critical role in supporting human and planetary health and are interconnected to the Sustainable Development Goals. To inform decision-making and guide actions towards sustainable food system transformations, the Food Systems Countdown to 2030 Initiative was launched in 2021 (Schneider et al. 2023).[1] This interdisciplinary collaboration aims to fill monitoring gaps, track progress annually until 2030, and provide evidence-based insights. The initiative currently tracks 50 indicators to assess progress toward global development, health, and sustainability goals (Schneider et al. 2023).

2.4.1 Traditional Systems

Traditional food systems are characterized by spatially short, fragmented value chains wherein consumers procure food from natural, predominately cultivated, food environments, as well as informal markets (Gómez and Ricketts 2013; HLPE 2017; Downs et al. 2020; Reardon et al. 2021a; Marshall et al. 2021). At the production level, smallholder farmers dominate and focus on producing staple crops for home consumption and local marketing (HLPE 2017; Marshall et al. 2021). Lack of adequate road, storage, and technology infrastructure limits the transportation and storage of perishable foods that are produced, restricting their distribution and reducing shelf life (Gómez and Ricketts 2013; Marshall et al. 2021). These constraints can discourage farmers from diversifying into the production of perishable, nutrient-rich crops, thus, limiting their availability in local markets (Gómez and Ricketts 2013; Marshall et al. 2021). Additionally, fresh food availability is often seasonal, with preservation techniques like drying used to extend shelf life (HLPE 2017). Although guidelines for staple food micronutrient fortification may exist (Marshall et al. 2021), there are few quality control or formal food safety standards for foods sold

[1] The Food Systems Countdown Initiative (https://www.foodcountdown.org/) "is a collaborative effort to monitor global food systems. It brings together indicators that span food systems and provides annual analysis to inform policy, business, and NGO priorities and actions. It supports the transformation of food systems, so they become equitable, sustainable, and resilient and positively contribute to achieving the 2030 SDGs and other global goals."

in the local informal markets (HLPE 2017). Only a small percentage of food produced and consumed is linked to international food trade (Marshall et al. 2021).

2.4.2 Transitional Systems

Transitional food systems are characterized by dynamic changes along different food value chains and food environments, as food production commercializes and incomes rise, coinciding with urbanization and shifts in consumer food behaviors. From the early to late transitional stages, there are generally more rapid transformations among shelf-stable products, including staple foods and processed foods, and slower transformations among fresh, perishable products, which require more developed storage and transport infrastructure (Reardon et al. 2003; Gómez and Ricketts 2013; HLPE 2017; Marshall et al. 2021).

In early transitional stages, the use of improved production inputs enables smallholders to become more productive, particularly, in staple grain production, which helps lower prices of staple foods and increase producer incomes (Pingali 2010). As supply chains lengthen and make food more accessible to growing urban areas, SMEs proliferate to coordinate storage and transport logistics (Reardon et al. 2021a). At the same time, food processing SMEs emerge and produce shelf-stable, convenience food products that can withstand transport and storage, meeting growing urban demand for foods that are ready-to-eat and faster to prepare. Transport and storage infrastructure is still limited to support of mostly shelf-stable products while value chains for perishable foods remain underdeveloped. Built food environments become increasingly dominant, with a mix of informal and formal markets. Formal markets, often concentrated in larger cities, are typically more accessible to higher- and middle-income consumers. Initially, formal markets primarily offer shelf-stable products, like staples and processed foods. As they expand, they may incorporate fresh produce, including fruits, vegetables, and ASF. However, informal markets continue to be important for many urban consumers, especially, in smaller cities, towns, and rural areas, for accessing fresh produce. Concurrently, food processors connect with informal retailers to offer affordable, shelf-stable packaged foods year round, expanding their reach into smaller cities, towns, and rural areas (Gómez and Ricketts 2013). Few food quality and safety standards exist for most foods, particularly, for unbranded, unpackaged foods.

In later transitional stages, diversified food production is undertaken by both smallholder farmers and larger producers, who rely more on mechanization and labor-saving technologies. Improved post-production and distribution infrastructure, including cold storage, facilitates the rapid development of value chains for fresh, perishable foods, extending their availability beyond seasonal limitations. However, infrastructure improvements often lack equity, with rural and lower income areas facing limitations (Reardon et al. 2021a).

Food supply chains become more spatially extensive, with third-party logistics providers assuming a greater role in distribution (Reardon et al. 2021a). Processed and packaged foods, produced by SMEs and larger enterprises, remain widely available. Formal markets expand, reaching more consumers in smaller cities and towns. Although formal markets may procure some products from smallholder farmers—especially, high value, perishable foods—informal markets continue to be a primary source of affordable fresh, perishable produce for lower income consumers. Therefore, differences in food quality between formal and informal markets can depend on the effectiveness of public food safety and quality standards. While formal markets often rely on private sector enforcement of quality standards, enforcement of public standards may be inconsistent.

2.4.3 Modern Systems

Modern food systems are characterized by consolidation and the dominance of formal markets. While value chains remain spatially long, food production is increasingly concentrated among both smallholder and large-scale, capital-intensive farms. Large enterprises dominate processing and distribution midstream stages, leveraging advanced technologies to optimize value chain efficiency. Developed infrastructure, including roads, storage facilities, and cold chains, increases the market connectedness between value chain actors and enables the distribution of fresh, perishable foods. Globalized food systems facilitate the year-round availability of a wide range of fresh and processed foods. Widespread household refrigeration further supports the purchase and consumption of perishable items.

Although formal markets offer a diverse range of products, including fresh and processed foods, access may vary across income groups and geographic locations. Lower income areas, both urban and rural, may have limited access to markets that carry a variety of fresh produce, necessitating longer travel distances or reliance on less diverse, often processed food offerings that are shelf-stable. Robust public food safety and quality standards are typically enforced in modern food systems. Additionally, private brands often implement their own standards, which may exceed public requirements.

2.5 Conclusion

In this chapter, we explored the complexities of food systems, describing their elements, transformations, and contribution to supporting nutrition and health outcomes. Food value chains, composed of upstream, midstream, and downstream stages, play a pivotal role in adding value and facilitating the flow of food products. Food environments, both natural and built, influence consumer choices and access to diverse food options. Consumer behavior, driven by a diverse set of factors, ultimately determines the demand for specific foods.

This chapter also highlighted the impact of structural transformation on food systems. As economies develop and urbanize, markets become more connected to meet changing consumer demands and shifting dietary patterns. Consumers increasingly demand diverse foods, including processed foods and high-value, perishable products. Although processed foods typically have greater shelf stability, perishable products generally consist of fresh, nutritious foods, such as fruits, vegetables, and animal-based products, which are inherently more susceptible to FLW. This evolving demand drives changes in the types of foods produced, the type of food processing, and the connections between rural producers and urban consumers. Monitoring food systems transformations across multiple domains, including nutrition and health, environment, livelihoods and equity, governance, and resilience, will be important for public and private stakeholders to make evidence-based decisions on minimizing FLW to support transformations toward healthy and sustainable food systems.

References

Alston JM, Pardey PG (2008) Public funding for research into specialty crops. HortScience 43(5):1461–1470. https://doi.org/10.21273/HORTSCI.43.5.1461. Accessed 21 Dec 2023

Ambikapathi R, Schneider KR, Davis B, Herrero M, Winters P, Fanzo JC (2022) Global food systems transitions have enabled affordable diets but had less favourable outcomes for nutrition, environmental health, inclusion and equity. Nat Food 3:764–779. https://doi.org/10.1038/s43016-022-00588-7. Accessed 06 Nov 2023

Anissa BP, Abate G, Bernard T, Bulte E (2021) Is the local wheat market a 'market for lemons'? Certifying the supply of individual wheat farmers in Ethiopia. Eur Rev Agric Econ 48:1162–1186. https://doi.org/10.1093/erae/jbab018. Accessed 01 Mar 2022

Beal T, Gardner CD, Herrero M, Iannotti LL, Merbold L, Nordhagen S, Mottet A (2023) Friend or foe? The role of animal-source foods in healthy and environmentally sustainable diets. J Nutr 153(2):409–425. https://doi.org/10.1016/j.tjnut.2022.10.016. Accessed 09 Dec 2024

Bradford KJ, Dahal P, Van Asbrouck J, Kunusoth K, Bello P, Thompson J, Wu F (2018) The dry chain: reducing postharvest losses and improving food safety in humid climates. Trends Food Sci Tech 71:84–93. https://doi.org/10.1016/j.tifs.2017.11.002. Accessed 24 Sep 2021

Carletto C, Corral P, Guelfi A (2017) Agricultural commercialization and nutrition revisited: empirical evidence from three African countries. Food Policy 67:106–118. https://doi.org/10.1016/j.foodpol.2016.09.020. Accessed 14 Mar 2022

Caspi CE, Sorensen G, Subramanian SV, Kawachi I (2012) The local food environment and diet: a systematic review. Health Place 18(5):1172–1187. https://doi.org/10.1016/j.healthplace.2012.05.006. Accessed 17 Feb 2022

Cui HD, Kuiper M, van Meijl H, Tabeau A (2018) Climate change and global market integration: implications for global economic activities, agricultural commodities and food security. In: The State of Agricultural Commodity Markets (SOCO) 2018: background paper. FAO, Rome. https://openknowledge.fao.org/server/api/core/bitstreams/94f8b7fd-d88a-45e1-b7f7-d171e043ff0b/content. Accessed 04 Mar 2022

de Bruin S, Dengerink J, van Vliet J (2021) Urbanisation as driver of food system transformation and opportunities for rural livelihoods. Food Secur 13:781–798. https://doi.org/10.1007/s12571-021-01182-8. Accessed 12 Apr 2022

Downs SM, Ahmed S, Fanzo J, Herforth A (2020) Food environment typology: advancing an expanded definition, framework, and methodological approach for improved characterization of wild, cultivated, and built food environments toward sustainable diets. Foods 9(4):532. https://doi.org/10.3390/foods9040532. Accessed 26 Oct 2021

Efficiency for Access Coalition (2019) The state of the off-grid appliance market. https://storage.googleapis.com/e4a-website-assets/Clasp-SOGAM-Report-final.pdf. Accessed 10 Mar 2022

Ericksen P, Stewart B, Dixon J, Barling D, Loring P, Anderson M, Ingram J (2010) The value of a food system approach. In: Ingram J, Ericksen P, Liverman D (eds) Food security and global environmental change. Earthscan, London, pp 25–45

FAO (Food and Agriculture Organization of the United Nations) (2003) The informal food sector: municipal support policies for operators. FAO, Rome. https://www.fao.org/sustainable-food-value-chains/library/details/en/c/266427/. Accessed 21 Apr 2022

FAO (Food and Agriculture Organization of the United Nations) (2020) The State of Agricultural Commodity Markets 2020: agricultural markets and sustainable development: global value chains, smallholder farmers and digital innovations. FAO, Rome. http://www.fao.org/documents/card/en/c/cb0665en. Accessed 04 Mar 2022

FAO (Food and Agriculture Organization of the United Nations) (2023) Global production. In: Fisheries and Aquaculture. https://www.fao.org/fishery/en/collection/global_production?lang=en. Accessed 04 Aug 2023

FAO (Food and Agriculture Organization of the United Nations) (2024) The State of Agricultural Commodity Markets 2024—trade and nutrition: policy coherence for healthy diets. FAO, Rome. https://openknowledge.fao.org/handle/20.500.14283/cd2144en. Accessed 10 Dec 2024

Fonseca JM, Njie DN (2014) Addressing losses of fresh produce grown in the Latin America and the Caribbean Region due to non-compliance with quality and safety requirements in export markets. Acta Hortic 1016:179–185. https://www.actahort.org/books/1016/1016_25.htm. Accessed 10 Apr 2022

Fonseca JM, Vergara N (2014) Logistics systems need to scale up reduction of produce losses in the Latin America and Caribbean Region. Acta Hortic 10173–179. https://doi.org/10.17660/ActaHortic.2014.1047.21. Accessed 10 Apr 2022

GAIN (Global Alliance for Improved Nutrition), The Columbia Climate School, Cornell University College of Agriculture and Life Sciences (2023) The Food Systems Dashboard. https://www.foodsystemsdashboard.org/. Accessed 11 Dec 2024

Global Panel (2017) Urban diets and nutrition: Trends, challenges and opportunities for policy action. Policy Brief No. 9. Global Panel on Agriculture and Food Systems for Nutrition, London. https://www.glopan.org/sites/default/files/Downloads/GlobalPanelUrbanizationPolicyBrief.pdf. Accessed 25 Feb 2022

Godfray HCJ (2011) Food for thought. P Natl Acad Sci USA 108(50):19845–19846. https://doi.org/10.1073/pnas.1118568109. Accessed 09 Mar 2022

Godfray HCJ, Aveyard P, Garnett T, Hall JW, Key TJ, Lorimer J, Pierrehumbert RT, Scarborough P, Springmann M, Jebb SA (2018) Meat consumption, health, and the environment. Science 361(3699):5324. https://doi.org/10.1126/science.aam5324. Accessed 02 Sep 2022

Gómez MI, Ricketts KD (2013) Food value chain transformations in developing countries: selected hypotheses on nutritional implications. Food Policy 42:139–150. https://doi.org/10.1016/j.foodpol.2013.06.010. Accessed 26 Aug 2021

Gouel C, Guimbard H (2019) Nutrition transition and the structure of global food demand. Am J Agr Econ 101(2):383–403. https://doi.org/10.1093/ajae/aay030. Accessed 21 Dec 2023

Hawkes C (2009) Identifying innovative interventions to promote healthy eating using consumption-oriented food supply chain analysis. J Hunger Environ Nutr 4(3–4):336–356. https://doi.org/10.1080/19320240903321243. Accessed 24 Feb 2022

Hawkes C, Grace D, Throw AM (2015a) Trade liberalization, food, nutrition and health. In: Smith R, Blouin

C, Mirza Z, Beyer P, Drager N (eds) Trade and health: towards building a national strategy. World Health Organization, Geneva, pp 92–116. https://apps.who.int/iris/handle/10665/183934. Accessed 18 Jan 2023

Hawkes C, Ruel MT (2012) Value chains for nutrition. In: Fan S, Pandya-Lorch R (eds) Reshaping agriculture for nutrition and health. International Food Policy Research Institute (IFPRI), Washington, DC, pp 73–81

Hawkes C, Smith TG, Jewell J, Wardle J, Hammond RA, Friel S, Thow AM, Kain J (2015b) Smart food policies for obesity prevention. Lancet 385(9985):P2410–P2421. https://doi.org/10.1016/S0140-6736(14)61745-1. Accessed 08 Mar 2022

Henson S, Reardon T (2005) Private Agri-food standards: implications for food policy and the Agri-food system. Food Policy 30(3):241–253. https://doi.org/10.1016/j.foodpol.2005.05.002. Accessed 22 Apr 2022

Herforth A, Arimond M, Álvarez-Sánchez C, Coates J, Christianson K, Muehlhoff E (2019) A global review of food-based dietary guidelines. Adv Nutr 10(4):590–605. https://doi.org/10.1093/advances/nmy130. Accessed 05 Oct 2022

Hernandez R, Belton B, Reardon T, Hu C, Zhang X, Ahmed A (2018) The "quiet revolution" in the aquaculture value chain in Bangladesh. Aquaculture 493:456–468. https://doi.org/10.1016/j.aquaculture.2017.06.006. Accessed 17 Aug 2023

Hernandez R, Berdegué JA, Reardon T (2015) Modernizing wholesalers and guava farmers in Mexico. Agric Econ 46(S1):41–52. https://doi.org/10.1111/agec.12197. Accessed 01 Mar 2022

HLPE (2014) Food losses and waste in the context of sustainable food systems. A report by the High Level Panel of Experts on Food Security and Nutrition of the Committee on World Food Security, Rome. https://openknowledge.fao.org/server/api/core/bitstreams/b1949fae-23d4-473c-8b87-8c4359b74d6c/content. Accessed 25 Aug 2021

HLPE (2017) Nutrition and food systems. A report by the High Level Panel of Experts on Food Security and Nutrition of the Committee on World Food Security, Rome. https://openknowledge.fao.org/server/api/core/bitstreams/4ac1286e-eef3-4f1d-b5bd-d92f5d1ce738/content. Accessed 02 Sep 2021

HLPE (2020) Food security and nutrition: building a global narrative towards 2030. A report by the High Level Panel of Experts on Food Security and Nutrition of the Committee on World Food Security, Rome. https://openknowledge.fao.org/server/api/core/bitstreams/8357b6eb-8010-4254-814a-1493faaf4a93/content. Accessed 02 Sep 2021

IIFIIR (International Institute of Refrigeration) (2019) The role of refrigeration in the global economy. 38th informatory note on refrigeration technologies. Paris. https://iifiir.org/en/fridoc/the-role-of-refrigeration-in-the-global-economy-2019-142028. Accessed 14 Apr 2022

Imamura F, Micha R, Khatibzadeh S, Fahimi S, Shi P, Powles J, Mozaffarian D (2015) Dietary quality among men and women in 187 countries in 1990 and 2010: a systematic assessment. Lancet Global Health 3(3):E132–E142. https://doi.org/10.1016/S2214-109X(14)70381-X. Accessed 14 Mar 2022

IFPRI (International Food Policy Research Institute) (2015) Global Nutrition Report 2015: actions and accountability to advance nutrition and sustainable development. IFPRI, Washington, DC. https://globalnutritionreport.org/reports/2015-global-nutrition-report/. Accessed 13 Apr 2022

Jaffee S, Henson S, Unnevehr L, Grace D, Cassou E (2019) The safe food imperative: accelerating Progress in low-and middle-income countries. World Bank, Washington, DC. https://openknowledge.worldbank.org/entities/publication/83a34062-c6c8-5edf-857c-62a643c16bb8. Accessed 01 Feb 2022

Kalmpourtzidou A, Eilander A, Talsma EF (2020) Global vegetable intake and supply compared to recommendations: a systematic review. Nutrients 12(6):1558. https://doi.org/10.3390/nu12061558. Accessed 20 Dec 2023

Kennedy E, Reardon T (1994) Shift to non-traditional grains in the diets of east and West Africa: role of women's opportunity cost of time. Food Policy 19(1):45–56. https://doi.org/10.1016/0306-9192(94)90007-8. Accessed 09 Mar 2022

Kitinoja L (2013) Use of cold chains for reducing food losses in developing countries. PEF White Paper No. 13-03. The Postharvest Education Foundation

Latino LR, Pica-Ciamarra U, Wisser D (2020) Africa: the livestock revolution urbanizes. Glob Food Secur 26:100399. https://doi.org/10.1016/j.gfs.2020.100399. Accessed 19 Dec 2023

Liverpool-Tasie S, Reardon T, Sanou A, Ogunleye W, Ogunbayo I, Omonona BT (2017) The transformation of value chains in Africa: evidence from the first large survey of maize traders in Nigeria. Feed the future innovation lab for Food Security Policy Research Paper 91, East Lansing, Michigan State University. https://www.canr.msu.edu/resources/the-transformation-of-value-chains-in-africa-evidence-from-the-first-large-survey-of-maize-traders-in-nigeria-dec. Accessed 01 Mar 2022

Marshall Q, Fanzo J, Barrett CB, Jones AD, Herforth A, R ML (2021) Building a global food systems typology: a new tool for reducing complexity in food systems analysis. Front Sustain Food Syst 5:746512. https://doi.org/10.3389/fsufs.2021.746512. Accessed 02 Mar 2022

Mason-D'Croz D, Bogard JR, Sulser TB, Cenacchi N, Dunston S, Herrero M, Wiebe K (2019) Gaps between fruit and vegetable production, demand, and recommended consumption at global and national levels: an integrated modelling study. Lancet Planetary Health 3(7):E318–E329. https://doi.org/10.1016/S2542-5196(19)30095-6. Accessed 02 Sep 2021

Miller DD, Welch RM (2013) Food system strategies for preventing micronutrient malnutrition. Food Policy 42:115–128. https://doi.org/10.1016/j.foodpol.2013.06.008. Accessed 10 Mar 2022

Minten B, Murshid KAS, Reardon T (2013) Food quality changes and implications: evidence from the rice value chain of Bangladesh. World Dev 42:100–113. https://doi.org/10.1016/j.worlddev.2012.06.015. Accessed 01 Mar 2022

Minten B, Reardon T, Das Gupta S, Hu D, Murshid KAS (2016) Wastage in food value chains in developing countries: evidence from the potato sector in Asia. In: Schmitz A, Kennedy PL, Schmitz TG (eds) Frontiers of economics and globalization. Emerald Group Publishing Limited, Bingley, pp 225–238. https://www.emerald.com/insight/content/doi/10.1108/S1574-871520150000016010/full/html. Accessed 19 Jan 2022

Monteiro CA, Cannon G, Moubarac J-C, Levy RB, Louzada MLC, Jaime PC (2018) The UN Decade of Nutrition, the NOVA food classification and the trouble with ultra-processing. Public Health Nutr 21(S1):5–17. https://doi.org/10.1017/S1368980017000234. Accessed 20 Sep 2021

Naylor RL, Hardy RW, Buschmann AH, Bush SR, Cao L, Klinger DH, Little DC, Lubchenco J, Shumway SE, Troell M (2021a) A 20-year retrospective review of global aquaculture. Nature 591:551–563. https://doi.org/10.1038/s41586-021-03308-6. Accessed 10 Aug 2023

Naylor RL, Kishore A, Sumaila UR, Issifu I, Hunter BP, Belton B, Bush SR, Cao L, Gelcich S, Gephart JA, Golden CD, Jonell M, Koehn JZ, Little DC, Thilsted SH, Tigchelaar M, Crona B (2021b) Blue food demand across geographic and temporal scales. Nat Commun 12:5413. https://doi.org/10.1038/s41467-021-25516-4. Accessed 11 Aug 2023

OECD/FAO (2023) OECD–FAO Agricultural Outlook 2023–2032. OECD Publishing, Paris. https://www.oecd-ilibrary.org/agriculture-and-food/oecd-fao-agricultural-outlook-2023-2032_08801ab7-en. Accessed 05 Dec 2023

Ohri-Vachaspati P, Leviton LC (2010) Measuring food environments: a guide to available instruments. Am J Health Promot 24(6):410–426. https://doi.org/10.4278/ajhp.080909-LIT-190. Accessed 08 Mar 2022

Pandey B, Reba M, Joshi PK, Seto KC (2020) Urbanization and food consumption in India. Sci Rep 10:17241. https://doi.org/10.1038/s41598-020-73313-8. Accessed 09 Mar 2022

Pingali P (2007) Westernization of Asian diets and the transformation of food systems: implications for research and policy. Food Policy 32(3):281–298. https://doi.org/10.1016/j.foodpol.2006.08.001. Accessed 25 Aug 2021

Pingali P (2010) Agriculture renaissance. Making "agriculture for development" work in the 21st century. In: Pingali P, Evenson R (eds) Handbook of agricultural economics, vol 4. Elsevier, Amsterdam, pp 3867–3894

Pingali P, Aiyar A, Abraham M, Rahman A (2019) Transforming food Systems for a Rising India. Springer, Cham. https://doi.org/10.1007/978-3-030-14409-8. Accessed 08 Sep 2021

Pingali P, Ricketts K, Sahn DE (2015) Agriculture for nutrition: getting policies right. In: Sahn DE (ed) The fight against hunger and malnutrition. Oxford, Oxford University Press, pp 165–191. Accessed 14 Mar 2022

Pingali PL, Rosegrant MW (1995) Agricultural commercialization and diversification: processes and policies. Food Policy 20(3):171–185. https://doi.org/10.1016/0306-9192(95)00012-4. Accessed 25 Feb 2022

Pounds A, Kaminski AM, Budhathoki M, Gudbrandsen O, Kok B, Horn S, Malcorps W, Mamun A-A, McGoohan A, Newton R, Ozretich R, Little DC (2022) More than fish—framing aquatic animals within sustainable food systems. Foods 11(10):1413. https://doi.org/10.3390/foods11101413. Accessed 18 Aug 2023

Reardon T (2015) The hidden middle: the quiet revolution in the midstream of agrifood value chains in developing countries. Oxford Rev Econ Pol 31:45–63. https://doi.org/10.1093/oxrep/grv011. Accessed 15 Apr 2022

Reardon T (2016) Growing food for growing cities: transforming food systems in an urbanizing world. Chicago Council on Global Affairs, Chicago. https://globalaffairs.org/sites/default/files/2021-02/report_growing-food-for-growing-cities.pdf. Accessed 17 Feb 2022

Reardon T, Liverpool-Tasie LSO, Minten B (2021a) Quiet revolution by SMEs in the midstream of value chains in developing regions: wholesale markets, wholesalers, logistics, and processing. Food Secur 13:1577–1594. https://doi.org/10.1007/s12571-021-01224-1. Accessed 07 Dec 2021

Reardon T, Timmer CP (2014) Five inter-linked transformations in the Asian agrifood economy: food security implications. Glob Food Sec 3:108–117. https://doi.org/10.1016/j.gfs.2014.02.001. Accessed 06 Jan 2022

Reardon T, Timmer CP, Barrett CB, Berdegué J (2003) The rise of supermarkets in Africa, Asia, and Latin America. Am J Agr Econ 85(5):1140–1146. https://www.jstor.org/stable/1244885. Accessed 07 Dec 2021

Reardon T, Timmer CP, Minten B (2012) Supermarket revolution in Asia and emerging development strategies to include small farmers. P Natl Acad Sci USA 109:12332–12337. https://doi.org/10.1073/pnas.1003160108. Accessed 20 Apr 2022

Reardon T, Tschirley D, Liverpool-Tasie LSO, Awokuse T, Fanzo J, Minten B, Vos R, Dolislager M, Sauer C, Dhar R, Vargas C, Lartey A, Raza A, Popkin BM (2021b) The processed food revolution in African food systems and the double burden of malnutrition. Glob Food Secur 28:100466. https://doi.org/10.1016/j.gfs.2020.100466. Accessed 01 Mar 2022

Rico-Sole R (2012) Economic impact of mycotoxins in nuts and dried fruit chain. Acta Hortic 963:155–172. https://doi.org/10.17660/ActaHortic.2012.963.26. Accessed 10 Apr 2022

Schneider KR, Fanzo J, Haddad L, Herrero M, Moncayo JR, Herforth A, Remans R, Guarin A, Resnick D, Covic N, Béné C, Cattaneo A, Aburto N, Ambikapathi R, Aytekin D, Barquera S, Battersby J, Beal T, Molina PB, Cafiero C, Campeau C, Caron P, Conforti P, Damerau K, Di Girolamo M, DeClerck F, Dewi D, Elouafi

I, Fabi C, Foley P, Frazier TJ, Gephart J, Golden C, Fischer CG, Hendriks S, Honorati M, Huang J, Kennedy G, Laar A, Lal R, Lidder P, Loken B, Marshall Q, Masuda YJ, McLaren R, Miachon L, Muñoz H, Nordhagen S, Qayyum N, Saisana M, Suhardiman D, Sumaila UR, Cullen MT, Tubiello FN, Vivero-Pol J-L, Webb P, Wiebe K (2023) The state of food systems worldwide in the countdown to 2030. Nat Food 4:1090–1110. https://doi.org/10.1038/s43016-023-00885-9. Accessed 16 Jan 2024

Senauer B, Sahn D, Alderman H (1986) The effect of the value of time on food consumption patterns in developing countries: evidence from Sri Lanka. Am J Agr Econ 68(4):920–927. https://doi.org/10.2307/1242138. Accessed 14 Mar 2022

Siegel KR, Ali MK, Srinivasiah A, Nugent RA, Venkat Narayan KM (2014) Do we produce enough fruits and vegetables to meet global health need? PLoS One 9(8):e104059. https://doi.org/10.1371/journal.pone.0104059. Accessed 02 Sep 2021

Swinburn B, Dominick CH, Vandevijvere S (2014) Benchmarking food environments: experts' assessments of policy gaps and priorities for the New Zealand government. University of Auckland, Auckland. https://www.fmhs.auckland.ac.nz/assets/fmhs/soph/globalhealth/informas/docs/Full%20Food-EPI%20report1.pdf. Accessed 08 Mar 2022

Tan KC (2001) A framework of supply chain management literature. Eur J Purch Supply Manag 7(1):39–48. https://doi.org/10.1016/S0969-7012(00)00020-4. Accessed 15 Dec 2021

Thornton PK (2010) Livestock production: recent trends, future prospects. Philos T Roy Soc B 365(1554):2853–2867. https://doi.org/10.1098/rstb.2010.0134. Accessed 22 Dec 2023

Tidwell JH (2012) Characterization and categories of aquaculture production systems. In: Tidwell JH (ed) Aquaculture production systems, 1st edn. Wiley, Ames, pp 64–78. https://onlinelibrary.wiley.com/doi/10.1002/9781118250105.ch4. Accessed 22 Dec 2023

Timmer CP (2009) A world without agriculture? The historical paradox of agricultural development. Development Policy Outlook No. 1, May. American Enterprise Institute for Public Policy Research, Washington, DC. Available at SSRN: https://www.aei.org/wp-content/uploads/2011/10/01%20DPO%20May%202009g.pdf. Accessed 24 Feb 2022

Timmer CP, Akkus S (2008) The structural transformation as a pathway out of poverty: analytics, empirics and politics. Center for Global Development Working Paper No. 150, Washington, DC. https://doi.org/10.2139/ssrn.1213154. Accessed 25 Aug 2021

Tonumaipe'a D, Cammock R, Conn C (2021) Food havens not swamps: a strength-based approach to sustainable food environments. Health Promot Int 36(6):1795–1805. https://doi.org/10.1093/heapro/daab021. Accessed 08 Mar 2022

Troell M, Jonell M, Crona B (2019) The role of seafood in sustainable and healthy diets. The EAT-Lancet Commission report through a blue lens. Stockholm Resilience Center, Stockholm University. https://eatforum.org/content/uploads/2019/11/Seafood_Scoping_Report_EAT-Lancet.pdf. Accessed 21 Aug 2023

UN (United Nations) (2018) 2018 revision of world urbanization prospects. Department of Economic and Social Affairs, Population Division, United Nations, New York. https://population.un.org/wup/. Accessed 10 Dec 2024

UN (United Nations) (2019) World urbanization prospects: the 2018 revision (ST/ESA/SER.A/420), Department of Economic and Social Affairs, Population Division, United Nations, New York. https://www.un-ilibrary.org/content/books/9789210043144#chapters. Accessed 19 Oct 2021

Willett W, Rockström J, Loken B, Springmann M, Lang T, Vermeulen S, Garnett T, Tilman D, DeClerck F, Wood A, Jonell M, Clark M, Gordon LJ, Fanzo J, Hawkes C, Zurayk R, Rivera JA, De Vries W, Majele Sibanda L, Afshin A, Chaudhary A, Herrero M, Agustina R, Branca F, Lartey A, Fan S, Crona B, Fox E, Bignet V, Troell M, Lindahl T, Singh S, Cornell SE, Srinath Reddy K, Narain S, Nishtar S, CJL M (2019) Food in the Anthropocene: the EAT–*Lancet* Commission on healthy diets from sustainable food systems. Lancet 393:447–492. https://doi.org/10.1016/S0140-6736(18)31788-4. Accessed 20 Aug 2021

Zhou Y, Staatz J (2016) Projected demand and supply for various foods in West Africa: implications for investments and food policy. Food Policy 61:198–212. https://doi.org/10.1016/j.foodpol.2016.04.002. Accessed 12 Apr 2022

A Globally Applicable Definition: A Framework for Food Loss and Waste

3

3.1 Introduction

In 2011, a study carried out by the Food and Agriculture Organization of the United Nations (FAO) reported that an estimated that one-third of all food produced for human consumption was lost or wasted globally (FAO 2011). Over 10 years later, United Nations (UN) agencies started reporting updated estimates. FAO estimated that 13.8% of food produced in 2016 was lost from the farm up to, but not including, before the retail stage (FAO 2019). The UN Environmental Programme (UNEP) estimated that 19% of food that reaches consumers is lost at the retailer, food service, or household levels (UNEP 2024). Each of these estimates are based on different food loss and waste (FLW) definitions. Therefore, the estimates are not comparable and, in the case of the most recent estimates from FAO and UNEP, cannot be combined into a complete value chain estimate from farm to consumer.

The first step to addressing FLW issues is to define the problem using consistent terminology for a common understanding of what is being measured, compared, and addressed. Unfortunately, there is no harmonized FLW definition. In 2014, the High Level Panel of Experts (HLPE) on Food Security and Nutrition recommended that stakeholders agree on a definition for FLW to improve data collection and knowledge sharing (HLPE 2014). At the same time, the FAO's Global Initiative on Food Loss and Waste Reduction sought to harmonize a global FLW definition and published a detailed definitional framework of food loss (FAO 2014).

Yet, there continue to be many, varied definitions and approaches to conceptualizing FLW. In this chapter, we summarize the current state of FLW definitions by examining major sources of FLW data and how FLW is defined according to an analytical framework specific to FLW definitions. We then explore definitional elements to identify opportunities for harmonization and present a comprehensive FLW pathways framework. We aim for a FLW pathways framework that is comprehensive and universally applicable, providing clarity and consistency as to how FLW is defined and how FLW impacts food systems outcomes and impacts.

3.2 Current State of Food Loss and Waste Definitions: Divergent and Inconsistent Approaches

Consistency in FLW definitions and terminology would allow for a common starting point from which FLW estimates can be compared and interpreted. However, the current lack of a harmonized FLW definition contributes to inconsistencies in terminology and data comparability, with terms used interchangeably, but not

J. Boiteau, P. Pingali, *Wasted Potential*, Sustainable Development Goals Series,
https://doi.org/10.1007/978 3 031 97411-3_3

referring to the same concepts (FAO 2014; Delgado et al. 2017; Kitinoja et al. 2018).

Food loss and waste may be sharply distinguished into *food loss* and *food waste*, based on the supply chain stage or segment where loss occurs (Parfitt et al. 2010; HLPE 2014; FAO 2019; Fabi et al. 2021; UNEP 2024). In contrast, *food loss and waste* may be used as a single term, with food waste considered to be an important part of food loss but with different underlying reasons for wasting food compared to unintended food loss (Lipinski et al. 2013; FAO 2014). *Food waste* can also refer to losses along the entire supply chain, from farm to consumer (Östergren et al. 2014; Bellemare et al. 2017).

Other terms have been proposed to be broad in scope, such as *potential food loss and waste* (Schuster and Torero 2016), or more specific to a supply chain stage—*preharvest loss, farm loss and waste, postharvest loss and waste,* or *consumer waste* (Xue et al. 2017; Kitinoja et al. 2018). Finally, *surplus food* may refer to food diverted to nonfood uses (ReFED 2021), whereas *food surplus* may refer to food that is redistributed for human consumption, used for animal feed, or used for bio-based materials or biochemical processing (UNEP 2024).

A critical barrier to arriving at a harmonized FLW definition is the overall recognition that FLW definitions reflect how stakeholders perceive the problem and their objectives for FLW measurement and management (Chaboud 2017). Adopting a definition is often seen as the first step in developing a measurement framework with FLW, with separate FLW definitions for individual objectives (Koester 2017; Fabi et al. 2021). Yet, different approaches to defining FLW contribute to the lack of clarity in exactly what the FLW data represents, limiting the interpretation and comparability of results.

Food loss and waste is typically approached by two different perspectives: a waste perspective that focuses on environmental sustainability objectives, and a food perspective that focuses on food security and nutrition objectives (HLPE 2014; FAO 2019). Ideally, a harmonized FLW definition is independent of stakeholder perspectives. Stakeholder objectives for FLW measurement and prevention are most useful in driving measurement and intervention frameworks.

Food loss and waste definitions are based on five major categories: (1) food value chain boundaries; (2) edibility and intended use; (3) quality dimensions, including nutrition, aesthetics, and shelf-life; (4) nature of food use (for example, nonfood use, productive use, etc.); and (5) food destination (Chauhan et al. 2021). An analytical framework presented by Chaboud and Daviron (2017) identifies key FLW definitional elements: timing, scope, terminology, criterion, perspective, and type. The framework is useful for comparing existing FLW definitions to understand where definitions agree or disagree and to identify challenges with comparing or aggregating loss estimates.

3.2.1 Major Food Loss and Waste Data Sources and Definitional Frameworks

To help track the state of FLW, FAO developed the Food Loss and Waste database, an online collection of FLW data across food products, value chain stages, and geographic areas (FAO 2022). The text-mining machine learning tool, used to identify relevant sources and extract and organize information, did not differentiate FLW definitions with respect to destinations and use of diverted food (Fabi et al. 2021). The database has sourced over 20,000 data points from over 480 openly accessible publications and reports (FAO 2022). Based on the dataset, 65% of FLW data come from the African Postharvest Losses Information System (APHLIS) online database, which is specific to FLW of cereal grains in sub-Saharan Africa (Fig. 3.1).

The major FLW data sources cover different geographic areas. Therefore, differences in how FLW is defined are associated with specific geographic areas. Most FLW data (75%) come from low-income and lower middle-income countries, following the APHLIS focus on sub-Saharan Africa (Table 3.1). Nearly all (96%) of data for low-income countries come from APHLIS. In lower middle-income countries, 74% of FLW

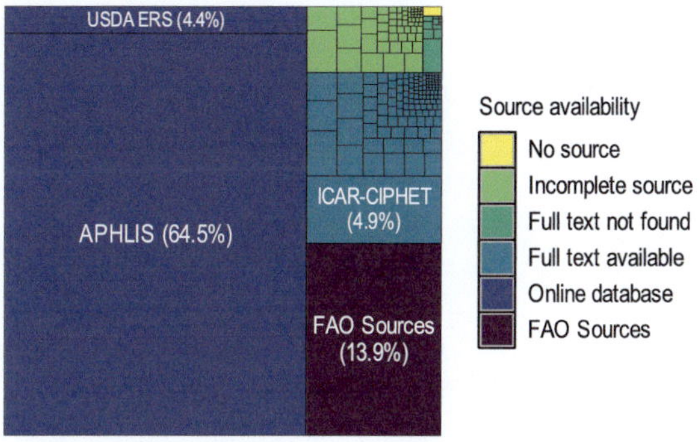

Fig. 3.1 Summary of data sources and source availability across food loss and waste data points (n = 21,786) reported in Food Loss and Waste Database from 2004 to 2021

Data source availability is indicated by colors, as shown in the legend. Each rectangle outlined in solid black lines represents a unique data source. The size of each rectangle corresponds to the share of FLW data that is contributed to the dataset. Data sources that contribute at least 4% of the data are indicated with white text and the percentage of data contributed shown in parentheses

No source, n = 21 (0.1%); Incomplete source, n = 896 (4.1%); Full text not found, n = 124 (0.6%); Full text available, n = 2696 (12.4%); Online database, n = 15,014 (68.9%); FAO Sources, n = 3035 (13.9%)

Abbreviations: APHLIS, African Postharvest Losses Information System; FAO, Food and Agriculture Organization of the United Nations; USDA ERS, United States Department of Agriculture's Economic Research Service; ICAR–CIPHET, Indian Council of Agricultural Research–Central Institute of Post-Harvest Engineering and Technology

Data from: FAO (2022), accessed 8 Dec 2021. Figure published in Boiteau and Pingali (2023)

Table 3.1 Data sources across all food loss and waste data points by country income group[a]

Data source	Low-income n (%)	Lower middle-income n (%)	Upper middle-income n (%)	High-income n (%)
APHLIS	9233 (95.7)	3945 (58.4)	884 (41.9)	0 (0.0)
FAO sources	83 (0.9)	319 (4.7)	994 (47.1)	1639 (50.2)
USDA ERS	0 (0.0)	0 (0.0)	0 (0.0)	952 (29.2)
ICAR–CIPHET	0 (0.0)	1063 (15.7)	0 (0.0)	0 (0.0)
Other, full text	192 (2.0)	809 (12.0)	147 (7.0)	485 (14.9)
Incomplete/no source	142 (1.5)	624 (9.2)	87 (4.1)	188 (5.8)

Data from FAO (2022), accessed on 08 Dec 2021
Note:
Abbreviations: APHLIS, African Postharvest Losses Information System; FAO, Food and Agriculture Organization of the United Nations; USDA ERS, United States Department of Agriculture's Economic Research Service; ICAR–CIPHET, Indian Council of Agricultural Research–Central Institute of Post-Harvest Engineering and Technology
[a] 2018 World Bank Income Classifications

data come from APHLIS and the Indian Council on Agricultural Research—Central Institute of Post-Harvest Engineering & Technology (ICAR–CIPHET) combined (still dominated by APHLIS). FAO sources become common FLW data sources in upper middle-income and high-income countries, contributing 47% and 50% of the data in each group, respectively. In high-income countries, 29% of data is sourced from the US Department of Agriculture Economic Research Service (USDA ERS).

3.2.2 Comparison of Definitional Elements Across Sources Relevant to Sustainable Development Goal Target 12.3

In 2014, the FAO's Global Initiative on Food Loss and Waste Reduction sought to harmonize a global FLW definition (FAO 2014). Using this FLW definitional framework as a starting point, Table 3.2 summarizes the comparison of cross-definitional elements for major FLW data sources

Table 3.2 Comparison of food loss and waste definitions[a]

Definitional element		FAO (2014)[b]	APHLIS[c]	USDA ERS[d]	ICAR–CIPHET[e]	Food Loss Index[f]	Food Waste Index[g]
Perspective		Food security	Food security, environmental, economic	Food security, environmental, economic	Food security, economic	Food security, environmental, economic	Food security, environmental, economic
Timing		Ready for harvest or slaughter[h]	Ready for harvest	Postharvest	Start of harvest or slaughter operations	Ready for harvest or slaughter[h]	Not specified
Scope		Food value chain: Intended for human consumption[i]	Human food value chain	Food available for human consumption	Human food value chain	Food value chain: Human-edible commodity	Food value chain: Intended for human consumption
Terminology:	Terms	Food loss and waste[j]	Post-harvest loss	Food loss; food waste	Harvest and post-harvest loss	Food loss	Food waste
	Stages	Ready for harvest or slaughter to the consumer.	Physiological maturity in the field to post-harvest activities from field to the consumer.	Food waste is part of food loss, occurs when food is discarded and unconsumed at retailer and consumer stages.	Farm operations, to market channels to kitchen and plate/table.	On-farm, postharvest up to, but not including, retail	Retail, food service (i.e., out-of-home consumption), household
Criterion:	Utilization	Any nonfood uses	Any nonfood uses	Any nonfood uses	Any nonfood uses	Limited nonfood uses[k]	Limited nonfood uses[l]
	Edibility	Edible portion	Not specified	Edible portion	Edible portion	Edible and inedible portion	Edible and inedible portion
Type		Quantitative and qualitative	Quantitative and qualitative	Quantitative	Quantitative and qualitative	Quantitative	Quantitative

Notes
[a]Adapted from Redlingshöfer (2015); Chaboud and Daviron (2017). Published in Boiteau and Pingali (2023)
[b]FAO (2014)
[c]Hodges (2012, 2013)

[d]Buzby et al. (2014)

[e]ICAR/ICAR–CIPHET (Nanda et al. 2012; Jha et al. 2015)

[f]Fabi and English (2019); FAO (2019)

[g]UNEP (2024)

[h]Examples of "Ready for harvest or slaughter" include crops that are harvest-mature or suitable to their purpose; animals that are ready for slaughter; milk that has been drawn from the udder; eggs laid by the bird; aquaculture fish that are mature in the pond; wild crops that are harvested; wild animals and fish that have been caught

[i]Food intended for human consumption depends on the food value chain, food system, and geographical and cultural contexts. Recommendations exist for assessing food loss if, at early stages of the supply chain, the intended use of the product is unknown

[j]Food waste is the removal of food, fit-for consumption, by choice or food spoilage due to negligence. Food waste is a part of food loss due to differences in the underlying reasons and motivations for wasting food, as compared to unintentional food loss

[k]Specific FLW destinations include: co/anaerobic digestion, incineration, landfill, discard, compost, or other waste use

[l]Waste destinations reported to the Food Waste Index include co/anaerobic digestion, compost/aerobic, controlled combustion, land application, landfill, refuse/discard, and sewer. Animal feed and biomaterial/processing are non-waste "surplus" destinations that are not reported to the FWI

Abbreviations: APHLIS, African Postharvest Loss Information System; FAO, Food and Agriculture Organization of the United Nations; FLW, Food Loss and Waste; FWI, Food Waste Index, USDA ERS, United States Department of Agriculture's Economic Research Service; ICAR–CIPHET, Indian Council of Agriculture Research- Central Institute of Post-Harvest Engineering & Technology

(APHLIS, USDA ERS, and ICAR–CIPHET), as well as the definitional frameworks for the current Sustainable Development Goal (SDG) Target 12.3 indicators, the Food Loss Index (FLI) and Food Waste Index (FWI).

APHLIS and FAO (2014) definitional frameworks both consider quantitative and qualitative FLW from the time that products are ready for harvest or slaughter to the time that food is either consumed or is removed from the food value chain. Food that entered the food value chain and was diverted to nonfood uses is considered to be FLW (Hodges 2013; FAO 2014).

The USDA ERS defines FLW as the edible portion of food starting at the postharvest stage (Buzby et al. 2014). Food diverted away from the supply chain to a nonfood use is considered to be FLW. Food waste is a part of food loss, and specifically, covers loss at the retail and consumer stages. The USDA ERS definition only considers quantitative losses (Buzby et al. 2014).

The ICAR–CIPHET survey (Nanda et al. 2012; Jha et al. 2015) indicates that food losses occur from the start of harvest operations through processing and marketing stages. Quantitative losses are defined as reductions in the weight of crops or commodities available for human consumption; thus, the scope is considered to be the human food value chain and criterion for loss are nonfood uses. Although the survey does not measure FLW beyond the marketing stages, the definitional framework recognizes losses at the consumer stage (that is, kitchen loss and plate/table loss). The ICAR–CIPHET approach also acknowledges quality deterioration and food value loss, aligning with qualitative types of FLW (Nanda et al. 2012; Jha et al. 2015).

At the time that SDG target 12.3 was included in the Global Indicator Framework, the target 12.3 indicator was classified as a Tier III indicator, meaning the measurement methodology needed to be developed and validated (Fabi and English 2019). After determining that the measurement methods would differ for supply-oriented and demand-oriented food chain stages, two sub-indicators were developed, the Food Loss Index (FLI) and the Food Waste Index (FWI), led by FAO and the United Nations Environment Programme (UNEP), respectively (Fabi and English 2019; UNEP 2021). The FLW definition used in the FLI builds on the FAO 2014 definitional framework (FAO 2014; Fabi and English 2019). While the conceptual framework strives for completeness, the FLI uses an operational framework, which is a modified FLW definition, aiming to produce consistent measurement (Fabi and English 2019). The FLW definition used in the FWI uses the Food Loss & Waste Protocol Accounting and Reporting Standard (FLW Standard) to specify the definitional elements (Hanson et al. 2016; UNEP 2021).

The FLI and FWI both take broad perspectives and associate FLW with food security, environmental, and economic outcomes. Using the operational definition, the FLI defines the scope as human-edible commodities (Fabi and English 2019), whereas the FWI defines the scope as intended for human consumption (UNEP 2024). The FLI and FWI align with supply chain stage breakdown specified in SDG target 12.3, using the terms *food loss* and *food waste*, respectively. The FLI considers food loss at the on-farm, postharvest up to, but not including, retail levels. The FLI also clarifies that preharvest stage losses are excluded from the SDG 12.3 target indicator, noting that policies to improve supply chain efficiencies cannot address unpredictable extreme events and natural disasters (Fabi and English 2019). Instead, these preharvest losses are captured in SDG indicator 1.5.2—direct agricultural losses attributed to disasters. Preharvest losses that are not due to extreme events or natural disasters remain, in effect, undefined and uncounted. The FLI and FWI both consider quantitative losses based on nonfood, noneconomically productive loss criteria. Destinations including animal feed and biomaterial and processing are beyond the scope of food loss and food waste. Unlike other definitions, the FLI and FWI consider the commodity as a whole and include both edible and inedible portions.

3.3 Clarifying Food Loss and Waste Pathways Using a Practical and Globally Applicable Framework

The FLW pathways framework proposed for this book and illustrated in Fig. 3.2 are based on the FLW definition published by the FAO's Global Initiative on Food Loss and Waste Reduction (FAO 2014), and summarized as: *Food loss and waste* is a reduction in the quantity or quality of the edible portion of food intended for human consumption when food is redirected to nonfood uses or when there is a decrease in the nutritional value, food safety, or other quality aspect from the time food is ready for harvest or slaughter to consumption.

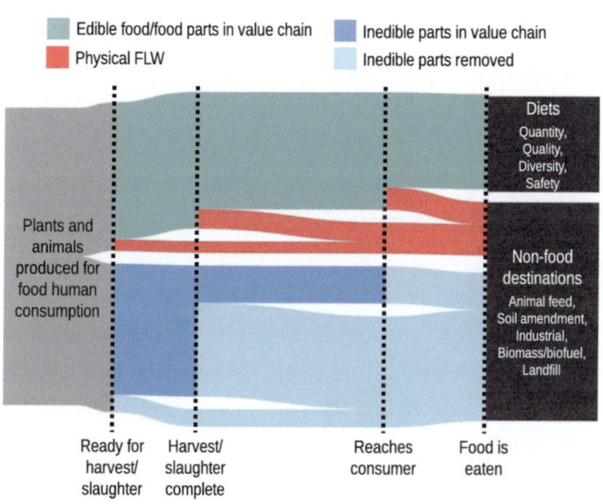

Fig. 3.2 Food loss and waste conceptual framework
FLW pathways begin after food production, once food intended for human consumption reaches market maturity and is ready for harvest or slaughter. As edible food moves through the value chain, FLW pathways involve FQL of edible food/food parts (green) and pFLW events (red). FLW pathways end when food is eaten, or when it is diverted to nonfood destinations. Inedible food parts (blue) that enter food value chains are directed to nonfood destinations at different points along the value chain (light blue). The proportion of edible to inedible parts and length and width of the pathways are illustrative, as each pathway will depend on the food commodity and value chain context

Based on the definition, we distinguish edible foods and food parts from inedible food parts (Fig. 3.2). Edible foods are intended to eventually be eaten by humans, thus contributing to diets. In contrast, inedible foods are directed to nonfood uses at a different point on the value chain. We use the overarching term FLW, which can occur at any point along the edible food value chain within a defined boundary (described in this section). To further distinguish the two types of FLW, (1) food redirected to nonfood use and (2) a decrease in quality aspects, we use the terms *physical food loss and waste (pFLW)* and *food quality loss (FQL)*, respectively. Acceptability thresholds of limiting food attributes connect FQL and pFLW (discussed in the next section).

FLW occurs within the context of food systems. Therefore, FLW pathways are affected by the same categories of drivers of food system changes: biophysical and environmental; innovation, technology, and infrastructure; political and economic, sociocultural, and demographic drivers (Ingram 2011; HLPE 2017). Since the structure of food value chains depend on the type of food system and food commodity, common stages anchor the general value chain flow, as depicted in Fig. 3.2: when food is ready for harvest, has been harvested, reaches consumers, and is eaten. The relative impact of drivers of food systems changes on FLW and food system outcomes depends on the food system context, type of FLW, and pathways. Food loss and waste pathways and drivers are discussed in more detail in Part II of this book.

This section focuses on the core elements of FLW pathways: pathway boundaries; food destinations; and food quality and acceptability thresholds. These elements are influenced by the food systems within which FLW pathways are embedded and contribute to final nutrition, health, environmental, economic, and social outcomes of diets and pFLW.

3.3.1 Food Loss and Waste Boundaries

The boundaries of FLW pathways denote which agricultural products are considered within the scope of FLW and at what points along food value chains FLW can occur. Agricultural products considered for FLW are those originally intended for human consumption, following the Codex Alimentarius definition of "food" (FAO and WHO 2019). Human-inedible agricultural products include nonfood parts that are not intended for human consumption and, therefore, fall outside the scope of FLW (FAO 2014). Human-edible agricultural products originally intended for nonfood uses (for example, animal feed, industrial uses) also fall outside the scope of FLW.

3.3.1.1 Food Intended for Human Consumption

Worldwide, on a mass basis, there are differences in the use allocation of different agricultural products according to food group (Fig. 3.3). Using data from FAO Food Balance Sheets, we divide the amount of food, by group, allocated to human food or animal feed by the sum of all productive uses for that food group (specifically, human food, animal feed, and other nonfood uses). Over 90% of fruits, vegetables, and animal-source foods (ASF) produced are allocated to human food value chains. In contrast, cereal grains and, to a small extent, starchy roots, currently have between 20–40% of production allocated to animal feed.

Early in the value chain, the intended use of human-edible agricultural products may yet to be determined. This is most relevant for cereals and, to a lesser extent, starchy roots (Fig. 3.3). Statistical information on the fraction of the product which, in a specific region and year, finally enters human food value chains can be used to determine the share of product that falls within the scope of FLW pathways (FAO 2014; Hanson et al. 2016). The decline in share of global supply of cereals allocated for human food and rising share allocated to animal feed coincides with diet transitions away from staple grains and increased dietary diversity, including ASF (Fig. 3.3).

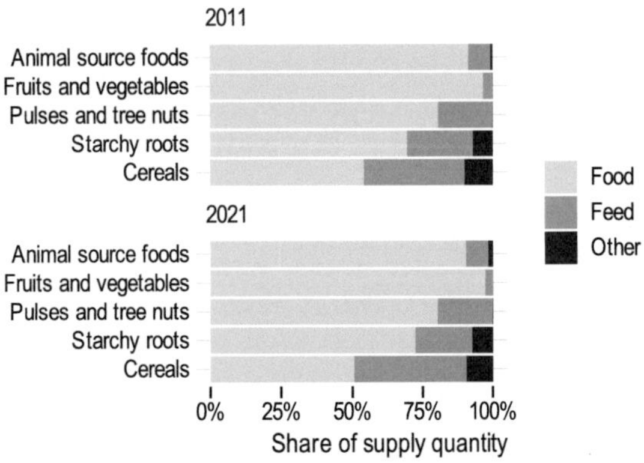

Fig. 3.3 Global share of commodity supply allocated to human food or animal feed, by food group (2011 and 2021)
Source: FAO Food Balance Sheets

3.3.1.2 Edible Foods and Food Parts

Conceptually, edibility is based on technical and cultural norms. Technical edibility refers to whether the food part is safe to eat. Cultural edibility refers to whether the food part is acceptable to eat in a given geography or culture (Muth et al. 2018; Nicholes et al. 2019; Moreno et al. 2020). Whole foods can be composed of edible, potentially edible, and inedible food parts based on technical and cultural criteria. Moreno et al. (2020) explained that food parts that are never edible include those that are considered unsafe to eat (for example, rhubarb leaves) or lack obvious examples of edible uses (for example, egg shells). Food parts are potentially edible if they are safe to eat and there is some consumption precedent in a given context (for example, melon rinds) (Moreno et al. 2020).

The composition of animal feed includes human-inedible and human-edible products (Mottet et al. 2017; Mottet and Tempio 2017). For example, in 2010, the majority of global livestock feed rations were composed of human-inedible products, including grass and leaves (46% of feed material) and crop residues (19% of feed material), compared to 14% of the feed ration from human-edible feed material (Mottet et al. 2017). In the same year, over half of global poultry feed rations were composed of human-edible material (58% of feed rations were from cereal grain) (Mottet and Tempio 2017).

3.3.1.3 Points along the Value Chain

The timing of when FLW is considered, up to when food is either consumed or removed from the human food value chain, marks the starting and ending boundaries of FLW pathways. The starting point of FLW pathways begins on the farm, when plants and animals intended for food reach market maturity and are ready for harvest or slaughter (Hodges 2013; Östergren et al. 2014; FAO 2014; Hanson et al. 2016; Delgado et al. 2017; Fabi and English 2019), setting FLW apart from production losses and yield gaps. The timing for some ASF begins at clear, specific moments, including when milk is drawn from the udder, eggs are laid, or wild fish or game are caught (FAO 2014). In contrast, the timing for crops and other ASF begins when crops or animals are considered ready for harvest or slaughter/catch (FAO 2014), which can be value chain-specific. Because individual stages are specific to each food value chain, the FLW pathways framework is anchored at three key moments once food is ready for harvest: when harvest is complete, when food reaches consumers, and when food is eaten. Therefore, FLW pathways cover all stages at the farmgate and post-farmgate (from preharvest through marketing), as well as the final consumer stage.

3.3.1.4 Worked Examples of Food Loss and Waste Pathways

Comparing Tomato and Rice: To illustrate the application of the FLW pathways framework to examine food-focused and waste-focused sustainability objectives, Fig. 3.4 provides a worked example of fresh tomatoes and rice, which are produced for human consumption and separated by edible food and inedible plant parts. Indicators of harvest maturity of both products are based on the edible food part (tomato fruit and rice grain, respectively). Inedible plant parts are removed from the food value chain at different value chain stages, requiring different management strategies. Physical FLW along each value chain reduces the final amount of edible food product that is eaten by consumers.

Using a hypothetical example of 100 kg edible food ready for harvest in India, data and

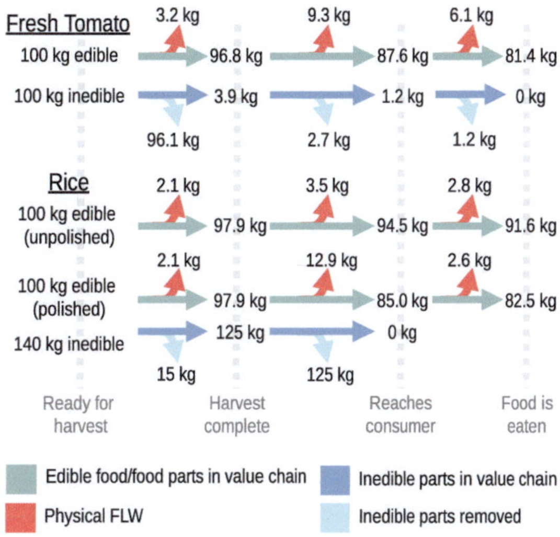

Fig. 3.4 Worked examples of simplified food loss and waste pathways: Fresh tomato and rice in India

Green straight arrows represent edible food parts remaining in the food value chain. Red curved arrows indicate physical FLW of edible food parts. Blue straight arrows indicate inedible food parts remaining in the food value chain. Light blue curved arrows indicate inedible food parts removed from the value chain. Following a hypothetical 100-kg edible food ready for harvest, the share of pFLW up to the point of "reaches consumer" is based on data from Jha et al. (2015). Physical FLW from "reaches consumer" to "food is eaten" stages are based on FAO (2011)

assumptions are applied for each FLW pathway described in Fig. 3.4. We use national-level data from India showing pFLW at harvest and at postharvest operations through retail storage channels for each food (Jha et al. 2015). There is a scarcity of consumer-level FLW data (Agarwal et al. 2021), and, to our knowledge, no recent study disaggregates consumer-level FLW by edible portion and food group (UNEP 2024). We therefore apply the FAO (2011) estimates of FLW to calculate the pFLW between when food reaches the consumer and when food is consumed.

Tomato fruit are the edible parts of a tomato plant. National-level data from India indicate pFLW of 3.2% at harvest and 9.3% at postharvest operations through retail storage channels (Jha et al. 2015). At the consumption end, pFLW of perishable fruits and vegetable are estimated at 7% (FAO 2011).

The majority of the inedible tomato plant parts remain on the field, unharvested with the edible part. Tomato plants each produce multiple tomatoes that grow and ripen at different times throughout a growing season. Therefore, each tomato plant is harvested several times and left intact after each harvest. Proper harvesting technique indicates that tomatoes should be harvested with the inedible pedicel and sepal attached (the small, secondary stalk holding the individual tomato along with modified leaves) (Naika et al. 2005). Postharvest and until tomatoes reach consumers, the pedicel and sepal plant parts may fall off or be intentionally removed. Otherwise, consumers remove these inedible parts prior to consumption.

Estimating the mass of inedible tomato plant parts required to produce tomatoes requires data on the mass of tomato plants after all tomatoes have been harvested; to our knowledge, this figure is not documented. Tomato plant growth and productivity are based on several factors, including the planting geometry, density, and fertilizer use (Mali et al. 2016). For illustrative purposes, we assume the ratio of total tomato fruit to one tomato plant is 1:1 by mass and that 3.9% of the inedible plant part is composed of the pedicel and sepal.

Rice grains (composed of germ, bran, and endosperm) are the edible food parts of the rice plant (also referred to as paddy). National-level data from India indicate a pFLW of 2.1% at harvest and 3.5% at postharvest operations through retail storage channels (Jha et al. 2015). Before reaching consumers, rice grains may be refined into white rice. When rice is polished into white rice, usually 10% by weight of edible portion (bran and germ) is removed during milling and is considered as pFLW (Juliano and Tuaño 2019). At the consumption end, pFLW of cereals in this region are estimated at 7% (FAO 2011).

The majority of inedible rice plant parts are harvested with the rice grains. Rice plants produce multiple grains of rice on each plant. The entire rice plant is harvested once rice grains reach harvest maturity. Rice is harvested by cutting the mature top part of the plant (containing the edible rice grains) as well as the straw above the ground (IRRI 2024). Some inedible plant parts are left in the ground, below where the cut was made for harvest. Postharvest, the rice grain is separated from the rest of the cut crop and cleaned to remove inedible material, including the grain husk (IRRI 2024). We assume a 1:1.05 ratio of rice grain to rice straw harvested (IRRI 2024) and a 4:1 ratio of rice grain to husk (Juliano and Tuaño 2019). For illustrative purposes, we also assume that for every 105 kg of rice straw harvested, 20 kg of inedible rice stubble is left unharvested. In-field rice straw management and off-field rice straw management have different environmental implications (IRRI 2024).

Box 3.1 Distinguishing Edible and Inedible Parts to Meet Sustainable Food System Objectives

The FLW pathways framework (Fig. 3.2) identifies two distinct pathways for edible and inedible plant/animal parts, emphasizing that FLW only applies to the edible components intended for human consumption while recognizing the significance of managing inedible parts.

Inedible plant/animal parts, never intended to be eaten by definition, are removed at different value chain stages. For example, tomatoes are harvested multiple times per plant, with inedible parts like the pedicel and sepal attached to the harvested tomato fruit. In contrast, rice plants are harvested once, but most of the plant is cut, including edible grains and inedible straw. In both cases, inedible plant parts remain unharvested, the tomato plant and rice stubble. The environmental impact of inedible food parts depends on where and how they are handled throughout the food supply chain.

Although the edible portions are relatively easy to identify, the accurate estimation of the inedible parts requires data on factors like plant mass, harvest methods, and processing techniques, which may not be available.

3.3.2 Food Destinations Determine What Counts as Physical Food Loss and Waste

Once ready for harvest, food intended for food value chains is either consumed as food (by humans) or diverted to nonfood destinations. Food diverted to nonfood destinations may follow two general pathways: recovered for nonfood uses or disposed of to landfills, sewers, or litter/discards (Bellemare et al. 2017). There is lack of consensus as to which nonfood destinations are considered pFLW (Hodges 2013; Buzby et al. 2014; FAO 2014; Hanson et al. 2016; Bellemare et al. 2017; Fabi and English 2019).

Based on the proposed FLW pathways framework (Fig. 3.2), pFLW pathways originate from edible food without quality loss or edible food with quality loss and lead to a nonfood destination. Food may be diverted accidentally or intentionally, the latter based on acceptability thresholds, described in the next section of this chapter. Food and/or associated inedible parts that are removed from the food value chain may be transformed, recycled, or valorized into nonfood products, including animal feed, soil amendments, products for industrial uses, and biomass for energy (FAO 2014, 2019; Hanson et al. 2016).

Some of the nonfood products are used as inputs in agricultural production. Although diverted food, which is valorized into agricultural inputs, "reenters" the food production system, the diverted food is still considered pFLW, because there is no direct feedback loop within the boundary of FLW pathways. Keeping track of nonfood destinations is important for assessing outcome trade-offs between keeping healthy and safe food in human food value chains and valorizing diverted food to productive uses.

For example, unmarketable food and inedible food parts are sometimes used as valuable alternative inputs for animal feed. Globally, feed materials from harvest wastes, second-grade crops, or other food discards are minor feed materials (Mottet et al. 2017; Mottet and Tempio 2017). However, animals not only serve as a food source, when ASFs enter the human food value chain to supply essential macro- and micronutrients, but animals also contribute to agricultural productivity (for example, manure and draft power) and can be used as a source of income generation (Mottet et al. 2017).

Box 3.2 Using Nonfood Use Criterion to Distinguish Prevention Versus Valorization Trade-Offs

The criteria used to define FLW must also differentiate between food and nonfood uses to inform measurement frameworks for estimating FLW to inform stakeholders of design FLW regulations and increase efficiencies (Buzby et al. 2014).

For example, in an analysis of global protein supply gap, Guo (2021) found that, when possible, FLW prevention is preferred because feed applications have a low protein conversion ratio and, from a climate change perspective, animal production is not favorable. When FLW is unavoidable, the chicken feed application strategy is recommended when FLW is suitable to be used as feed (Guo 2021).

Dou et al. (2018) examined the cost–benefit of different management options for consumption-stage FLW and found that animal feed is the least costly option, compared to composting, biogas generation, incineration, and landfill use. Although incorporating feeds derived from consumption-stage FLW is cost-effective for livestock producers, the management strategy may not be economically self-sustaining (Dou et al. 2018).

Using panel survey data of rural Chinese households from 1991 to 2009, Qi et al. (2021) examined the linkages between household-level FLW and livestock systems over a period of market liberalization and government policies that encouraged livestock intensification. They found that households raising livestock—with many of the households using uneaten food for animal feed—generated more FLW than non-raising households during the period prior to these policies, but experienced significantly greater FLW reductions after the onset of livestock intensification (Qi et al. 2021).

3.3.3 Food Attributes and Acceptability Thresholds

For food to successfully move through the value chain from farm to consumer stages, it must be acceptable to each value chain actor. People generally evaluate food based on the assigned value, quality, and utility of the product (Tijskens and Schouten 2014). Although individuals may perceive value, quality, and utility attributes differently, there is a shared process of evaluating these elements and determining which foods are acceptable (Fig. 3.5).

Food acceptance hinges on value chain actor awareness and prioritization of different food attributes. Food with FQL continues moving through food value chains and is either eaten or diverted to nonfood destinations, the latter linking FQL to pFLW pathways based on food acceptability decisions (Fig. 3.2). The search, experience, credence (SEC) framework is useful for distinguishing food attributes based on observability (Nelson 1970, 1974; Darby and Karni 1973):

- *Search attributes*—those that are easily identified before consumption or use.
- *Experience attributes*—those that are identified at consumption or use.
- *Credence attributes*—those that cannot be immediately identified by direct experience.

3.3.3.1 Value Attributes Based on Food Availability and Accessibility

Market conditions determine the cost of making food available and physically accessible to consumers, as well as the price that value chain actors must be able and willing to pay and the price value chain actors are willing to sell food for to consumers. Value attributes are typically either search or credence attributes. For example, prices are known prior to use or consumption, but costs in terms of environmental impacts are not immediately observable. Value chain actors assess food value based on food attributes, such as accessibility, price, cost, and prior experience, with respect to the market conditions and weigh these attributes against quality and utility attributes (Tijskens and Schouten 2014).

For value chain actors upstream to consumers, value attributes are linked to their livelihoods and success in their given business. At the start of food value chains, producers evaluate food value attributes to make decisions at the farmgate that affect the availability of food moving post-farmgate. Post-farmgate value chain actors operate as both food buyers and sellers to assess food value attributes that affect food accessibility in terms of which foods reach certain markets and the food price for consumers.

For end consumers, value attributes are linked to purchase and consumption behaviors. Consumers assess food affordability based on

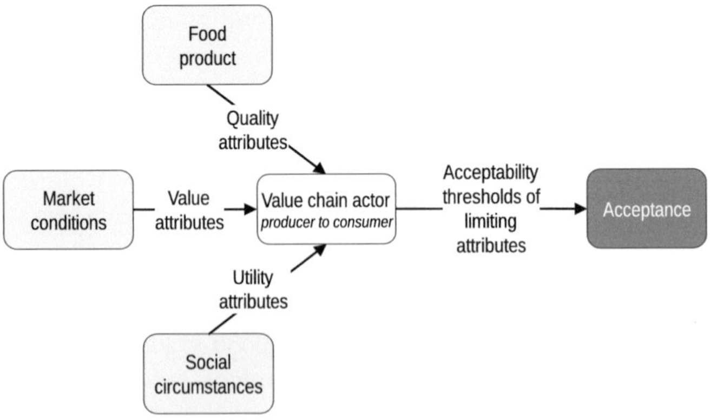

Fig. 3.5 Food acceptance based on quality, value, and utility attributes
Food acceptance is determined by how value chain actors (from producer to consumers) perceive quality attributes of the food product, value attributes based on market conditions, and utility attributes based on social circumstances, and set acceptability thresholds of limiting attributes
Adapted from Tijskens and Schouten (2014)

value attributes within market context, including the physical food accessibility and food price. At the consumer level, food acceptance begins at the point of purchase when consumers determine their ability to afford and their willingness to pay for the food product. Post-purchase, when food reaches the consumer, value attributes are also weighted with quality and utility attributes to determine final food acceptance, when food is eaten.

From farm to retail, value chain actors can enhance value attributes through value addition practices that target quality and utility attributes. Value addition can be used to enhance or preserve quality attributes, such as food safety and attributes associated with perishability, as well attributes related to how food was produced, processed, packaged, or marketed. Value addition can also address market conditions directly, such as formalizing quality standards.

3.3.3.2 Quality Attributes Based on Food Properties

Food quality is a complex human construct in which individuals perceive food properties (physical, chemical, and sensorial) and interpret them as quality attributes (Abbott 1999; Tijskens and Schouten 2014). The relationship between food quality attributes and properties is complex, particularly when food quality attributes are based on several food properties.

Although the SEC framework is often used to explore end-consumer preferences and behaviors (Kapoor and Kumar 2015; Ariyawardana et al. 2017; Uribe et al. 2020), each value chain actor behaves as a consumer with respect to the previous actor and sets their own acceptable food quality attributes and criteria, which vary by context, personal expectations, and needs (Abbott 1999). The extent of observability of each type of quality attribute affects an individual's ability to confirm information about the product. Generally, the level of information asymmetry between a supplier and consumer is low for search attributes, high for credence attributes, and initially high for experience attributes but declines with experience over time (Comyns et al. 2013).

Food search quality attributes are easy to confirm and primarily include appearance attributes, such as shape, size, color, odor, firmness, and evidence of injury, damage, or contamination (for example, blemishes, bruises, insect parts). Appearance attributes are particularly important for foods that are sold to consumers loose or in transparent packaging. Search attributes are often used to determine acceptable quality limits, under which food products are either redirected to lower quality food market channels or rejected and diverted to nonfood channels (Kyriacou and Rouphael 2018). Food grade and quality labels based on search attributes may be used for high-value foods as a marketing strategy to add value through product differentiation (USDA n.d.).

Food experience quality attributes include taste, texture, convenience, and preparation, among others. Although experience attributes are those identified at the time of product consumption or use, producers and retailers may signal experience quality attributes through labels and advertising. Consumers make repeat food purchases and can draw upon their previous experiences and update their perceptions on product quality, perhaps favoring the same brands or retailers (Caswell and Mojduszka 1996).

Finally, credence quality attributes may fall under health attributes, including nutritious and safe foods, which are related to nutrient content and food hazard properties of the food. Credence attributes, such as ethical and sustainable, relate to product properties determined by production and marketing activities (Moser et al. 2011). Unique to credence attributes, consumers cannot verify these attributes, even after consuming or using the product. Therefore, the level of information asymmetry is high, and consumers must rely on supplier-provided information. Added value comes from reputable private or public certification mechanisms to communicate accurate information to consumers (Caswell and Mojduszka 1996).

Food safety is a subset of food quality and is distinguished from other food quality attributes because of the potential effects of food hazards on human health (Henson 2008). In some situations, food safety can be considered an experience attribute when consumption of a food hazard causes acute illness that is traceable to the specific unsafe food source (Adalja et al. 2023). Most often, food safety is a credence attribute often arising from the uncertainty regarding specific food hazards and unsafe food sources.

Quality attributes are either intrinsic or extrinsic attributes. Intrinsic quality attributes refer to those that change based on changes to physical, chemical, and sensorial product properties (Olson and Jacoby 1972; Linnemann et al. 2006; Fernqvist and Ekelund 2014). Extrinsic quality attributes are fixed, product-related attributes, often related to how food was produced and marketed, that are meant to add value to the product without directly affecting the product properties (Olson and Jacoby 1972; Linnemann et al. 2006; Schreiner et al. 2013).

Box 3.3 Examples of Intrinsic and Extrinsic Quality Attributes Using the SEC Framework

Intrinsic Quality Attributes

- *Search attributes:* Color, firmness, shape.
- *Experience attributes*: Flavor, texture.
- *Credence attributes:* Nutritional value, food safety, ingredients (for example, natural, artificial).

Extrinsic Quality Attributes

- *Search attributes*: Packaging, branding, labeling.
- *Experience attributes*: Convenience, preparation.
- *Credence attributes*: Organic, fair trade, designation of origin.

Desirable and undesirable changes to intrinsic food attributes occur throughout the production and post-production value chain. Food perishability describes the potential for undesirable quality changes that lead to food spoiling, deteriorating, or becoming unsafe for consumption. Fruits, vegetables, and ASF are among the more perishable food groups and are, therefore, more susceptible to FQL.

Within the FLW boundary, FQL occurs once food is ready for harvest. Preharvest, intrinsic quality attributes are important for determining when food reaches commercial maturity and is ready for harvest or slaughter (Watada et al. 1984). Undesirable quality changes occurring during the production stage fall outside the boundary of FLW. However, production-side quality changes can lead to pFLW when lower quality food is left unharvested or harvested and immediately diverted to nonfood destinations.

Edible food may also be diverted to nonfood uses due to undesirable extrinsic quality attributes. Food may be diverted to nonfood destinations when supply of foods with specific extrinsic quality attributes exceeds demand, for example, during food recalls of food from a specific origin due to risk of food safety. Undesirable changes in extrinsic attributes, which do or do not affect the food itself, may also cause food to be diverted to nonfood destinations—for example, in the potential for damaged packaging to affect intrinsic food quality attributes.

3.3.3.3 Utility Attributes

Within a social context, value chain actors consider the usefulness or benefit of food products, based on utility attributes, including preferences, intended use, and social status (Tijskens and Schouten 2014). As food moves along the value chain, actors at each stage consider trade-offs between which attributes are most useful for their activities and how their processes and procedures impact favorable attributes-related consumer preferences.

Food quality attributes may be considered useful attributes for multiple value chain actors. For example, fruit size consistency for protective packaging can prevent bruising and accelerated decay. However, some quality attributes are not necessarily useful, but may have a perceived benefit at the consumer stage. For example, cosmetic attribute standards may be used for value addition and product differentiation (De Hooge et al. 2018), which consumers may perceive as a useful benefit.

Cosmetic attributes generally refer to search attributes that are not essential for maintaining experience and credence quality attributes (for example, maintaining taste, performance, nutritive value, or food safety), and often include natural variations in the size, shape, and color of foods (De Hooge et al. 2018). Food standards based on cosmetic attributes can narrow the variation appearances presented to consumers as "normal." Suboptimal food, also referred to as imperfect or "ugly" food, describes food that has cosmetic attributes deviating from average or common cosmetic attributes (Xu et al. 2021).

Along the value chain, utility attributes also include the potential to upcycle foods that are below marketable grades and food by-products of other food processes. Before unmarketable foods and by-products are removed from human food value chains, they may be upcycled into food ingredients and final food products for human consumption (Upcycled Foods Definition Task Force 2020). The transformation of edible foods and food parts (including edible by-products) into upcycled foods means food has not left the edible food pathways. Upcycled foods can also be made out of inedible food parts (including inedible by-products), and the now edible products enter the food intended for human consumption pathway (Fig. 3.2).

Box 3.4 The Definition and Scope of Upcycled Foods

In 2020, the Upcycled Foods Definition Task Force[1] developed a comprehensive definition of upcycled foods and identified definitional elements. According to the definition, "upcycled foods use ingredients that otherwise would not have gone to human consumption, are procured and produced using verifiable supply chains, and have a positive impact on the environment" (Upcycled Foods Definition Task Force 2020). To facilitate a common understanding of the definition, five elements further elaborate on the definition. Upcycled foods:

1. are made from ingredients that otherwise would have ended up in a food waste destination;
2. are value-added food products;
3. are for human consumption;
4. have an auditable supply chain; and,
5. indicate which ingredients are upcycled on their labels.

Upcycled foods are inherently novel foods because they are made from ingredients that would have otherwise been diverted away from food value chains. Therefore, upcycling specifically refers to the process of transforming an unusable food or food part into a usable ingredient. As an upcycled food becomes common or widespread, value chains for that product become more established and inputs are no longer considered unmarketable foods or by-products (Upcycled Foods Definition Task Force 2020).

3.3.3.4 Attributes Limiting Product Acceptance

Although there are seemingly infinite individual food attributes, the impact of FLW on food systems outcomes are based on a smaller number of attributes and their acceptability thresholds. Attributes that limit food acceptance are context-dependent, may be predefined or depend on specific conditions, and can shift from one set of attributes to another (Tijskens and Polderdijk 1996). Therefore, value chain actors use one or a combination of quality, value, and utility food attributes to decide on the acceptance of a food product.

Which food attributes are prioritized as limiting attributes generally follows structural transformation and food systems transformations (see

[1]The Task Force included the Upcycled Food Association, Natural Resources Defense Council, ReFED, Harvard Law School Food Law and Policy Clinic, Drexel University Food Lab, and the World Wildlife Foundation.

Chap. 2). Value chain actors assess trade-offs between food attributes to select limiting attributes that optimize each of their objectives related to livelihoods and diets. At earlier stages of food systems transformation, limited market connections and lower consumer incomes influence limiting attributes that tend to focus on value attributes (for example, cost and price). As food systems elongate and connections become more complex, quality attributes are increasingly prioritized, particularly for high value, perishable foods. At later stages of food systems transformations, food is more affordable, and limiting attributes focus on those related to value addition and product differentiation, tapping into consumers' preferences and status. Limiting attributes may also suddenly shift in response to shocks, such as collapse of food prices or foodborne disease outbreaks.

Acceptability thresholds delineate the point at which food no longer meets formal or informal requirements for a limiting attribute, depending on which limiting attribute first becomes unacceptable. Food with or without FQL that fails to meet acceptability thresholds is intentionally diverted to nonfood destinations (Fig. 3.2). Alternatively, food that fails to meet acceptability thresholds for limiting attributes in one value chain may be redirected to other food value chains with different limiting attributes and/or acceptability thresholds. Food that reaches consumers and is eventually eaten represents food with or without FQL that has been accepted by value chain actors (Fig. 3.2). Depending on which attributes are limiting in a certain context, food with FQL may reach consumers and be eaten—for example, when food becomes unsafe to eat but is still consumed because food safety attributes were either not detected or not prioritized.

Food quality standards emerge, set by both public and private entities and implemented throughout the value chain. As food retail becomes more concentrated, public and private standards focus on food quality attributes (Lee et al. 2012). Public standards emerge when governments set minimum requirements that are typically focused on food safety (Lee et al. 2012; Unnevehr and Ronchi 2014). Private voluntary standards for food safety are used to protect the reputation of private brands (Fulponi 2006; Lee et al. 2012). As food production and manufacturing become more concentrated, private standards tend to go beyond the minimum public food quality standards and focus on value-added quality attributes (for example, appearance, taste, and convenience), following consumer demand and preferences for higher quality and diversified products in particular market channels (Fulponi 2006; Lee et al. 2012; Unnevehr and Ronchi 2014).

The extent to which public or private food quality standards impact FLW pathways depends on the quality attributes defined in the requirement, as well as the implementation and enforcement of the standard. Compared to low- and middle-income countries, high-income countries typically have more rigorous standards that overemphasize cosmetic food quality attributes, thus removing foods from the value chain that could be safely consumed (HLPE 2014; Kader 2010).

Governments can set and enforce labeling standards to help ensure messaging is truthful and credible, particularly, for those that directly impact public health and nutrition (Caswell and Mojduszka 1996). Private brands, third party certifications, and other labels may also exist outside of government standards and rely on consumer confidence in the certification process and oversight. Brands, themselves, may signal product credence attributes, such as food safety, as they meet consumer needs and establish trust (Lassoued and Hobbs 2015).

3.4 Outcomes and Impacts of FLW Pathways: Diets and Nonfood Destinations

Attention to FLW has gained momentum as a threat to sustainable and healthy food systems because of potential negative impacts on food security, environmental, and livelihood outcomes. FLW pathways end once food is either eaten or once food is diverted to a nonfood destination (Fig. 3.2), contributing to broader food system nutrition and health outcomes and social, economic, and environmental impacts (HLPE 2017).

Diets consist of individual foods that people consume and can be characterized based on the quantity, diversity, quality, and safety of different foods (Hu 2002; HLPE 2017). As part of the food system, FLW pathways influence the quantity, diversity, quality, and safety of foods that reach consumers and are eaten by consumers. Diets are directly linked with nutrition and health outcomes, as healthy diets are required for preventing malnutrition and diet-related diseases. Additionally, demand shifts for certain diets impact environmental outcomes related to food production and consumption practices, economic outcomes related to income and employment for value chain actors, and social equity outcomes related to food distribution (HLPE 2017).

Nonfood destinations of FLW pathways also contribute to food system impacts, depending on where and how food is diverted from food value chains. Sending food intended for human consumption to nonfood destinations may impact the economic livelihoods of value chain actors. Different nonfood destinations can have different environmental impacts, which have been ranked from most preferred to least preferred, based on benefits to the environment and to a circular economy (US EPA 2023). Impacts to social equity outcomes relate to who stands to benefit or lose out most when food is diverted to nonfood destinations and how diverted food is handled.

Although preventing FLW may lead to multiple outcomes, these may not all be beneficial. Therefore, it is important to understand the cascading effects and feedback loops involving FLW pathways and food systems outcomes to evaluate potential synergies, trade-offs, and negative effects that may result from FLW reduction strategies.

3.5 Conclusion

Estimating FLW at the global level requires aggregation of FLW data from many sources reaching across value chain stages, food commodities, and geographies. The lack of a harmonized FLW definition presents a major challenge to interpreting, comparing, and aggregating FLW data. The majority of data on quantitative FLW (65%) comes from the APHLIS database, which covers staple foods in sub-Saharan Africa. Yet, the FLW definition used in APHLIS does not align with FLW definitions used in the SDG 12.3 indicators (Food Loss Index and Food Waste Index). After examining definitional elements, we find that the FLW definition from FAO's Global Initiative on Food Loss and Waste Reduction is comprehensive and globally applicable. In summary, FLW is a reduction in the quantity or quality of the edible portion of food intended for human consumption when food is redirected to nonfood uses or when there is a decrease in the nutritional value, food safety, or other quality aspect from the time food is ready for harvest or slaughter to consumption.

In this chapter, we use this definition as the basis to describe FLW pathways, which serves as a guiding framework for the book. FLW is an overarching term that encompasses two types of FLW—FQL and pFLW. Distinguishing FLW that is based on the type rather than the value chain stage is important for identifying the pathways and linkages, as food flows from farm to consumer. Within this framework, FLW pathways only involve the edible portion of food, because once food reaches the consumer, only the edible portion is reasonably eaten.

The FLW pathways framework also sets clear boundaries of when pathways begin and end, aligning with the FLW definition. FLW pathways begin once food is ready for harvest or slaughter and ends once food is either eaten or is diverted to a nonfood destination. This clear boundary becomes important for demarcating pFLW as any food that is diverted to nonfood uses. Even when nonfood uses are considered productive, the diverted food remains outside the FLW pathways boundary. For example, when food is diverted to be used as animal feed, this productive use is considered an input either for food production (that is, ASF) or for another value chain involving animals (for example, animal labor).

This chapter also details how food value, quality, and utility attributes are an essential component of FLW pathways. Food quality attributes have different levels of observability and either change over time or remain fixed. FQL pathways involve undesirable changes in intrinsic food quality attributes. Physical FLW pathways are based on acceptability thresholds for limiting attributes. Once foods fail to meet a minimum acceptability level for a given attribute, food may be diverted to nonfood destinations. The limiting attribute and acceptability thresholds can each change, depending on the value chain stage, actor, food commodity, and food system context.

FLW is one of many factors that impact food systems nutrition, health, environmental, and social outcomes. FLW pathways contribute to diet outcomes, based on the food that becomes available and accessible to and eaten by consumers. Additionally, where, when, and how food is diverted to nonfood uses have different environmental, economic, and social outcomes. The FLW pathways framework presented in this chapter offers a tool to understand, compare, and address FLW problems and evaluate trade-offs between food systems outcomes.

References

Abbott JA (1999) Quality measurement of fruits and vegetables. Postharvest Biol Tec 15:207–225. https://doi.org/10.1016/S0925-5214(98)00086-6

Adalja A, Lichtenberg E, Page ET (2023) Collective investment in a common pool resource: grower associations and food safety guidelines. Am J Agr Econ 105(1):144–173. https://doi.org/10.1111/ajae.12291

Agarwal M, Agarwal S, Ahmad S, Singh R, Jayahari KM (2021) Food loss and waste in India: the knowns and the unknowns. World Resources Institute Working Paper. https://doi.org/10.46830/wriwp.20.00106. Accessed 7 Aug 2024

Ariyawardana A, Ganegodage K, Mortlock MY (2017) Consumers' trust in vegetable supply chain members and their behavioural responses: a study based in Queensland, Australia. Food Control 73:193–201. https://doi.org/10.1016/j.foodcont.2016.08.006

Bellemare MF, Çakir M, Peterson HH, Novak L, Rudi J (2017) On the measurement of food waste. Am J Agr Econ 99:1148–1158. https://doi.org/10.1093/ajae/aax034

Boiteau JM, Pingali P (2023) Can we agree on a food loss and waste definition? An assessment of definitional elements for a globally applicable framework. Glob Food Secur 37:100677. https://doi.org/10.1016/j.gfs.2023.100677

Buzby JC, Farah-Wells H, Hyman J. (2014) The estimated amount, value, and calories of postharvest food losses at the retail and consumer levels in the United States. USDA ERS Economic Information Bulletin 121. https://www.ers.usda.gov/webdocs/publications/43833/43680_eib121.pdf. Accessed 7 Aug 2024

Caswell JA, Mojduszka EM (1996) Using informational labeling to influence the market for quality in food products. Am J Agri Econ 78:1248–1253. https://doi.org/10.2307/1243501

Chaboud G (2017) Assessing food losses and waste with a methodological framework: insights from a case study. Resour Conserv Recy 125:188–197. https://doi.org/10.1016/j.resconrec.2017.06.008

Chaboud G, Daviron B (2017) Food losses and waste: navigating the inconsistencies. Glob Food Secur 12:1–7. https://doi.org/10.1016/j.gfs.2016.11.004

Chauhan C, Dhir A, Akram MU, Salo J (2021) Food loss and waste in food supply chains. A systematic literature review and framework development approach. J Clean Prod 295:126438. https://doi.org/10.1016/j.jclepro.2021.126438

Comyns B, Figge F, Hahn T, Barkemeyer R (2013) Sustainability reporting: the role of "search", "experience" and "credence" information. Account Forum 37:231–243. https://doi.org/10.1016/j.accfor.2013.04.006

Darby MR, Karni E (1973) Free competition and the optimal amount of fraud. J Law Econ 16:67–88. https://doi.org/10.1086/466756

De Hooge IE, Van Dulm E, Van Trijp HCM (2018) Cosmetic specifications in the food waste issue: supply chain considerations and practices concerning suboptimal food products. J Clean Prod 183:698–709. https://doi.org/10.1016/j.jclepro.2018.02.132

Delgado L, Schuster M, Torero M (2017) Reality of food losses: a new measurement methodology. IFPRI Discussion Paper 1686. International Food Policy Research Institute (IFPRI), Washington, DC

Dou Z, Toth JD, Westendorf ML (2018) Food waste for livestock feeding: feasibility, safety, and sustainability implications. Glob Food Secur 17:154–161. https://doi.org/10.1016/j.gfs.2017.12.003

Fabi C, Cachia F, Conforti P, English A, Rosero MJ (2021) Improving data on food losses and waste: from

theory to practice. Food Policy 98:101934. https://doi.org/10.1016/j.foodpol.2020.101934
Fabi C, English A (2019) Methodological proposal for monitoring SDG target 12.3. Sub-indicator 12.3.1.A the food loss index design, data collection methods and challenges. FAO, Rome
FAO (Food and Agriculture Organization of the United Nations) (2022) Technical platform on the measurement of and reduction of food loss and waste. Food Loss and Waste Database. https://www.fao.org/platform-food-loss-waste/flw-data/en/. Accessed 14 Jun 2022
FAO (Food and Agriculture Organization of the United Nations) (2019) The state of food and agriculture 2019. Moving forward on food loss and waste reduction. FAO, Rome
FAO (Food and Agriculture Organization of the United Nations) (2014) Definitional framework of food loss. Working paper of the Global Initiative on Food Loss and Waste Reduction. FAO, Rome
FAO (Food and Agriculture Organization of the United Nations) (2011) Global food losses and food waste—extent, causes and prevention. FAO, Rome
FAO (Food and Agriculture Organization of the United Nations) and WHO (World Health Organization). (2019) Codex Alimentarius Commission—procedural manual, 27th edn. Rome
Fernqvist F, Ekelund L (2014) Credence and the effect on consumer liking of food—a review. Food Qual Prefer 32:340–353. https://doi.org/10.1016/j.foodqual.2013.10.005
Fulponi L (2006) Private voluntary standards in the food system: the perspective of major food retailers in OECD countries. Food Policy 31:1–13. https://doi.org/10.1016/j.foodpol.2005.06.006
Guo X (2021) Policy brief: food loss and waste reduction strategies to fill the protein gap. Wageningen University & Research, November 2. https://cgspace.cgiar.org/server/api/core/bitstreams/b4104a7e-6915-414c-92ef-0fac02c7d806/content. Accessed 5 Aug 2024
Hanson C, Lipinski B, Robertson K, Dias D, Gavilan I, Gréverath P, Ritter S, Fonseca J, van Otterdijk R, Timmermans T, Lomax J, O'Connor C, Dawe A, Swannell R, Berger V, Reddy M, Somogyi D, Tran B, Leach B, Quested T (2016) Food loss and waste accounting and reporting standard. World Resources Institute, Washington, DC
Henson S (2008) The role of public and private standards in regulating international food markets. J Int Agr Trade Dev 4:63–81
HLPE (High Level Panel of Experts) (2017) Nutrition and food systems. Report of HLPE on Food Security and Nutrition, Committee on World Food Security, Rome. https://openknowledge.fao.org/server/api/core/bitstreams/4ac1286e-eef3-4f1d-b5bd-d92f5d1ce738/content. Accessed 7 Aug 2024
HLPE (High Level Panel of Experts) (2014) Food losses and waste in the context of sustainable food systems. Report of HLPE on Food Security and Nutrition, Committee on World Food Security, Rome. https://openknowledge.fao.org/server/api/core/bitstreams/b1949fae-23d4-473c-8b87-8c4359b74d6c/content. Accessed 7 Aug 2024
Hodges R (2013) How to assess postharvest cereal losses and their impact on grain supply: rapid weight loss estimation and the calculation of cumulative cereal losses with the support of APHLIS (African Postharvest Losses Information System)., July 2013 (Version 1.1). https://www.aphlis.net/downloads/APHLIS%20Loss%20Assessment%20Manual.pdf. Accessed 4 Aug 2024
Hodges RJ (2012) Postharvest quality losses of cereal grains in Sub-Saharan Africa. APHLIS (African Postharvest Losses Information System). http://archive.aphlis.net/downloads/Postharvest_quality_losses%20review_Sept_2012_revised.pdf. Accessed 4 Aug 2024
Hu FB (2002) Dietary pattern analysis: a new direction in nutritional epidemiology. Curr Opin Lipidol 13:3–9. https://doi.org/10.1097/00041433-200202000-00002
Ingram J (2011) A food systems approach to researching food security and its interactions with global environmental change. Food Secur 3:417–431. https://doi.org/10.1007/s12571-011-0149-9
IRRI (International Rice Research Institute) (2024) Harvesting. Rice Knowledge Bank. http://www.knowledgebank.irri.org/step-by-step-production/postharvest/harvesting#harvesting-processes. Accessed 7 Jul 2024
Jha SN, Vishwakarma RK, Ahmad T, Rai A, Dixit AK (2015) Report on Assessment of Quantitative Harvest and Post-Harvest Losses of Major Crops/Commodities in India. ICAR-All India Coordinated Research Project on Post-Harvest Technology, ICAR-CIPHET, Ludhiana. https://www.researchgate.net/profile/Rajesh-Vishwakarma-3/publication/289637983_Assessment_of_Quantitative_Harvest_and_Post-Harvest_Losses_of_Major_CropsCommodities_in_India/links/5690c83b08aed0aed811bae6/Assessment-of-Quantitative-Harvest-and-Post-Harvest-Losses-of-Major-Crops-Commodities-in-India.pdf. Accessed 7 Aug 2024
Juliano BO, Tuaño APP. (2019) Gross structure and composition of the rice grain. In: Bao J (ed) Rice: chemistry and technology. 4th edn. AAC International Press, pp 31–53. https://doi.org/10.1016/B978-0-12-811508-4.00002-2
Kader AA (2010) Handling of horticultural perishables in developing vs. developed countries. In Erkan M (ed) Acta Hortic 877, 121–126. Proceedings of the VI International postharvest symposium, Antalya, Turkey, Acta Horticulturae, 8 Apr 2009
Kapoor S, Kumar N (2015) Fruit and vegetable consumers' behavior: implications for organized retailers in emerging markets. J Int Food Agribusiness Mark 27:203–227. https://doi.org/10.1080/08974438.2014.940118
Kitinoja L, Tokala VY, Brondy A (2018) Challenges and opportunities for improved postharvest loss measure-

ments in plant-based food crops. J Postharvest Tec 6:16–34
Koester U (2017) Food loss and waste as an economic and policy problem. Front Econ Global 17:275–288. https://doi.org/10.1108/S1574-871520170000017018
Kyriacou MC, Rouphael Y (2018) Towards a new definition of quality for fresh fruits and vegetables. Sci Hortic 234:463–469. https://doi.org/10.1016/j.scienta.2017.09.046
Lassoued R, Hobbs JE (2015) Consumer confidence in credence attributes: the role of brand trust. Food Policy 52:99–107. https://doi.org/10.1016/j.foodpol.2014.12.003
Lee J, Gereffi G, Beauvais J (2012) Global value chains and agrifood standards: challenges and possibilities for smallholders in developing countries. P Natl Acad Sci 109:12326–12331. https://doi.org/10.1073/pnas.0913714108
Linnemann AR, Benner M, Verkerk R, van Boekel MAJS (2006) Consumer-driven food product development. Trends Food Sci Tech 17:184–190. https://doi.org/10.1016/j.tifs.2005.11.015
Lipinski B, Hanson C, Lomax J, Kitinoja L, Waite R, Searchinger T (2013) Reducing food loss and waste. Installment 2 of creating a sustainable food future. World Resources Institute Working Paper, Washington, DC. http://pdf.wri.org/reducing_food_loss_and_waste.pdf. Accessed 7 Aug 2024
Mali SS, Jha BK, Naik SK, Singh AK, Kumar A (2016) Effect of fertigation pattern and planting geometry on growth, yield and water productivity of tomato (Solanum lycopersicum). Indian J Agr Sci 86(9). https://doi.org/10.56093/ijas.v86i9.61522
Moreno LC, Tran T, Potts MD (2020) Consider a broccoli stalk: how the concept of edibility influences quantification of household food waste. J Environ Manag 256:109977. https://doi.org/10.1016/j.jenvman.2019.109977
Moser R, Raffaelli R, Thilmany-McFadden D (2011) Consumer preferences for fruit and vegetables with credence-based attributes: a review. Int Food Agribus Man 14(22):121–142
Mottet A, de Haan C, Falcucci A, Tempio G, Opio C, Gerber P (2017) Livestock: on our plates or eating at our table? A new analysis of the feed/food debate. Glob Food Secur 14:1–8. https://doi.org/10.1016/j.gfs.2017.01.001
Mottet A, Tempio G (2017) Global poultry production: current state and future outlook and challenges. World Poultry Sci J 73:245–256. https://doi.org/10.1017/S0043933917000071
Muth MK, Giombi KC, Bellemare M, Ellison B, Roe B, Smith T (2018) Expert panel on technical questions and data gaps for the ERS Loss-Adjusted Food Availability (LAFA) Data Series Agreement No. 59-4000-6-0069 final report (No. RTI Project Number 0215523.000.001). RTI International, Research Triangle Park. https://www.ers.usda.gov/webdocs/publications/92409/ccr-70.pdf?v=9728.9. Accessed 7 Aug 2024
Naika S, van Lidt de Jeude J, de Goffau M, Hilmi M, van Dam B (2005) Cultivation of tomato: production, processing and marketing. Agromisa Foundation and CTA, Wageningen. http://www.agromisa.org/wp-content/uploads/Agrodok-17-Cultivation-of-tomato_sample.pdf. Accessed 7 Aug 2024
Nanda SK, Vishwakarma RK, Bathla HVL, Rai A, Chandra P (2012) Harvest and post harvest losses of major crops and livestock produce in India. All India Coordinated Research Project on Post Harvest Technology, Council of Agricultural Research (ICAR), Ludhiana. https://www.mofpi.gov.in/sites/default/files/phl_2010_ciphet.pdf. Accessed 7 Aug 2024
Nelson P (1974) Advertising as information. J Polit Econ 82:729–754. https://doi.org/10.1086/260231
Nelson P (1970) Information and consumer behavior. J Polit Econ 78:311–329. https://doi.org/10.1086/259630
Nicholes MJ, Quested TE, Reynolds C, Gillick S, Parry AD (2019) Surely you don't eat parsnip skins? Categorising the edibility of food waste. Resour Conserv Recy 147:179–188. https://doi.org/10.1016/j.resconrec.2019.03.004
Olson JC, Jacoby J (1972) Cue utilization in the quality perception process. In: Venkatesan M (ed) SV - proceedings of the third annual conference of the Association for Consumer Research. Association for Consumer Research, Chicago, pp 167–179
Östergren K, Gustavsson J, Ole-Jørgen H, Møller H, Anderson G, O'Connor C, Soethoudt H, Quested T, Easteal S, Politano A, Bellettato C, Canali M, Falasconi L, Gaiani S, Vittuari M, Schneider F, Moates G, Waldron K, Redlingshöfer B (2014) FUSIONS definitional framework for food waste. 3 July. isbn:978-91-7290-331-9
Parfitt J, Barthel M, Macnaughton S (2010) Food waste within food supply chains: quantification and potential for change to 2050. Philos T R Soc B 365:3065–3081. https://doi.org/10.1098/rstb.2010.0126
Qi D, Lai W, Roe BE (2021) Food waste declined more in rural Chinese households with livestock. Food Policy 98:101893. https://doi.org/10.1016/j.foodpol.2020.101893
Redlingshöfer B (2015) La méthodologie utilisée dans l'étude INRA pour l'analyse des pertes alimentaires dans les filières. Innov Agronomiques 48:11–22. 10.15454/1.4622704219926074E12. hal-02634690. https://hal.inrae.fr/hal-02634690/document
ReFED (2021) Roadmap to 2030: reducing U.S. Food Waste by 50% and the ReFED insights engine. At-A-Glance. https://refed.org/uploads/refed_roadmap2030-FINAL.pdf. Accessed 6 Aug 2024
Schreiner M, Korn M, Stenger M, Holzgreve L, Altmann M (2013) Current understanding and use of quality characteristics of horticulture products. Sci Hortic 163:63–69. https://doi.org/10.1016/j.scienta.2013.09.027
Schuster M, Torero M (2016) Reducing food loss and waste. In: 2016 Global Food Policy Report, Chapter 3.

International Food Policy Research Institute (IFPRI), p 22–31
Tijskens LMM, Polderdijk JJ (1996) A generic model for keeping quality of vegetable produce during storage and distribution. Agric Syst 51:431–452. https://doi.org/10.1016/0308-521X(95)00058-D
Tijskens LMM, Schouten RE (2014) Modeling quality attributes and quality related product properties. In: Florkowski WJ, Shewfelt RL, Brueckner B, Prussia SE (eds) Postharvest handling: a systems approach. Elsevier Academic Press, Amsterdam/Boston, pp 411–448
UNEP (United Nations Environment Programme) (2024) Food Waste Index Report 2024. Nairobi. https://www.unep.org/resources/publication/food-waste-index-report-2024. Accessed 7 Aug 2024
UNEP (United Nations Environment Programme) (2021) Food Waste Index Report 2021. Nairobi. https://www.unep.org/resources/report/unep-food-waste-index-report-2021. Accessed 7 Aug 2024
Unnevehr LJ, Ronchi L. (2014) Food safety and developing markets: research findings and research gaps. IFPRI Discussion Paper 1376. International Food Policy Research Institute, Washington, DC. https://cgspace.cgiar.org/server/api/core/bitstreams/40539cd0-79b9-4ef9-885c-d395ad2964b9/content. Accessed 7 Aug 2024
Upcycled Foods Definition Task Force (2020) Defining upcycled foods: a definition for use across industry, government, and academia. Center for Health Law and Policy Innovation, Harvard Law School. https://chlpi.org/wp-content/uploads/2013/12/Upcycled-Food_Definition.pdf. Accessed 5 Aug 2024
Uribe R, Infante R, Kusch C, Contador L, Pacheco I, Mesa K (2020) Do consumers evaluate new and existing fruit varieties in the same way? Modeling the role of search and experience intrinsic attributes. J Food Products Mark 26:521–534. https://doi.org/10.1080/10454446.2020.1821419
US EPA (United States Environmental Protection Agency) (2023) Wasted Food Scale. https://www.epa.gov/sustainable-management-food/wasted-food-scale. Last updated 25 Apr 2024
USDA (United States Department of Agriculture) (n.d.) Understanding Food Quality Labels. https://www.ams.usda.gov/sites/default/files/media/AMSProductLabelFactsheet.pdf. Accessed 7 Aug 2024
Watada AE, Herner RC, Kader AA, Romani RJ, Staby GL (1984) Terminology for the description of developmental stages of horticultural crops. HortScience 19:20–21. https://doi.org/10.21273/HORTSCI.19.1.20
Xu Y, Jeong E, Jang S(S), Shao X (2021) Would you bring home ugly produce? Motivators and demotivators for ugly food consumption. J Retailing Cons Ser 59:102376. https://doi.org/10.1016/j.jretconser.2020.102376
Xue L, Liu G, Parfitt J, Liu X, Van Herpen E, Stenmarck Å, O'Connor C, Östergren K, Cheng S (2017) Missing food, missing data? A critical review of global food losses and food waste data. Environ Sci Technol 51:6618–6633. https://doi.org/10.1021/acs.est.7b00401

From Data to Action: Building Evidence on Food Loss and Waste

4

4.1 Introduction

Since the 2030 Agenda for Sustainable Development recognized food loss and waste (FLW) as a global issue (UN 2015), efforts have been initiated to identify and define indicators, as well as collect, compile, and share data to effectively address this issue. A comprehensive FLW data value chain is needed for understanding the extent of FLW at different scales and for transforming information into actionable decisions. Establishing such a data value chain requires setting clear priorities, identifying critical data gaps, and determining the platforms and institutions that will provide and use the information (Piwoz et al. 2019).

Data plays a pivotal role in driving informed decision-making and effective interventions to reduce FLW. By collecting, analyzing, and interpreting relevant data, stakeholders can gain valuable insights into the factors contributing to FLW, identify areas for improvement, and monitor progress toward reduction goals. A robust data ecosystem, composed of government agencies, regional and international organizations, civil society, and the private sector, is essential for ensuring that data are collected, shared, and used effectively, based on harmonized standards and methods (Global Partnership for Sustainable Development Data 2016).

Open-access FLW platforms that gather and share data are required for enhancing information accessibility and enabling evidence-based decision-making. However, a significant challenge lies in the coordination of the multitude of modeled estimates of primary data in existing databases, including the Food Balance Sheets (FBS) of the Food and Agriculture Organization of the United Nations (FAO), the Food Loss and Waste Database (FAO), and the Food Waste Atlas developed by a multi-stakeholder partnership that includes United Kingdom's Waste and Resources Action Programme (WRAP), the World Resources Institute (WRI), and Walmart Foundation. The lack of coordinated efforts for data collection raises concerns about the quality and comparability of new primary data, which is often resource-intensive to obtain. While facing significant data gaps, FAO has used a modeling approach to calculate the Food Loss Index (FLI), assessing loss trends (Fabi and English 2019; Mingione et al. 2021). In contrast, the United Nations Environment Programme (UNEP) has deemed a modeling approach insufficient for calculating the Food Waste Index (FWI) to assess waste trends and is awaiting the availability of sufficient primary data before proceeding with the index calculation (UNEP 2024).

This chapter examines the data requirements for addressing FLW and identifies outstanding knowledge gaps. It explores the existing global indicators for measuring physical FLW (pFLW), including the strengths and limitations of the FLI and the FWI, and identifies gaps in current

J. Boiteau, P. Pingali, *Wasted Potential*, Sustainable Development Goals Series,
https://doi.org/10.1007/978-3-031-97411-3_4

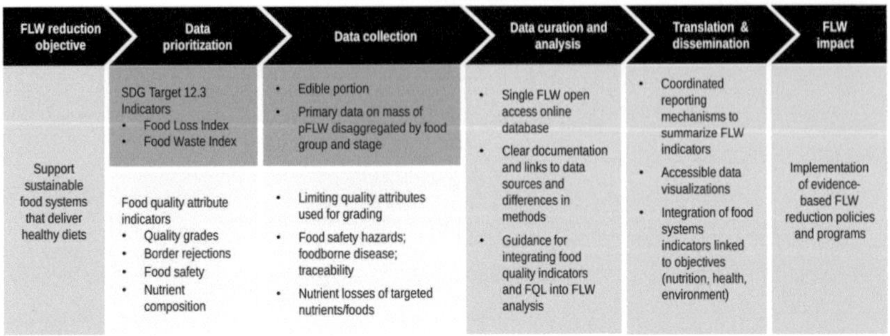

Fig. 4.1 **Food loss and waste data value chain roadmap**

measurement approaches. Additionally, the chapter examines the need for a broader range of indicators to capture the complex nature of FLW, such as food quality loss (FQL) and associated factors in determining FLW, and the challenges for standardized measurement methods. Finally, we discuss the limitations of existing databases and the difficulties in translating and disseminating accurate FLW estimates using current data and statistical approaches. To address these challenges, we propose a data value chain roadmap (Fig. 4.1) to guide the use of data for informing policies, interventions, and monitoring efforts. This chapter aims to contribute to a more informed and effective data-driven response to the FLW problem.

4.2 Data Prioritization

At the global level, Sustainable Development Goal (SDG) Target 12.3 calls for reducing per capita food waste at the retail and consumer levels by 50% and reducing food losses along production and supply chains. Stakeholders are motivated to reduce FLW based on the potential linkages with other SDGs, which are focused on food security, natural resources, and the environment (FAO 2019). Sustainable food systems objectives influence FLW data priorities, including food products targeted, stages of the food chain considered, geography, and measurement and reporting methodologies (Chaboud 2017; Cattaneo et al. 2021). Identifying data priorities is the first step in the data value chain to define indicator categories and operational guidance needed to understand the causes of FLW, formulate policies, design interventions, and monitor progress (Piwoz et al. 2019).

Quantifying pFLW has been, and continues to be, the indicator category prioritized for FLW reduction attention and efforts. The SDG Target 12.3 indicators, the FLI and the FWI monitor trends in pFLW at national and global levels. Without global FQL indicators, efforts to fill knowledge gaps primarily focus on measuring and estimating pFLW, and few studies measure FQL (Kitinoja et al. 2018a; Parfitt et al. 2021). Information on pFLW and FQL is needed to link FLW pathways to other food systems indicators and to inform comprehensive policies on FLW reduction to achieve desired sustainable food systems outcomes. Narrowly prioritizing pFLW data may ignore broader policy objectives and fail to allocate resources in the most cost-effective way to reach overarching goals to improve nutrition, enhance food safety, and minimize environmental impacts (Rutten 2013; Sheahan and Barrett 2017).

4.2.1 Food Loss Index

As the custodian agency, FAO is responsible for global monitoring and reporting of the FLI, the SDG Target 12.3 indicator used to track food losses along production and supply chains from farm to retail. The FLI is a country-level index calculated as the ratio of the Food Loss Percentage in the current year and that in the base year. The Global Food Loss Index (GFLI) is calculated based on the weighted average of country-level FLIs and is used for global monitoring. The FLI and GFLI summarize the complexities of food loss, indicating trends in the share of agriculture production that does not reach the retail stage compared to the base year (Fabi and English 2019). The Food Loss Percentage is an indicator of the extent of loss and informs the magnitude of the problem.

Calculating the Food Loss Percentage requires data on pFLW and production (both expressed in kgs and tonnes). Loss percentages for each commodity are calculated using cumulative loss to account for losses at each value chain stage from postharvest to retail (with an option to include the harvest stage at the country level) (Fabi and English 2019). The Food Loss Percentages can be disaggregated by commodity and value chain stage and used for comparison across countries. FAO guidance on data collection calls for increasing data availability of primary food loss data and imputing losses using an estimation model when primary data is unavailable (Fabi and English 2019). Baseline estimates require two or three consecutive years of data to avoid bias of using a single year. Then, data at critical loss points should be collected every three to five years (Fabi and English 2019).

4.2.2 Food Waste Index

As the custodian agency, UNEP is responsible for global monitoring and reporting of the FWI, the SDG Target 12.3 indicator used to track per capita food waste at retail and consumer levels. The FWI is a country-level index calculated as the ratio of Per Capita Food Waste in the current year and that in the base year. Per Capita Food Waste is disaggregated individually at retail, food service, and household stages. The country-level FWIs are kept disaggregated at each stage and are not combined into a single, country-level FWI (UNEP 2024). To our knowledge, the UNEP has not published an approach for calculating a Global FWI, but has referred to the potential to aggregate the FWI at the subregional, regional, and global levels, provided sufficient country-level data are available (UNEP 2021a).

Calculating Per Capita Food Waste requires data on the total mass of food waste (edible and inedible portions), rather than specific food commodities. The requirement is a major limitation to using the FWI to assess diet and environmental outcomes that require disaggregation at the food group level. UNEP guidance on data collection calls for national-level data collection every two years and prioritizes direct measurement of food waste for reporting on the FWI to accurately form national baselines, inform reduction policies, and track progress (UNEP 2021b; UNEP 2024). Additional data on the share of edible and inedible food waste, waste destination, and food losses at the manufacturing stage not captured by the FLI are not included in the FWI, but provide additional information for policy and program decision-making at the country level (UNEP 2024).

Box 4.1 Comparing the Food Loss Index and the Food Waste Index Indicators

Although the FLI and FWI appear to be complementary, the former covers farm to retail stages and the latter covers retail and consumer stages. The two indices cannot be directly compared or summed due to differences how the indicators are composed:

Food Loss Index (FLI) is a country-level, fixed-base index (base 100) that aggregates losses of ten key commodities in five main food groups (two commodities per food group), using economic weights (value of production in the base year). Lower values indicate a decrease in losses from the farm level to retail.

Global Food Loss Index (GFLI) is a weighted average of countries' Food Loss Indices.

Food Loss Percentage is the average share of production that is lost before reaching the retail stage, based on a weighted sum across commodities.

Food Waste Index (FWI) is a country-level, fixed-base index (base 100), based on total mass of food waste per capita, disaggregated at the retail, food service, and household stages. Lower values indicate a decrease in waste at each specific sector.

Per Capita Food Waste is the total mass of food waste (edible and inedible parts) per capita, disaggregated at the retail, food service, and household stages.

4.2.3 Food Quality Indicators Linked with Food Loss and Waste

Food quality aspects of FLW are not captured in SDG Target 12.3. Food quality loss indicators are important to identify causes of pFLW related to acceptability and to identify quality losses of food eaten that negatively affect health and nutrition outcomes. Although there are no global indicators for FQL, value chain actors use different quality indicators that typically summarize a group of food quality attributes. Most important is how food quality indicators communicate accurate and useful information with consensus between value chain actors to make marketing, purchase, and consumption decisions that determine which foods remain in the food value chain and are eaten. In this section, we review food quality indicators that have been used to assess food quality in the context of FLW.

Grades and standards are specific systems for sorting food into different categories defined by sets of quality attributes, including food safety, cosmetic and experiential attributes, and other extrinsic quality attributes, based on established classification and measurement procedures set either by convention or regulation (Doumeizel 2019; Giovannucci and Reardon 2000). Standards for food safety are directly connected to consumer health and welfare, whereas standards for other quality attributes are used to describe a product for marketing purposes (Giovannucci and Reardon 2000). To assess FLW, several approaches have been used to create quality grade indicators based on groups of quality attributes and intended end uses (Compton et al. 1992; Hodges 2013; Ali et al. 2021; Delgado et al. 2021b). Using grades as a FQL indicator not only requires consistent terminology and consensus in selection and assessment of quality attributes, but also that intrinsic quality attributes, which change over time, are used to define categories.

Food safety standards can be mandatory or voluntary, with several standard-setting systems developed by different entities including international organizations, national governments, nongovernment organizations, and the private sector (Doumeizel 2019). Foodborne hazards are not synonymous with foodborne disease risk. Hazards may cause harm in the absence of control, whereas risks are the likelihood that foodborne disease will occur with public health consequences dependent on several population factors (Roesel et al. 2015). Although international and national sets of food safety indicators have been developed (for example, food hazard occurrence, foodborne disease burden), there is still a lack of harmonized food safety indicators, particularly that can be adapted to the informal sector (GAIN 2020; Schneider et al. 2023).

Date labels serve as a signal of food quality, indicating FQL as the labeled date approaches or passes. Food date labels are often based on ambiguous terminology, which causes confusion in the marketplace and for consumers (Newsome et al. 2014). Misinterpretation of date labels often involves the distinction between food safety and other food quality attributes, sometimes leading retailers and consumers to discard food that is still consumable but is past a date label (Leib et al. 2013; Gruber et al. 2016; Wilson et al. 2018; Neff et al. 2019). The Codex Alimentarius Commission (2018) sets international food standards for food labeling of prepackaged foods, including date labeling reference guides, which are not enforceable (see Box 4.2). Individual countries and regions may have labeling standards, either voluntary or mandatory, which often follow similar date labeling phrases as the Codex. When public requirements do not exist, date labeling is typically left to the private sector. The lack of consensus and understanding of food date labels make date labeling a poor FQL indicator.

Box 4.2 Lack of Consensus on Date Labeling at the International and National Levels

The Codex Standards for the Labeling of Prepackaged Foods offers guidance on several date marking approaches to prepackaged foods (Codex Alimentarius 2018). **"Date of Manufacture"** and **"Date of**

Packaging" do not indicate the food product durability, but rather provide information about when the ingredients were transformed into the food product and when it was packaged, respectively. **"Best Before Date"** and **"Best Quality Before Date"** are meant to provide information on the date by which the product should retain its intended food quality attributes, excluding food safety (as the product should be acceptable to consume after the date). **"Use-by Date"** or **"Expiration Date"** are meant to signal the date by which the product should not be consumed because of food quality degradation, including food safety (Codex Alimentarius 2018).

Food date labeling in the United States. The United States' federal law only requires that infant formula be labeled with a **"Use-By"** date to indicate safety and nutrient content; all other labels are left to the discretion of manufacturers or fall under individual US state food labeling laws (USDA FSIS 2019). Legislation first proposed in 2021 and most recently reintroduced in 2023 intends to create two distinct labels: **"Best if Used By"** to indicate food is at peak quality before the date, and **"Use By"** to indicate potential food safety concerns if used beyond the date (US Congress 2023).

Food date labeling in the United Kingdom. In the United Kingdom, packaged products must either display a **"best before"** or **"use by"** date, where "use by" indicates a safety issue with eating the food after the date indicated (Government of the United Kingdom 2022). Thus, it is illegal for retailers to sell products past their "use by" dates (Government of the United Kingdom 2022).

Nutritional value FQL can occur without correlated pFLW, for example, when nutrients degrade or leach from food that will be eaten or when specific edible food parts that are nutrient-rich are removed (HLPE 2014). Indicators of nutrient adequacy and diversity often rely on food composition tables (Remons et al. 2014; Gustafson et al. 2016; Ahmed et al. 2022). However, changes in nutritional value, based on nutrient composition, are not indicative of other changes related to nutritional qualities, such as processing, bioavailability, and other food matrix effects (Delgado et al. 2021a). A comprehensive nutritional value FQL indicator should incorporate nutrient composition as well as other nutritional quality attributes to inform policy decisions, including food labels, health claims, and diet recommendations.

Finally, **food price** has been used as a FQL indicator, based on the assumption that price differences within a food commodity reflect quality differences (Delgado et al. 2021b). Using price as an indicator for quality differences requires that everything else is kept equal (for example, market conditions and structure, time and season, etc.) (Delgado et al. 2021b). However, quality attributes reflected in price are not necessarily well-defined, and this approach relies on constant variance over time in the difference between the best price for ideal quality foods and the actual price. For example, consumers may interpret price-reduced foods as food with inferior health or taste qualities, even when the price reduction is based on cosmetic attributes (Aschemann-Witzel et al. 2017). Price is best considered as an indicator for food value attributes that, in addition to food quality and utility attributes, influences an individual's acceptability of and willingness to pay for a certain food product (see Chap. 3).

Box 4.3 Identifying Food Quality Loss Indicators Remains an Outstanding Priority

Food quality indicators are an important but missing component in FLW data prioritization. Although the SDG Target 12.3 focuses on physical FLW, FQL is also necessary for understanding FLW and its impact on delivery of healthy and sustainable diets.

Food quality indicators help identify causes of FLW related to acceptability and assess quality losses in consumed food. While these indicators often rely on grades and standards, date labels, nutritional value, and price, their application to comprehensive FQL indicators presents several challenges. These challenges include the need for standardization, adaptability to diverse contexts, the integration of various quality attributes, and the identification of which attributes limit food acceptability. By addressing these barriers and developing robust FQL indicators, complex FLW pathways can be more effectively measured and managed.

4.3 Data collection

Generating high quality data using standardized approaches to reduce FLW requires robust, time-relevant data on the extent and causes of FLW across different contexts, commodities, and value chain stages. The lack of a consistent FLW definition, inconsistent data collection methods, opaque reporting practices, and the limitations of secondary data sources often lead to gaps in FLW data, making it difficult to compare, aggregate, and interpret (Chaboud 2017; Kitinoja et al. 2018b). To ensure widespread acceptance, a unified definition of FLW must be accompanied by consistent methods for applying the definition to standardized data collection and reporting. This will help address concerns about data gaps and ensure compatibility with other statistical definitions (FAO 2019). Comprehensive FLW measurement and reporting approaches should clearly outline the specific methods used to collect data, including the representativeness of the data, standardized measurement techniques, explanations of the chosen indicators, the nature of the data (units, etc.), and the timeframe (Östergren et al. 2014; Hanson et al. 2016a; Chaboud 2017).

Developing standardized FLW data collection approaches is central to the FLW data value chain, providing evidence for the design, implementation, and monitoring of targeted and informed policies and programs aimed at addressing key FLW challenges. Operationalizing a common framework for measuring and reporting FLW must be aligned with current FLW targets, especially SDG Target 12.3 (FAO 2019). However, operational realities can limit data collection and impact the generalizability of FLW estimates (Delgado et al. 2017). Three key challenges must be addressed to ensure consistent FLW measurement and production of adequate data for prioritized indicators:

1. Systematically distinguishing food safety and cultural dimensions of edibility;
2. Standardizing measurement and reporting methods for pFLW to generate comparable primary data across value chain stages, food groups, and time periods; and
3. Identifying the key quality attributes that limit food acceptance and the data sources that use standardized methods to assess intrinsic food quality attributes.

4.3.1 Edibility is a Complex Concept That Involves Food Safety and Cultural Dimensions

To define the scope of FLW, as discussed in Chap. 3, it is necessary to distinguish between edible and inedible food parts. Defining edibility involves examination of both food safety and the cultural dimensions of what is considered fit for human consumption, which can vary across time and context (Rutten 2013; HLPE 2014). Edible food that becomes unsafe due to FQL and associated foodborne hazards should be removed from the food value chain as pFLW. Culturally inedible food parts are more challenging to identify. Inedible parts that are technically safe to eat could refer to a range of food parts, such as stems, cores, and peels of fruits and vegetables, to bones, skin, and separable fat of meat, fish, and poultry (Muth et al. 2018). For example, in a study of pFLW in tomato sauce processing, tomato skin and seeds were considered inedible (Secondi et al. 2019).

The importance of classifying foods as edible or inedible depends on context-specific information about culturally inedible parts or those typically not consumed at the start of the food value chain. Differences in edibility categorizations, as broad as the institutional level and as narrow as the individual level, can impact FLW quantification. For example, Moreno et al. (2020) compared the edibility categorization of 69 food items and found that categorizations used in the United States (USDA National Nutrient Database for Standard Reference) are more restrictive and consider more food parts as inedible than those used in the United Kingdom's WRAP. The generalizability of these classifications to other contexts, particularly in LMICs, remains uncertain. Questionnaire-based methods that rely on self-reported consumption and perceptions of edibility can help address the data gap in classifying food parts as edible or inedible (Nicholes et al. 2019).

Box 4.4 Edibility Classification Approach

Nicholes et al. (2019) developed a reproducible classification method, using a scoring system and a threshold score, to determine food items that are perceived as technically edible and are eaten often or always by the majority participants. The authors averaged the perception of edibility and self-reported consumption scores to arrive at the final classification. Among consumer participants in the United Kingdom, 10 out of 16 food items were classified as edible, based on consumer perceptions that 15 food items were technically edible and 5 food items were often or always consumed. This approach can be replicated and adjusted (for example, threshold scores, food proxies, food list subsamples) to determine location- or cultural-specific edibility classifications (Nicholes et al. 2019).

4.3.2 Standardizing Measurement and Reporting Methods for Physical Food Loss and Waste Across the Food Value Chain

Estimates for pFLW, measured in mass or volume, are derived from (1) direct measurement, (2) approximations based on primary data, or (3) indirect measurement and calculations using secondary data. Most available pFLW estimates rely on secondary data (Xue et al. 2017). Comprehensive FLW reporting should include the units of measure (for example, mass, volume) and details about the reported figures (for example, cumulative percentages, proportions, means, standard deviations, medians, interquartile ranges) for clear interpretation (Kitinoja et al. 2018b). Guidance documents for the SDG Target 12.3 indicators provide only broad recommendations for measurement tools by value chain stage, and do not include standardized measurement methods (Fabi and English 2019; UNEP 2024).

Standardized data collection methods and measurement practices are essential for generating reliable primary pFLW data. These data can help establish baseline estimates, identify effective policies and interventions, track progress toward FLW reduction goals, and inform future planning. Several primary data collection methods exist to directly measure or approximate pFLW, as highlighted in Box 4.5. Beyond specific measurement limitations, operational and structural challenges within a given context and the stage of development can further affect the quality of data collection (Trienekens 2011; Hanson et al. 2016a; Soethoudt et al. 2021). Using mixed methodologies can help to overcome the weaknesses of specific methods and contextual limitations (Kitinoja et al. 2018b).

Standardized reporting is necessary for accurate interpretation and comparability of pFLW data. Reporting should clearly outline how estimates are summarized across value chain stages. For example, reporting total pFLW as a simple sum of estimates across value chain stages differs from reporting cumulative loss, which accounts for pFLW from each preceding stage. (Kitinoja et al. 2018b). While the *FLW Standard* offers guidance on practical and consistent accounting and reporting (for example, time frame, material type, destination, and boundary), and describes the quantification method used (Hanson et al. 2016a), there are still underlying limitations due to the lack of a unified FLW definition and standardized data collection approaches.

Box 4.5 Primary Data Collection Methods for Physical Food Loss and Waste

Primary FLW estimation methods are wide-ranging in terms of data collection time, cost, accuracy, objectivity, and reliability (Hanson et al. 2016b; Chaboud 2017; Xue et al. 2017). Direct measurement methods, while highly accurate and objective, are less frequently used, compared to approximations, due to their time and resource demands (Hanson et al. 2016b; Xue et al. 2017; Kitinoja et al. 2018a). The FLW Protocol offers guidance on selection and implementation of data collection methods, but falls short of providing systematic approaches for each method (Hanson et al. 2016b; Kitinoja et al. 2018b).

Methods that are resource-intensive, with high objectivity, accuracy, and reliability include:

- **Direct weighing** using a calibrated scale.
- **Waste composition analysis** of food physically separated from other material in waste streams.
- **Volume assessment** to determine weight of a product in a specified container or physical space based on a density factor.

Methods that largely depend on personal perceptions and subjectivity of observers or respondents include:

- **Surveys** collecting information on perceptions, behaviors, and approximations of FLW through questionnaires, interviews.
- **Diaries** in which respondents keep a log of FLW, as it occurs in real time, with or without weighing.
- **Records** of data collected for other purposes used to estimate FLW.
- **Observation** by visual method or counting to approximate volume and convert to weight.

The transformation of single food commodities into processed foods complicates pFLW measurement. Single food commodities are handled between the time the food is ready for harvest to when harvest is completed. Postharvest, processing can occur at different stages, from minimal to ultra-processing, before reaching consumers. Households may also process and mix food as part of food preparation before consumption. The type and extent of processed foods accessible to the consumer are influenced by the stage of food systems transformation (see Chap. 2) and affect the amount of consumer preparation needed. Therefore, measuring pFLW of individual food commodities becomes more difficult as foods move farther down the value chain.

Physical FLW measurement approaches must disaggregate individual food groups and their edible and inedible parts to understand which foods are removed from the value chain, especially for evaluating nutrition and environmental outcomes. The FLI methodology uses Technical Conversion Factors to convert data between primary and secondary equivalents (for example, live animal to meat) for processed foods (Fabi and English 2019). However, no such conversion factors exist for retail, food service, or household stages covered by the FWI. To avoid duplication, pFLW measurement methodologies might consider expanding and adapting existing repositories of food lists, recipes, and ingredients (often used in dietary data collection) to serve as conversion factors at these stages. The Global Food Matters Database is an example of such an open access online repository (Intake 2022).

Technological innovations can reduce the cost and complexity of implementing standardized measurement and reporting methods for pFLW. Smartphone apps can expedite crowdsourced data collection at different value chain stages, and implementation must include training and data quality assurance. The FAO Food Loss App (FLAPP) is a recent tool designed to collect crowdsourced data from farmers to analyze where and why farm-level FLW occurs (FAO 2024a). On the consumer side, smartphone apps that allow users to share photographs of food selection, pFLW, waste reasons, and destinations are promising tools for measuring pFLW during food preparation, eating, and discarding stored food (Roe et al. 2020). When deciding to use technology for data collection, it is important to consider the potential for excluding certain populations, which can introduce biases and limitations in the data.

Box 4.6 Standardized Physical Food Loss and Waste Data Collection Approaches in India

The ICAR–CIPHET studies conducted between 2005 and 2012 contributed methodologies for quantifying pFLW nationwide in India (Nanda et al. 2012; Jha et al. 2015). These studies evaluated major crops across all food groups, including cereals, pulses, oilseeds, fruits, vegetables, plantation crops, spices, and animal-source foods (ASFs). The studies covered food value chain stages from harvest to on-farm postharvest activities, transportation, and storage channels (for example, farm, godown/warehouse, wholesale, retail, and processing) at a national scale, across all agroclimatic zones. Data were collected using surveys and measurements (direct weighing and counting) over a one-year crop cycle. FLW was estimated as the percent loss at each value chain stage, with overall pFLW calculated as the sum of these losses (Nanda et al. 2012; Jha et al. 2015). Jha et al. (2015) compared their 2012 estimates with those from the previous 2005–2007 study (Nanda et al. 2012), finding significant increases and decreases in FLW for specific products.

4.3.3 Identifying Limiting Quality Attributes and Sourcing Standardized Data

The diversity and complexity of food quality indicators requires different approaches to identify appropriate types and sources of food quality data for assessing FQL. While individual food quality attributes, food safety analysis, and food nutrient analysis are broad fields, this section focuses on data collection methods and data sources most relevant to FLW.

Food quality indicators that consider multiple attributes, such as grades, standards, and date labels, require data on which attributes limit product acceptance and the methods used to assess each attribute. Human perception and instrumental methods are used to assess individual food quality attributes. Human perception, while complex, cannot be perfectly replicated by instruments (Lawless and Heymann 2010). However, instrumental methods can objectively detect quality attributes, especially credence attributes (for example, food safety and nutrient content) that are imperceptible to humans. Measurement objectives should guide the selection of measurement methods to quantify important quality attributes at critical points in the value chain (Barrett et al. 2010; Shewfelt 2014).

The relative importance of individual quality attributes varies depending on the food product, the value chain stage, and intended use (for example, fresh or processed) (Kitinoja and Kader 2002; Barrett et al. 2010; Shewfelt 2014). Standardized surveys can help identify which attributes are prioritized in assessing food quality across different food commodities and value chains. Delgado et al. (2021b) proposed an attribute method (A-method) in which value chain actors define the percentage of their produce affected by FQL for specific appearance, texture, and flavor attributes. Additionally, published grades and standards or documented reasons for rejection (Box 4.7) can provide data on prioritized quality attributes. For date labels, primarily used for packaged foods, shelf-life studies focused on food safety or other food quality attributes (for example, color, aroma, taste, texture) determine how long the product can last under specified conditions before falling below food safety or other food quality standards.

Intrinsic search and experience attributes of food products fall under three main categories: appearance (visual), texture (feel), and flavor (taste and aroma) (Pearson 1994; Kader 2002; Barrett et al. 2010; Su et al. 2017; Delgado et al. 2021b). Although there is extensive literature on measuring these attributes in controlled environments (Lawless and Heymann 2010; Nicolaï et al. 2014; Shewfelt 2014; Su et al. 2017), assessing them in real-world settings varies, based on available resources and value chain requirements. The combined knowledge can inform surveys on specific quality attributes of intrinsic quality attributes to develop FQL indicators.

High quality, timely data are crucial for assessing food safety-related FQL to identify where, when, and how food safety hazards arise along the food value chain. Efforts to address data gaps include improving accessibility of analytical testing, identifying common foodborne disease sources, and identifying high-risk foods and environments. Analytical testing methods determine the presence and extent of food hazards (Ismail and Nielsen 2017). Rapid screening methods are often used initially, followed by lab-based methods (Ismail and Nielsen 2017). High costs associated with lab-based analyses can be a barrier in resource-limited contexts (Choi et al. 2019; Saloni Sharma et al. 2020). Emerging point-of-care technologies, using a device integrated with a smartphone app, offer promising alternatives for food safety monitoring in such settings (Choi et al. 2019; Saloni Sharma et al. 2020). Source attribution methods determine where in the food value chain the human disease burden of foodborne infections can be attributed to specific sources (Pires et al. 2009; Hald et al. 2016). The WHO Foodborne Disease Burden Epidemiology Reference Group (FERG) provides estimates of the burden of foodborne disease using food source attribution methods (Hoffmann et al. 2017). Identifying high-risk foods and environments can guide additional data collection efforts. For example, the US Food and Drug Administration (FDA) Food Traceability List identifies foods requiring additional traceability records (US FDA 2024).

The nutritional value of food changes over time due to storage and handling conditions and due to processing. Lab-based methods are used to measure changes in specific nutrients as food is stored fresh, frozen, or canned, and then prepared for eating in controlled settings (Rickman et al. 2007a; Rickman et al. 2007b). However, less is known about how nutrient content changes in fresh foods under real world storage and handling conditions in different value chain contexts. Food composition databases provide information on nutrient content of foods, but they vary in data sources and update frequency (Delgado et al. 2021a). Current challenges in evaluating food composition for evidence-based decision-making include reproducibility and standardization, representation and quantification, and accessibility (Ahmed et al. 2022). The *Periodic Table of Food Initiative* aims to apply standardized tools for mapping food composition across the food system and sharing this information in an accessible database (Ahmed et al. 2022).

Box 4.7 Using Food Rejection Data to Identify Limiting Quality Attributes

Food can be rejected at various stages of the value chain due to noncompliance with quality indicators. Data on rejected food can identify limiting quality attributes and critical loss points.

Border rejection data provide a broad overview of noncompliance trends across traded food products, exporting countries, and destination markets. High-income countries collect disaggregated border rejection data across agri-food products and time (Henson and Olale 2011). However, only a small portion of imported food is inspected at the border, and there may be nonrandom subjectivity in border inspections (Henson and Olale 2011). Rejected food may be returned to the country of origin, destroyed, or diverted to another destination (Henson and Olale 2011).

Data on rejection rates between domestic value chain actors is less available. As a leading food waste organization in the United States, ReFED (https://refed.org/) uses expert interviews to determine buyer rejection rates, the breakdown of final destination for rejected food, and the reasons for rejection (Powell and Curtis 2020). Expert interviews could also collect data on specific noncompliance with requirements.

4.4 Data Curation and Analysis

Given the complexity of FLW indicators, data curation is important for creating useful datasets that can be analyzed for evidence-based decision-making. It involves aggregating, structuring, and synthesizing data to create analytical tools and models that generate insights. This section focuses on data curation and analysis for estimating pFLW, as global goals and data priorities primarily center on this area.

Calculation-based FLW estimation methods using secondary data sources include mass-balance modeling and proxy and literature data (Hanson et al. 2016b; Chaboud 2017; Xue et al. 2017). Mass-balance methods estimate FLW by measuring inputs, outputs, and stock changes to food weight (Hanson et al. 2016b). Modeling can generate provisional data that can be updated with more primary data, but it risks oversimplifying complex food value chains and their contexts (Affognon et al. 2015). Estimation methods using proxy and literature data are less costly but may have limited scope and representativeness (for example, focusing on certain crops, regions, and production processes), potentially overestimating country averages (Fabi et al. 2021).

The quality and accuracy of FLW estimates depend on the quality and comprehensiveness of primary data sources, statistical tools, and estimation models. Inconsistent definitions, measurements, and reporting of FLW create challenges in data curation and analysis. A review of 202 publications reporting FLW data found that over half used only secondary data, highlighting uncertainties in the global FLW data landscape and the need for more consistent data that are interoperable across geographies, food groups, and food value chain stages (Xue et al. 2017). Due to differences in estimation approaches, global estimates from FAO (2011), FAO (2019), and UNEP (2024) are not comparable.

4.4.1 Open Access Food Loss and Waste Data Sources: Food Balance Sheets, Food Loss and Waste Database, and the Food Waste Atlas

Food loss and waste estimation typically requires multiple calculation-based methods, which draw upon data from different sources. For example, the widely cited FLW estimates from FAO (2011) are based on mass-balance estimates, using data from FAOSTAT (food volumes produced and mass flows) and literature data on loss weight percentages (Gustavsson et al. 2013). The accuracy of calculation-based estimation methods depends on the quality and comprehensiveness of the data used for pFLW quantities and model parameters (Xue et al. 2017). Since the FAO (2011) report, efforts to increase FLW data curation and analysis have emerged, including two open access databases that are specific to FLW: the Food Loss and Waste Database and the Food Waste Atlas.

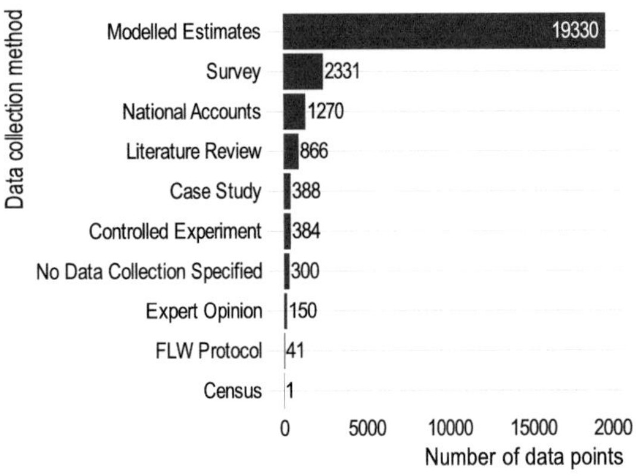

Fig. 4.2 Share of food loss and waste data from 2000–2022 by data collection method, Food Loss and Waste Database
Data from FAO (2024b), accessed on June 25, 2024

The Food Loss and Waste database (https://www.fao.org/platform-food-loss-waste/flw-data/en/) was developed by FAO, using machine learning to automatically identify relevant sources and extract information on FLW (Fabi et al. 2021). However, the database does not distinguish studies with different FLW definitions (Fabi et al. 2021), which could affect the quality of data, if definitions are not compatible. As of November 2021, the database contained over 29,000 data points from over 700 openly accessible publications and reports (FAO 2022). The vast majority of data in the FLW database comes from modeled estimates using secondary data (Fig. 4.2). As discussed in Chap. 3, data from the African Postharvest Losses Information System (APHLIS) dataset (see Box 4.8 for details) contribute significantly (more than half of data points) to the FLW Database. The FLW Database offers various filter options to visualize data by region, food categories, commodities, stages in the food value chain, and data collection methods. Users can also choose the reporting year range. Importantly, data can be downloaded for further analysis.

Box 4.8 African Postharvest Loss Information System

The APHLIS database, launched in 2009, and openly available online, focuses on pFLW estimates for nine cereal crops in sub-Saharan Africa and is expanding to include pulses, roots, tubers, and bananas (Rembold et al. 2011; Stathers et al. 2018; APHLIS 2022a). Loss estimates are based on data from peer-reviewed literature (for postharvest loss profiles) and local experts (for contextual factors) (Rembold et al. 2011; Hodges et al. 2014). Table 4.1 summarizes the literature used for 2021 estimates by supply chain stage.

APHLIS provides guidance to scientists for rapid, systematic approaches, based on visual scales and questionnaires, to collect new FLW data for inclusion in the default loss profiles (Hodges 2013). The quality of secondary data is assessed based on the data collection method (Hodges et al. 2014). To estimate losses when data are missing, APHLIS models available data, clustered by climate and crop (Hodges et al. 2014). APHLIS provides a postharvest loss calculator, by which users can calculate FLW with their own figures to estimate FLW under different contextual factors (Rembold et al. 2011; APHLIS 2022b).

Table 4.1 Number of references contributing to the African Postharvest Losses Information System's postharvest loss profiles across all countries for all crops in 2021, by value chain stage

Year	Number of publications
1970–79	3
1980–89	17
1990–99	18
2000–09	6
2010–19	6

Data from aphlis.net (https://perma.cc/H8S5-CNDD)

Box 4.9 Food Loss and Waste Data in the United States

The Economic Research Service (ERS) of the United States Department of Agriculture (USDA) monitors FLW at the retail and consumer levels using ERS's Loss-Adjusted Food Availability (LAFA) data series, which is open access and downloadable. LAFA data are derived from food availability data after adjusting for food spoilage, plate waste, and other losses (USDA ERS 2020). The primary purpose of LAFA data is to estimate the loss-adjusted amount of food available for consumption (Buzby et al. 2014). Therefore, FLW estimates using LAFA data are considered preliminary, pending further improvements to food loss assumptions and estimates. More details of the methodology used to create the LAFA data series can be found at USDA ERS (2021).

Starting in 2018, the US Environmental Protection Agency (EPA), US Food and Drug Administration (FDA), and US Department of Agriculture (USDA) began a formal agreement with the NGO partner, ReFED, to collaborate on efforts to cut food loss and waste in the United States by 50 percent by 2030 (US EPA 2019; US EPA et al. 2020). To measure progress in reducing FLW, federal agencies will partner with ReFED. ReFED's Insights Engine, an online tool, provides data and solutions related to FLW. The data is based on modeling estimates. For more details on ReFED's methodology, including definitions, sources, and data quality, see Powell and Curtis (2020).

The Food Waste Atlas (https://thefoodwasteatlas.org/), a global online database cofounded by the UK's WRAP and the WRI, with funding from WRAP and the Walmart Foundation, and support from UNEP, tracks global FLW across food types, value chain stages, and geographies. Data for the database come from several sources, including FAO, academic literature, businesses, and national governments. Entities interested in submitting data to the Food Waste Atlas are encouraged to follow the *FLW Standard*. However, the Food Waste Atlas does not provide a guarantee that submitted data are in full compliance with all reporting requirements.

The Food Waste Atlas contains over 150,000 records from organizations worldwide. However, the platform currently limits users to viewing only 10,000 entries at a time. Without a download option, users can only analyze a small portion of the data. While the SDG Target 12.3 uses the term "food waste" to focus on retail, food service, and household consumer levels, only 1,400 records in the Food Waste Atlas cover these stages, representing less than 1% of the total data (Food Waste Atlas 2024). To identify the primary data sources for individual countries, records were filtered by country and FAO's FBS data. Table 4.2 lists the total entries for several countries, revealing that FBS data constitutes the majority of available information. However, the database's search limitations hinder users to fully assess data availability and the unique contributions of different sources.

Table 4.2 Share of data entries on the Food Waste Atlas sourced from the Food and Agriculture Organization's Food Balance Sheets in select countries

Country	Total records	Food Balance Sheets records	Share of records from Food Balance Sheets source
United States	2495	1777	71.2%
Mexico	1860	1857	99.8%
China	5233	5124	97.9%
India	2258	2253	99.8%
Australia	1250	1216	97.3%
UK	1716	1412	82.3%
Kenya	1475	1474	99.9%
Brazil	2011	1986	98.8%

Data retrieved from thefoodwasteatlas.org/records (Food Waste Atlas 2024)

Box 4.10 A Closer Look at Loss Data on Food Balance Sheets

The FBS provide open access agrifood data for nearly all countries, with annual data dating back to 1961. Data are available on the FAOSTAT website, allowing public access to explore global, regional, and country trends in the supply and utilization of hundreds of primary and processed food items. The FBS methodology was updated in 2020, creating two datasets: *Food Balances (2010–)* and *Food Balances (−2013, old methodology and population)*. Accounting and adjusting for potential inconsistencies across datasets is important when combining data from these datasets to avoid skewed trends (Vonderschmidt et al. 2024).

FAO collects loss data, primary or estimated, from countries using the Agriculture Production Questionnaire and the newer Food Losses from Production to the Retail Stages Questionnaire (FAO 2024c). These data are used to create FBS. Countries are encouraged to carry out targeted surveys to gather official loss data. When official data is unavailable, loss data are imputed or estimated using several approaches (FAO 2020).

The FLI is designed to align with FAO's annual Agriculture Production Questionnaire and the FBS framework (Mingione et al. 2021; FAO 2024d). The index defines food loss as "human-edible commodity quantities" (Fabi and English 2019), which is consistent with the "Loss" variable in the Agriculture Production Questionnaire. This variable encompasses "all quantity losses along the supply chain for all utilizations (food, feed, seed, industrial, other)" until the retail stage, including edible and inedible parts (FAO 2024c). However, the FBS loss data may not be exclusively related to food intended for human consumption. The proportion of human-edible food commodities used for food versus feed varies by food group (for example, fruits and vegetables compared to cereals, see Chap. 3).

4.4.2 Existing Data and Modeling Approaches for SDG Target 12.3 Indicators

Due to limited primary data, FAO calculates the FLI using modeled estimates based on the FBS framework (Fabi and English 2019). The current modeling approach supplements the 7% of officially reported loss data (from FBS) with data from literature review, data from the FLW Database, and a dataset of over 200 potential explanatory variables (Mingione et al. 2021). FAO has estimated global food loss percentages for 2016, 2020, and 2021, comparing them to the base year of 2015 to calculate the GFLI. Although improvements to the modeling approach have been identified, they require more primary data for most commodity groups (Mingione et al. 2021).

Unlike the FLI, the FWI emphasizes the need for primary data on food waste at the national level for tracking changes over time and informing policy decisions (UNEP 2024). Currently, UNEP has only reported modeled estimates of per capita food waste at retail, food service, and household levels for 233 countries and areas. These estimates are based on literature review and extrapolations from estimates in other countries (UNEP 2024). Due to the uncertainty of this evidence and the limitations of modeled estimates for tracking food waste over time, UNEP has not published an FWI (UNEP 2024).

Box 4.11 Data Gaps in Physical Food Loss and Waste Data Curation

Open access databases have revealed significant gaps in primary pFLW data. The FLW Database and the Food Waste Atlas, both online repositories, aggregate FLW data. Despite containing thousands of data points, most of the data in both databases are calculated estimates rather than primary data. FAO's FLW Database, while extensive, relies heavily on modeled estimates and may have data quality issues, due to varying definitions. A large share of data on the FLW Database come from the APHLIS database. The Food Waste Atlas offers a vast amount of data but is limited by its user viewing capacity and lack of download functionality. Among data that can be viewed, it appears that most data on the Food Waste Atlas are sourced from FAO FBS.

SDG Target 12.3 Indicators use different methods to calculate their respective indices. The FLI uses the food loss percentage, a proportion of food production lost. Due to data limitations, FAO used a modeling approach to calculate the FLI for 2016, 2020, and 2021, using 2015 as the baseline (Mingione et al. 2021). Conversely, the FWI focuses on per capita food waste. Although UNEP has used modeling for estimations, this approach is not suitable for tracking changes over time, and no FWI has been reported yet (UNEP 2024). UNEP recommends countries measure food waste and plans to collect primary data through the United Nations Statistical Division (UNSD)/UNEP Questionnaire on Environmental Statistics.

4.4.3 Developing an Integrated Approach to Analyze Food Loss and Waste

Addressing FLW to support sustainable food systems requires a robust data infrastructure. Currently, limitations exist. To improve the availability and accessibility of FLW data, UN agencies must take a more coordinated approach that includes increasing primary data collection, establishing comprehensive data reporting mechanisms, and standardizing statistical methods used to estimate FLW. While FAO and UNEP currently serve as custodian agencies for monitoring and reporting on SDG 12.3, broader collaboration among UN agencies will be necessary as they look beyond the 2030 Agenda for Sustainable Development.

Global initiatives focused on FLW need to prioritize data curation and accessibility, by promoting data interoperability, so that data from different sources can be easily integrated and analyzed together. Open access platforms that complement each other, avoiding redundancy, are also key. Finally, establishing clear and consistent global guidelines on when estimated FLW data is sufficient and when primary data is necessary for informing policy and strategy decisions would promote support for an integrated approach (Xue et al. 2017).

Effective decision-making requires robust analytical tools and clear guidance. User-friendly tools and guides can empower decision-makers to analyze FLW data effectively. Furthermore, prioritizing data collection that allows analysis by food type, value chain stage, and final destination of FLW can offer valuable insights into the specific context of FLW issues. Integrating FQL indicators into FLW analysis is also essential for a holistic understanding of the problem. Structured guidance on how to do this is needed.

Exploring the potential of technology to improve food traceability offers encouraging possibilities for increased transparency within the food chain (Resende-Filho and Hurley 2012; Tripoli and Schmidhuber 2018). Strengthening global capacity to analyze FLW data and translate insights into actionable information for policymakers at all levels is another key priority. Initiatives like the ReFED Insights Engine (https://insights.refed.org/), a comprehensive open access online portal for FLW data and solutions in the United States, serve as useful examples of how collaboration and data sharing build platforms for informed decision-making.

4.5 Translation and Dissemination

The final stage of the FLW data value chain involves translating insights into actionable programs and policies to inform evidence-based decision-making, requiring effective knowledge. While reports like the *State of Food and Agriculture* 2019 (FAO 2019) and the Food Waste Index Report 2024 (UNEP 2024) track progress on SDG Target 12.3 indicators, the limitations in

FLW data estimation are often not adequately communicated (Box 4.12).

This can be further compounded by outdated information on online platforms. For instance, the SDG 12 Hub, managed by the One Planet Network currently displays progress on the Food Loss Index and Food Waste Index from 2020 (One Planet Network 2024). The One Planet Network implements the 10-Year Framework on Programmes on Sustainable Consumption and Production and works toward achieving SDG 12: ensuring sustainable patterns of consumption and production. Serving as the secretariat of the 10-Year Framework on Programmes, the UNEP facilitates the One Planet network. The SDG 12 Hub lags behind the FAO SDG Indicators Data Portal with 2021 FLI data (FAO 2024e), and even behind UNEP's FWI reports from 2021 and 2024 (UNEP 2021b; UNEP 2024).

To enhance the FLW data value chain and enable more effective policy decisions, UN agencies must prioritize ensuring consistent access to up-to-date FLW indicator data across various platforms. Additionally, integrating evidence and data on food quality into policy-oriented communications on FLW and food systems outcomes, particularly nutrition and health, will help to highlight the critical role of food quality in addressing FLW and its broader implications for food systems. Addressing these areas to build a more robust and actionable FLW data value chain will effectively support the development and implementation of programs and policies to reduce FLW.

Box 4.12 Data Influencers: Oversized Statements with Undersized Discussions

Since FAO's 2011 report revealed that approximately one-third of global food production was lost or wasted (FAO 2011), headline-grabbing statistics on FLW have become common. For nearly a decade, the "one-third" figure dominated discussions on FLW. However, in recent years, this narrative has evolved with the release of updated estimates by both FAO and UNEP. In 2019, FAO reported revised figures for FLW from postharvest up to retail stages (FAO 2019), and in 2021, UNEP provided updated estimates from the retail to consumer stages (UNEP 2021b).

Headlines now feature these updated number. For example, UN News reported: "Some 19 percent of food available to consumers was lost overall at retail, food service, and household levels. That is in addition to around 13 per cent of food lost in the supply chain, as estimated by the UN Food and Agriculture Organization, from post-harvest up to the point of sale" (UN 2024).

The translation of current FLW estimates from FAO and UNEP often overlooks crucial distinctions between estimation approaches and the quality and availability of underlying data, which can lead to misleading interpretations and comparisons. For instance, while the terms "food loss and waste," "food loss," and "food waste" are often used interchangeably, they are defined using slightly different concepts, making it challenging to combine or compare estimates. Moreover, the representativeness of global FLW estimates is limited by data gaps across value chain stages, food groups, and geographies. Effective knowledge dissemination on FLW should not be confined to high-level figures but must provide a comprehensive understanding, including the limitations and assumptions inherent in the data, to facilitate accurate interpretation and informed decision-making.

4.6 Conclusions

Since the recognition of FLW as a global challenge in the 2030 Agenda, significant strides have been made in developing indicators, collecting data, and sharing information to address this issue. However, a comprehensive data value chain is still needed to fully understand the extent of FLW and inform effective interventions.

A robust data ecosystem is essential for collecting, sharing, and using data effectively. Open access FLW platforms can enhance information accessibility and enable evidence-based decision-making. However, the reliance on modeled estimates at the global and national levels, with the lack of coordinated data collection efforts, pose significant challenges.

In this chapter, we examined the data requirements for addressing FLW and identified key knowledge gaps. By developing a data value chain roadmap and addressing these challenges, broader collaboration among UN agencies can improve stakeholder understanding of FLW and inform more effective policies and interventions beyond 2030.

One critical area that requires further attention is the development of robust FQL indicators. Although existing indicators can provide valuable insights, a more comprehensive approach is needed to capture the complex nature of FQL, including factors such as food safety, nutrient degradation, sensory attributes, and consumer acceptability. Developing and implementing a core set of standardized FQL indicators will enable governments to better understand the factors contributing to FLW and identify targeted policy interventions and investments to create enabling environments for loss reduction.

Strengthening international cooperation and collaboration is essential for addressing the global challenges of FLW. This includes supporting partnerships between governments, international organizations, academic institutions, and the private sector to develop and implement effective data-driven strategies. A comprehensive approach to reducing FLW requires sufficient data that is relevant and reliable, with clearly defined indicators that collectively address complex FLW pathways, along with improved cooperation across data stakeholders. Investment in these areas will improve the understanding of FLW, inform effective interventions, and ultimately deliver healthier and more sustainable diets for all.

References

Affognon H, Mutungi C, Sanginga P, Borgemeister C (2015) Unpacking postharvest losses in sub-Saharan Africa: A meta-analysis. World Dev 66:49–68. https://doi.org/10.1016/j.worlddev.2014.08.002. Accessed 12 Sep 2024

Ahmed S, de la Parra J, Elouafi I, German B, Jarvis A, Lal V, Lartey A, Longvah T, Malpica C, Vázquez-Manjarrez N, Prenni J, Aguilar-Salinas CA, Srichamnong W, Rajasekharan M, Shafizadeh T, Siegel JB, Steiner R, Tohme J, Watkins S (2022) Foodomics: a data-driven approach to revolutionize nutrition and sustainable diets. Front Nutr 9:874312. https://doi.org/10.3389/fnut.2022.874312. Accessed 13 Sep 2024

Ali A, Xia C, Ouattara N, Mahmood I, Faisal M (2021) Economic and environmental consequences' of post-harvest loss across food supply chain in the developing countries. J Clean Prod 323:129146. https://doi.org/10.1016/j.jclepro.2021.129146. Accessed 13 Sep 2024

APHLIS (2022a) The APHLIS+ Project. African Postharvest Loss Information System. https://perma.cc/F633-54P7. Accessed 27 Jun 2022

APHLIS (2022b) How APHLIS estimates loss. African Postharvest Loss Information System https://perma.cc/76TY-BK5X. Accessed 28 Jun 2022

Aschemann-Witzel J, Jensen JH, Jensen MH, Kulikovskaja V (2017) Consumer behaviour towards price-reduced suboptimal foods in the supermarket and the relation to food waste in households. Appetite 116:246–258. https://doi.org/10.1016/j.appet.2017.05.013. Accessed 13 Sep 2024

Barrett DM, Beaulieu JC, Shewfelt R (2010) Color, flavor, texture, and nutritional quality of fresh-cut fruits and vegetables: desirable levels, instrumental and sensory measurement, and the effects of processing. Crit Rev Food Sci 50:369–389. https://doi.org/10.1080/10408391003626322. Accessed 13 Sep 2024

Buzby JC, Farah-Wells H, Hyman J (2014) The estimated amount, value, and calories of postharvest food losses at the retail and consumer levels in the United States. Economic Information Bulletin No. 121, Economic Research Service, United States Department of Agriculture. https://www.ers.usda.gov/webdocs/publications/43833/43680_eib121.pdf. Accessed 14 Sep 2024

Cattaneo A, Sánchez MV, Torero M, Vos R (2021) Reducing food loss and waste: five challenges for policy and research. Food Policy 9:101974. https://www.sciencedirect.com/science/article/pii/S0306919220301780. Accessed 9 Sep 2024

Chaboud G (2017) Assessing food losses and waste with a methodological framework: insights from a case study. Resour Conserv Recy 125:188–197. https://doi.org/10.1016/j.resconrec.2017.06.008. Accessed 9 Sep 2024

Choi JR, Yong KW, Choi JY, Cowie AC (2019) Emerging point-of-care technologies for food safety analysis. Sensors 19(4):817. https://doi.org/10.3390/s19040817. Accessed 14 Sep 2024

Codex Alimentarius (2018) Codex general standard for the labelling of prepackaged foods CODEX STAN 1-1985. https://www.fao.org/4/y2770e/y2770e02.htm. Accessed 13 Sep 2024

Compton JAF, Wright M, Gay C, Stabrawa A (1992) A rapid method for loss assessment in stored maize and dried cassava. Working paper, University of Greenwich, London. https://gala.gre.ac.uk/id/eprint/10762/. Accessed 13 Sep 2024

Delgado L, Schuster M, Torero M (2017) The reality of food losses: a new measurement methodology. IFPRI Discussion Paper 1686. International Food Policy Research Institute, Washington, DC. https://ebrary.ifpri.org/digital/collection/p15738coll2/id/131530. Accessed 13 Sep 2024

Delgado A, Issaoui M, Vieira MC, Saraiva de Carvalho I, Fardet A (2021a) Food composition databases: does it matter to human health? Nutrients 13(8):2816. https://doi.org/10.3390/nu13082816. Accessed 13 Sept 2024

Delgado L, Schuster M, Torero M (2021b) Quantity and quality food losses across the value chain: a comparative analysis. Food Policy 98:101958. https://www.sciencedirect.com/science/article/pii/S0306919220301627?via%3Dihub. Accessed 13 Sep 2024

Doumeizel V (2019) Foresight review of food safety: feeding the world safely and sustainably [online] Lloyd's Register Foundation. https://www.lrfoundation.org.uk/496740/siteassets/pdfs/foresight_review_of_food_safety_web.pdf. Accessed 13 Sep 2024

Fabi C, English A (2019) Methodological Proposal for Monitoring SDG Target 12.3. Sub-Indicator 12.3.1.a The Food Loss Index Design, Data Collection Methods and Challenges. Working Paper Series ESS/18-13. FAO, Rome. https://openknowledge.fao.org/server/api/core/bitstreams/6603bc2f-7105-434b-baaa-f440e3505a88/content. Accessed 9 Sep 2024

Fabi C, Cachia F, Conforti P, English A, Rosero Moncayo J (2021) Improving data on food losses and waste: from theory to practice. Food Policy 98:101934. https://doi.org/10.1016/j.foodpol.2020.101934. Accessed 13 Sep 2024

FAO (2011) Global food losses and food waste—extent, causes and prevention. Food and Agriculture Organization of the United Nations, Rome. https://www.fao.org/4/mb060e/mb060e00.pdf. Accessed 10 Sep 2024

FAO (2019) The State of Food and Agriculture 2019. Moving forward on food loss and waste reduction. Food and Agriculture Organization of the United Nations, Rome. https://openknowledge.fao.org/server/api/core/bitstreams/11f9288f-dc78-4171-8d02-92235b8d7dc7/content. Accessed 9 Sep 2024

FAO (2020) New food balances: description of utilization variables. https://files-faostat.fao.org/production/FBS/New%20FBS%20methodology.pdf. Accessed 13 Sep 2024

FAO (2022) Food Loss and Waste Database. https://www.fao.org/platform-food-loss-waste/flw-data/en/. Accessed 14 Jun 2022

FAO (2024a) FAO Food Loss App (FLAPP). In: Technical Platform on the Measurement of Reduction and Food Loss and Waste. https://www.fao.org/platform-food-loss-waste/news/news-detail/fao-food-loss-application-(flapp)-virtual-introduction/en. Accessed 21 Aug 2024

FAO (2024b) Food Loss and Waste Database. https://www.fao.org/platform-food-loss-waste/flw-data/en/. Accessed 25 Jun 2024

FAO (2024c) Data collection. https://www.fao.org/statistics/data-collection/agriculture/en#navbarNavWebsite. Accessed 19 Aug 2024

FAO (2024d) SDG Indicators Data Portal. Metadata. https://www.fao.org/sustainable-development-goals-data-portal/methods-and-tools/metadata/en. Accessed 13 Sep 2024

FAO (2024e) SDG Indicators Data Portal. https://perma.cc/RD4B-2RWN. Accessed 9 Aug 2024

Food Waste Atlas (2024) The Food Waste Atlas. https://thefoodwasteatlas.org/. Accessed 19 Aug 2024

GAIN (2020) Indicators of food safety performance: a review. A USAID EatSafe Project Report

Giovannucci D, Reardon T (2000) Understanding grades and standards. The guide to developing agricultural markets and agro-enterprises. https://papers.ssrn.com/sol3/papers.cfm?abstract_id=996761. Accessed 13 Sep 2024

Global Partnership for Sustainable Development Data (2016) The State of Development Data Funding 2016. An initiative of the Resource Mobilization and Alignment Working Group of the Global Partnership for Sustainable Development Data. https://www.data4sdgs.org/resources/state-development-data-funding. Accessed 11 Sep 2024

Government of the United Kingdom (2022) Food labelling: Giving food information to consumers. https://www.gov.uk/guidance/food-labelling-giving-food-information-to-consumers#show-the-best-before-or-use-by-date. Accessed 1 Aug 2022

Gruber V, Holweg C, Teller C (2016) What a waste! Exploring the human reality of food waste from the store manager's perspective. J Public Policy Mark 35(1):3–25. https://doi.org/10.1509/jppm.14.095. Accessed 13 Sep 2024

Gustafson D, Gutman A, Leet W, Drewnowski A, Fanzo J, Ingram J (2016) Seven food system metrics of sustainable nutrition security. Sustainability 8(3):196. https://doi.org/10.3390/su8030196. Accessed 13 Sep 2024

Gustavsson J, Cederberg C, Sonesson U, Emanuelsson A (2013) The methodology of the FAO study: "Global Food Losses and Food Waste—extent, causes and prevention"– FAO, 2011. The Swedish Institute for Food and Biotechnology (SIK). https://www.diva-portal.org/smash/get/diva2:944159/FULLTEXT01.pdf. Accessed 13 Sep 2024

Hald T, Aspinall W, Devleesschauwer B, Cooke R, Corrigan T, Havelaar AH, Gibb HJ, Torgerson PR, Kirk MD, Angulo FJ, Lake RJ, Speybroeck N, Hoffmann S (2016) World Health Organization estimates of the relative contributions of food to the burden of disease due to selected foodborne hazards: a structured expert elicitation. PLoS One 11:e0145839. https://doi.org/10.1371/journal.pone.0145839. Accessed 13 Sep 2024

Hanson C, Lipinski B, Robertson K, Dias D, Gavilan I, Gréverath P, Ritter S, Fonseca J, van Otterdijk R, Timmermans T, Lomax J, O'Connor C, Dawe A, Swannell R, Berger V, Reddy M, Somogyi D, Tran B, Leach B, Quested T (2016a) Food Loss and Waste Accounting and Reporting Standard. World Resources Institute. https://www.wri.org/research/food-loss-and-waste-accounting-and-reporting-standard. Accessed 14 Sep 2024

Hanson C, Lipinski B, Robertson K, Dias D, Gavilan I, Gréverath P, Ritter S, Fonseca J, van Otterdijk R, Timmermans T, Lomax J, O'Connor C, Dawe A, Swannell R, Berger V, Reddy M, Somogyi D, Tran B, Leach B, Quested T (2016b) Guidance on FLW Quantification Methods. Supplement to the Food Loss and Waste (FLW) Accounting and Reporting Standard, Version 1.0. https://flwprotocol.org/wp-content/uploads/2017/06/FLW-Protocol_Guidance-on-FLW-Quantification-Methods.pdf. Accessed 14 Sep 2024

Henson S, Olale E (2011) What do border rejections tell us about trade standards compliance of developing countries? Analysis of EU and US data 2002–2008. UNIDO Working Paper. Vienna, United Nations Industrial Development Organization. https://www.unido.org/sites/default/files/2011-12/rejection_analysis_0.PDF. Accessed 14 Sep 2024

HLPE (2014) Food losses and waste in the context of sustainable food systems. Report by the High Level Panel of Experts on Food Security and Nutrition, Rome. https://openknowledge.fao.org/server/api/core/bitstreams/b1949fae-23d4-473c-8b87-8c4359b74d6c/content. Accessed 13 Sep 2024

Hodges R (2013) How to assess postharvest cereal losses and their impact on grain supply: rapid weight loss estimation and the calculation of cumulative cereal losses with the support of APHLIS. African Postharvest Losses Information System https://www.aphlis.net/downloads/APHLIS%20Loss%20Assessment%20Manual.pdf. Accessed 13 Sep 2024

Hodges R, Bernard M, Rembold F, European Commission Joint Research Centre (2014) APHLIS: postharvest cereal losses in Sub-Saharan Africa, their estimation, assessment and reduction. Publications Office, Luxembourg

Hoffmann S, Devleesschauwer B, Aspinall W, Cooke R, Corrigan T, Havelaar A, Angulo F, Gibb H, Kirk M, Lake R, Speybroeck N, Torgerson P, Hald T (2017) Attribution of global foodborne disease to specific foods: findings from a World Health Organization structured expert elicitation. PLoS One 12:e0183641. https://doi.org/10.1371/journal.pone.0183641. Accessed 10 Sep 2024

Intake (2022) Global Diet Quality Score toolkit. Intake—Center for Dietary Assessment/FHI Solutions, Washington, DC. https://www.intake.org/sites/default/files/2022-10/GDQS_Toolkit_Final_6Oct2022_2.pdf. Accessed 14 Sep 2024

Ismail BP, Nielsen SS (2017) Chapter 33 Analysis of food contaminants, residues, and chemical constituents of concern. In: Nielsen SS (ed) Food analysis. Springer, Cham, pp 573–597

Jha SN, Vishwakarma RK, Ahmad T, Rai A, Dixit AK (2015) Report on assessment of quantitative harvest and post-harvest losses of major crops/commodities in India. ICAR-All India Coordinated Research Project on Post-Harvest Technology, ICAR–CIPHET, PO: PAU, Ludhiana. https://www.researchgate.net/publication/289637983_Assessment_of_Quantitative_Harvest_and_Post-Harvest_Losses_of_Major_CropsCommodities_in_India. Accessed 14 Sep 2024

Kader AA (2002) Quality and safety factors: definition and evaluation for fresh horticultural crops. In: Kader AA (ed) Postharvest technology of horticultural crops, 3rd edn. University of California Agriculture and Natural Resources, Davis, pp 279–286

Kitinoja L, Kader AA (2002) Small-scale postharvest handling practices: a manual for horticultural crops, 4th edn. University of California, Davis Postharvest Technology Research and Information Center

Kitinoja L, Tokala VY, Brondy A (2018a) A review of global postharvest loss assessments in plant-based food crops: recent findings and measurement gaps. J Postharvest Technol 6:1–15

Kitinoja L, Tokala VY, Brondy A (2018b) Challenges and opportunities for improved postharvest loss measurements in plant-based food crops. J Postharvest Technol 6:16–34

Lawless HT, Heymann H (2010) Sensory evaluation of food: principles and practices. Springer, New York

Leib EB, Gunders D, Ferro J, Nielsen A, Nosek G, Qu J (2013) The dating game: How confusing food date labels lead to food waste in America. NRDC Report R:13-09-A, September, Natural Resources Defense Council. https://www.nrdc.org/sites/default/files/dating-game-report.pdf. Accessed 13 Sep 2024

Mingione M, Fabi C, Lasinio GJ (2021) Measuring and modeling food losses. J Off Stat 37:171–211. https://doi.org/10.2478/jos-2021-0008. Accessed 9 Sep 2024

Moreno LC, Tran T, Potts MD (2020) Consider a broccoli stalk: how the concept of edibility influences quantification of household food waste. J Environ Manag 256:109977. https://doi.org/10.1016/j.jenvman.2019.109977. Accessed 14 Sep 2024

Muth MK, Giombi KC, Bellemare M, Ellison B, Roe B, Smith T (2018) Expert Panel on Technical Questions and Data Gaps for the ERS Loss-Adjusted Food Availability (LAFA) Data Series. Agreement No. 59-4000-6-0069 Final Report, RTI International, Research Triangle Park, North Carolina, prepared for United States Department of Agriculture, Economic Research Service. https://permanent.fdlp.gov/gpo229667/ccr-70.pdf. Accessed 14 Sep 2024

Nanda SK, Vishwakarma RK, Bathla HVL, Rai A, Chandra P (2012) Harvest and post harvest losses of major crops and livestock produce in India. All India Coordinated Research Project on Post Harvest Technology, Indian Council of Agricultural Research (ICAR), Ludhiana. https://www.mofpi.gov.in/sites/default/files/phl_2010_ciphet.pdf. Accessed 14 Sep 2024

Neff RA, Spiker M, Rice C, Schklair A, Greenberg S, Leib EB (2019) Misunderstood food date labels and reported food discards: a survey of U.S. consumer attitudes and behaviors. Waste Manag 86:123–132. https://doi.org/10.1016/j.wasman.2019.01.023. Accessed 13 Sep 2024

Newsome R, Balestrini CG, Baum MD, Corby J, Fisher W, Goodburn K, Labuza TP, Prince G, Thesmar HS, Yiannas F (2014) Applications and perceptions of date labeling of food: date labeling of food. Compr Rev Food Sci Food Saf 13(4):745–769. https://doi.org/10.1111/1541-4337.12086. Accessed 13 Sep 2024

Nicholes MJ, Quested TE, Reynolds C, Gillick S, Parry AD (2019) Surely you don't eat parsnip skins? Categorising the edibility of food waste. Resour Conserv Recy 147:179–188. https://doi.org/10.1016/j.resconrec.2019.03.004. Accessed 14 Sep 2024

Nicolaï B, Bulens I, De Baerdemaeker J, De Ketelaere B, Lammertyn J, Saeys W, Verboven P, Hertog MLATM (2014) Non-destructive evaluation: detection of external and internal attributes frequently associated with quality and damage. In: Florkowski WJ, Shewfelt RL, Brueckner B, Prussia SE (eds) Postharvest handling: a systems approach, 3rd edn. Elsevier Academic Press, Amsterdam, pp 363–385

One Planet Network (2024) SDG 12 Hub: food loss and waste. https://www.oneplanetnetwork.org/sdg-12-hub. Accessed 9 Aug 2024

Östergren K, Gustavsson J, Brouwers HB, Timmermans T, Hansen O-J, Møller H, Anderson G, O'Connor C, Soethoudt H, Quested T, Easteal S, Politano A, Bellettato C, Canali M, Falasconi L, Gaiani S, Vittuari M, Schneider F, Moates G, Waldron K, Redlingshöfer B (2014) FUSIONS definitional framework for food waste. Swedish Institute for Food and Biotechnology (SIK)

Parfitt J, Croker T, Brockhaus A (2021) Global food loss and waste in primary production: a reassessment of its scale and significance. Sustainability 13:12087. https://doi.org/10.3390/su132112087. Accessed 9 Sep 2024

Pearson AM (1994) Appendix 1.1. Major quality traits of meat, poultry and fish products—some factors associated with them and their effects on acceptability of different products. In: Pearson AM, Dutson TR (eds) Quality attributes and their measurement in meat, poultry and fish products. Springer, Boston, pp 18–24

Pires SM, Evers EG, van Pelt W, Ayers T, Scallan E, Angulo FJ, Havelaar A, Hald T (2009) Attributing the human disease burden of foodborne infections to specific sources. Foodborne Pathog Dis 6(4):417–424. https://doi.org/10.1089/fpd.2008.0208. Accessed 10 Sep 2024

Piwoz E, Rawat R, Fracassi P, Kim D (2019) Strengthening the nutrition data value chain for accountability and action. Sight Life 33(1):38–43. https://doi.org/10.52439/HDXT8911. Accessed 11 Sep 2024

Powell C, Curtis P (2020) Insights Engine Food Waste Monitor. 2020 Methodology. ReFED. https://insights.refed.org/uploads/documents/refed-insights-engine-food-waste-monitor-methodology-vfinal-2022-03-11.pdf. Accessed 13 Sep 2024

Rembold F, Hodges R, Bernard M, Knipschild H, Léo O (2011) The African Postharvest Losses Information System (APHLIS): An innovative framework to analyse and compute quantitative postharvest losses for cereals under different farming and environmental conditions in East and Southern Africa. https://core.ac.uk/download/pdf/38621127.pdf. Accessed 13 Sep 2024

Remons R, Wood SA, Saha N, Anderman TL, DeFries RS (2014) Measuring nutritional diversity of national food supplies. Glob Food Secur 3(3–4):174–182. https://doi.org/10.1016/j.gfs.2014.07.001. Accessed 13 Sep 2024

Resende-Filho MA, Hurley TM (2012) Information asymmetry and traceability incentives for food safety. Int J Prod Econ 139(2):596–603. https://doi.org/10.1016/j.ijpe.2012.05.034. Accessed 15 Sep 2024

Rickman JC, Barrett DM, Bruhn CM (2007a) Nutritional comparison of fresh, frozen and canned fruits and vegetables. Part 1. Vitamins C and B and phenolic compounds. J Sci Food Agric 87(6):930–944. https://doi.org/10.1002/jsfa.2825. Accessed 13 Sep 2024

Rickman JC, Bruhn CM, Barrett DM (2007b) Nutritional comparison of fresh, frozen, and canned fruits and vegetables II. Vitamin A and carotenoids, vitamin E, minerals and fiber. J Sci Food Agric 87(7):1185–1196. https://doi.org/10.1002/jsfa.2824. Accessed 13 Sep 2024

Roe BE, Qi D, Beyl RA, Neubig KE, Martin CK, Apolzan JW (2020) The validity, time burden, and user satisfaction of the FoodImage™ smartphone app for food waste measurement versus diaries: A randomized crossover trial. Resour Conserv Recy 160:104858. https://doi.org/10.1016/j.resconrec.2020.104858. Accessed 14 Sep 2024

Roesel K, Grace D, Makita K, Bonfoh B, Kang'ethe E, Kurwijila L, Hendrickx S, McCrindle C, Tano-Debrah K, Zewde G, Matusse H (2015) Hazards do not always translate to risks. In: Roesel K, Grace D (eds) Food safety and informal markets: animal products in Sub-Saharan Africa. Routledge, Taylor & Francis Group, London, pp 33–44

Rutten MM (2013) What economic theory tells us about the impacts of reducing food losses and/or waste: implications for research, policy and practice. Agr Food Secur 2:Article No. 13. https://doi.org/10.1186/2048-7010-2-13. Accessed 10 Sep 2024

Schneider KR, Fanzo J, Haddad L, Herrero M, Moncayo JR, Herforth A, Remans R, Guarin A, Resnick D, Covic N, Béné C, Cattaneo A, Aburto N, Ambikapathi R, Aytekin D, Barquera S, Battersby J, Beal T, Molina PB, Cafiero C, Campeau C, Caron P, Conforti P, Damerau K, Di Girolamo M, DeClerck F, Dewi D, Elouafi I, Fabi C, Foley P, Frazier TJ, Gephart J, Golden C, Fischer CG, Hendriks S, Honorati M, Huang J, Kennedy G, Laar A, Lal R, Lidder P, Loken B, Marshall Q, Masuda YJ, McLaren R, Miachon L, Muñoz H, Nordhagen S, Qayyum N, Saisana M, Suhardiman D, Sumaila UR, Cullen MT, Tubiello FN, Vivero-Pol J-L, Webb P, Wiebe K (2023) The state of food systems worldwide in the countdown to 2030. Nat Food 4:1090–1110. https://doi.org/10.1038/s43016-023-00885-9. Accessed 14 Sep 2024

Secondi L, Principato L, Ruini L, Guidi M (2019) Reusing food waste in food manufacturing companies: The case of the tomato-sauce supply chain. Sustainability 11(7):2154. https://doi.org/10.3390/su11072154. Accessed 14 Sep 2024

Sharma S, Kaur S, Kumar A, Sharma N, Kapoor P, Chunduri V, Kumari A, Garg M (2020) Chapter 13 Diagnostic tools for food safety. In: Sharma TR, Deshmukh R, Sonah H (eds) Advances in agri-food biotechnology. Springer, Singapore, pp 333–365

Sheahan M, Barrett CB (2017) Review: food loss and waste in Sub-Saharan Africa. Food Policy 70:1–12. https://doi.org/10.1016/j.foodpol.2017.03.012. Accessed 10 Sep 2024

Shewfelt RL (2014) Measuring quality and maturity. In: Florkowski WJ, Shewfelt RL, Brueckner B, Prussia SE (eds) Postharvest handling: a systems approach, 3rd edn. Elsevier Academic Press, Amsterdam, pp 387–410

Soethoudt JM, Pedrotti M, Bos-Brouwers HEJ, Castelein RB (2021) Adoption of food loss and waste-reducing interventions in low- and middle-income countries. Wageningen Food & Biobased Research, Wageningen. https://research.wur.nl/en/publications/adoption-of-food-loss-and-waste-reducing-interventions-in-low-and. Accessed 14 Sep 2024

Stathers T, Ognakossan KE, Priebe J, Mvumi BM, Tran B (2018) Counting losses to cut losses: quantifying legume postharvest losses to help achieve food and nutrition security. Proceedings of the 12th International Working Conference on Stored Product Protection (IWCSPP), Berlin, October 7–11, Vol. 1. Julius-Kühn-Archiv Nr 463:8–18. https://doi.org/10.5073/JKA.2018.463.004. Accessed 13 Sep 2024

Su W-H, He H-J, Sun D-W (2017) Non-destructive and rapid evaluation of staple foods quality by using spectroscopic techniques: A review. Crit Rev Food Sci 57(5):1039–1051. https://doi.org/10.1080/10408398.2015.1082966. Accessed 14 Sep 2024

Trienekens JH (2011) Agricultural value chains in developing countries: a framework for analysis. Int Food Agribus Man Rev 14(2):51–82. https://edepot.wur.nl/189057. Accessed 14 Sep 2024

Tripoli M, Schmidhuber J (2018) Emerging opportunities for the application of blockchain in the agri-food industry. Issue Paper. Food and Agriculture Organization of the United Nations (FAO), Rome, and International Centre for Trade and Sustainable Development (ICTSD), Geneva. http://www.fao.org/3/ca1335en/CA1335EN.pdf. Accessed 15 Sep 2024

UN (2024) With 783 million people going hungry, a fifth of all food goes to waste. UN News

UN (United Nations) (2015) A/RES/70/1 Transforming our world: The 2030 Agenda for Sustainable Development

UNEP (2021a) SDG indicator metadata (Indicator 12.3.2). https://sdg12hub.org/sites/default/files/2021-06/Metadata-12-03-01B.pdf. Accessed 22 Aug 2022

UNEP (2021b) Food Waste Index Report 2021. United Nations Environment Programme, Nairobi. https://www.unep.org/resources/report/unep-food-waste-index-report-2021. Accessed 10 Sep 2024

UNEP (2024) Food Waste Index Report 2024. United Nations Environment Programme, Nairobi.. https://www.unep.org/resources/publication/food-waste-index-report-2024. Accessed 9 Sep 2024

US Congress (2023) Food Date Labeling Act of 2023. https://www.congress.gov/bill/118th-congress/house-bill/3159. Accessed 26 Sept 2024

US EPA (2019) Winning on Reducing Food Waste. FY 2019-2020 Federal Interagency Strategy. April. https://19january2021snapshot.epa.gov/sites/static/files/2019-08/documents/interagency_strategy_on_reducing_food_waste_final.pdf. Accessed 15 Sep 2024

US EPA, US FDA, USDA (2020) Formal agreement among the United States Environmental Protection Agency and the United States Food and Drug Administration and the Unites States Department of Agriculture Relative to a Federal—Nongovernmental Organization Formal Agreement with ReFED, Inc. in Order to Better Evaluate and Improve Upon Strategies to Reduce Food Loss and Waste. https://19january2021snapshot.epa.gov/sites/static/files/2019-04/documents/refed-epa-usda-fda_formal_agreement.pdf. Accessed 15 Sep 2024

US FDA (2024) Food Traceability List. United States Food and Drug Administration. https://www.fda.gov/food/food-safety-modernization-act-fsma/food-traceability-list. Accessed 8 Aug 2024

USDA ERS (2020) Loss-Adjusted Food Availability Documentation. United States Department of Agriculture/Economic Research Service. https://perma.cc/2LT2-QT7S. Accessed 29 Jun 2022

USDA ERS (2021) Food Availability (Per Capita) Data System. https://catalog.data.gov/dataset/food-availability-per-capita-data-system. Accessed 15 Sep 2024

USDA FSIS (2019) Food product dating. Food Safety and Inspection Service, United States Department of Agriculture. https://www.fsis.usda.gov/food-safety/safe-food-handling-and-preparation/food-safety-basics/food-product-dating. Accessed 19 Jul 2022

Vonderschmidt A, Arendarczyk B, Jaacks LM, Bellows AL, Alexander P (2024) Analysis combining the multiple FAO food balance sheet datasets needs careful treatment. Lancet Planet Health 8(2):e69–e71. https://doi.org/10.1016/S2542-5196(23)00276-0. Accessed 13 Sep 2024

Wilson NLW, Miao R, Weis C (2018) Seeing is not believing: Perceptions of date labels over food and attributes. J Food Prod Mark 24(5):611–631. https://doi.org/10.1080/10454446.2018.1472700. Accessed 13 Sep 2024

Xue L, Liu G, Parfitt J, Liu X, Van Herpen E, Stenmarck Å, O'Connor C, Östergren K, Cheng S (2017) Missing food, missing data? A critical review of global food losses and food waste data. Environ Sci Technol 51(12):6618–6633. https://doi.org/10.1021/acs.est.7b00401. Accessed 14 Sep 2024

5 Protecting the Harvest: Farmgate Loss Pathways Linked to Food Availability

5.1 Introduction

Diversifying food production and increasing the supply of nutritious foods are essential first steps toward transforming food systems to deliver healthy diets. However, production alone will not be enough. Maintaining the quality and quantity of nutritious foods throughout the postproduction process is equally crucial. For food systems to deliver healthy diets, sufficient quantities of nutrient-rich and staple foods must be readily available to enter food value chains and become accessible to consumers (Global Panel 2020a; HLPE 2020). This requires both the production of a diverse variety of nutritious foods and the prevention of food loss and waste (FLW) at the farmgate.

FLW pathways at the preharvest and harvest stages determine the food supply composition available for post-farmgate stages. Physical food loss and waste (pFLW) before and at the time of harvest reduces the mass or volume of food entering value chains. Foods that have food quality loss (FQL) at the farmgate stages may be diverted as pFLW or may enter food value chains, lowering the food supply quality. FLW feedback loops occurring throughout the value chain are linked to food availability when perceived FLW risks influence production decisions and when quality tolerance limits influence producers' capacity to invest in and manage FLW.

FLW at the farmgate impacts the composition of the food supply available for post-farmgate stages. Physical FLW reduces the quantity of food entering value chains. Global pFLW of total agricultural production intended for human consumption at the farmgate was estimated at 8.3% at harvest and an additional 7.0% at postharvest stages (Parfitt et al. 2021). FQL is a factor in food acceptability, which contributes to agency in food preferences. Food acceptability is influenced by the perceived quality of a food, its value in a given market, and its usability and intended use, all of which are shaped by social and market conditions (Tijskens and Schouten 2014). Therefore, FQL of limiting attributes contributes to farmgate pFLW when quality falls below acceptable thresholds. Alternatively, FQL can result in lower quality food entering food value chains. The impact of this result depends on the specific quality attributes affected and the standards of the relevant value chain.

This chapter focuses on identifying FLW pathways at the farm level that significantly impact the availability of nutritious, perishable foods entering both fresh and processed food value chains. We begin by examining the current state of global FLW data at the farm level, highlighting persistent measurement challenges and knowledge gaps. Subsequently, we present evidence from the literature to describe the factors and processes that influence FQL and pFLW, and the availability and accessibility of technologies,

J. Boiteau, P. Pingali, *Wasted Potential*, Sustainable Development Goals Series,
https://doi.org/10.1007/978-3-031-97411-3_5

knowledge, and infrastructure that shape producer decision-making. When discussing different food commodity groups, we use the terms "preharvest" and "harvest" to refer to the farm-level stages, as defined in Chap. 3. Additionally, we include postharvest processing and storage activities that occur on-farm.

5.2 Farmgate Food Loss and Waste in the Context of Regional Food Production and Diet Recommendations

Food systems must prioritize a diverse, nutritious food supply, ensuring sufficient production and efficient distribution to support healthy diets. However, a significant gap exists between the global availability of foods and those required for meeting diet recommendations (KC et al. 2018; Global Panel 2020b). Optimizing food production and minimizing FLW to close the gap requires sufficient and timely data on food production and pFLW once food is ready for harvest. Existing data approaches often overlook key farmgate stages, limiting our knowledge of food production and FLW (Flanagan et al. 2019; Parfitt et al. 2021).

Determining the exact point where food production ends and the FLW boundary begins can be challenging (see Chap. 3). Classifying FLW at the farm level depends on when foods are considered ready for harvest (that is, the preharvest stage), when foods are harvested, and the specific postharvest activities performed on-farm versus off-farm, all of which change with food system transformation (Parfitt et al. 2021). The diversification of food production systems and the growing complexity of food value chains affect the timing of commercial maturity and harvest, particularly, for perishable foods. Additionally, postharvest activities, such as processing and sorting, shift away from on-farm operations and contribute to growing off-farm rural employment (Parfitt et al. 2021; Reardon et al. 2021).

Although farm-level activities are important contributors to global FLW, existing measurement frameworks often start at the postharvest stages, overlooking the preharvest and harvest stages. National production data frequently only report on harvested products, excluding any edible food that remained unharvested or discarded during harvest. For example, many FLW estimates, including the Food Loss Index (FLI), rely on data from the Food Balance Sheets (FBS) of the Food and Agriculture Organization of the United Nations (FAO) (Box 5.1), with different approaches to using the data based on the FLW definition and estimation approach (Kummu et al. 2012; Gustavsson et al. 2013; Aragie et al. 2023). The FLI framework focuses on the postharvest stages, in alignment with the FAO FBS framework (Fabi and English 2019). Therefore, there is a lack of data, particularly, in-field measurements of how much edible food is lost at the preharvest and harvest stages (Johnson et al. 2018a, 2018b), posing a challenge for assessing farmgate FLW pathways (Parfitt et al. 2021; Aragie et al. 2023; Boiteau and Pingali 2023).

Box 5.1 Spotlight on Food Production and Loss Data from FAO Food Balance Sheets

FAO Food Balance Sheets (FBS) provide open access agrifood data, including losses, for most countries. Production volumes are reported for total domestic production of primary crops, livestock items, primary fish items, and production of processed commodities (FAO 2001). Production at the farm level includes primary crops, excluding harvesting losses, and livestock items. Losses are reported to include all quantity losses along the supply chain up to retail and to exclude losses during preharvest and harvesting stages.

FBS data report on agricultural commodities, including those intended for human consumption and those not (for example, feed, seed, industrial uses) (FAO 2001). This means that losses reported in the FBS encompass a broader range than just loss of food intended for human consumption. Consequently, FLW estimation approaches that rely on FBS data, like the

Food Loss Index, may include losses of commodities not strictly intended for human consumption.

Another limitation of using FBS data is the scarcity of officially reported loss data. Although countries are encouraged to conduct specific surveys to collect loss information, FAO often relies on estimated or imputed data when official figures are missing. As a result, only a small fraction (approximately 7%) of the loss data in the FBS is officially reported (Mingione et al. 2021).

A comprehensive assessment of the postproduction food supply at the farmgate requires considering preharvest and harvest losses. In Fig. 5.1, we analyze the balance between regional food production (before any pFLW) and the components of healthy diets, expanding upon the work of KC et al. (2018). Using FAOSTAT production data from 2016 to 2020, we average the figures to account for annual variation (Kummu et al. 2012; Aragie et al. 2023). We apply an allocation factor to estimate the portion of production intended for human consumption, following FAO (2011), and using Kummu et al. (2012) as a reference. To address in-field and harvest losses, we adjust FAO production statistics using regional loss factors (FAO 2011). Similar to Kummu et al. (2012), we do not convert agricultural production to edible weight as this does not affect calorie content. Finally, we calculate the total food production for each food category in terms of dietary servings, as in KC et al. (2018). Comparing this distribution to the Healthy Diet Bundle (Herforth et al. 2022), we can evaluate the alignment between food production and the nutritional requirements of a healthy diet.

Globally, production of starchy staples and oils and fats surpasses the dietary recommendations of the Healthy Diet Bundle, while production of nutrient-rich, perishable foods (fruits, vegetables, protein-rich foods) falls short (Fig. 5.1). The imbalance persists across most regions, despite variations in per capita production levels. Consequently, all global regions struggle to produce enough nutritious food to support balanced, healthy diets—even before losses occur. As countries modernize, consumer

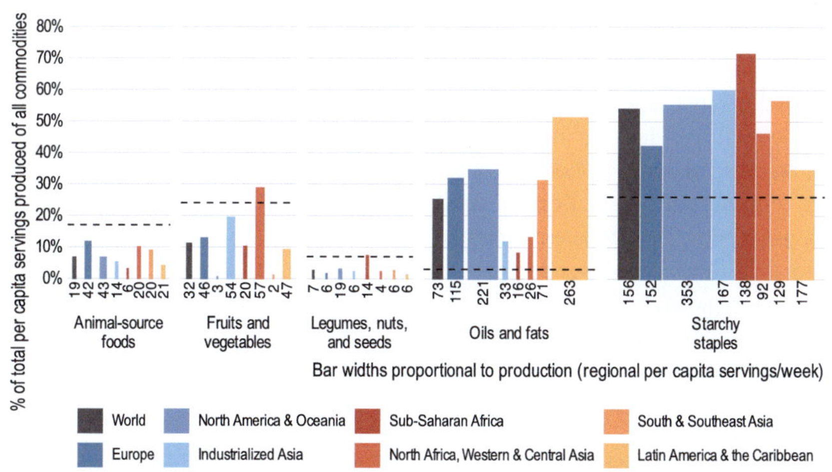

Fig. 5.1 Global and regional food production of food intended for human consumption

Bar widths are proportional to the number of servings/per capita/week of food produced that is intended for human consumption. Bar heights represent the percent of total food production of all commodities for the indicated region, before any physical FLW. The dashed line represents Healthy Diet Basket recommendations (Herforth et al. 2022), from KC et al. (2018). Cool colors (blues) are used for high- and middle-income regions; warm colors (reds and oranges) are used for low-income regions according to FAO (2011). Production data (harvested weight) was calculated from FAOSTAT 2016–2020 (FAOSTAT 2024)

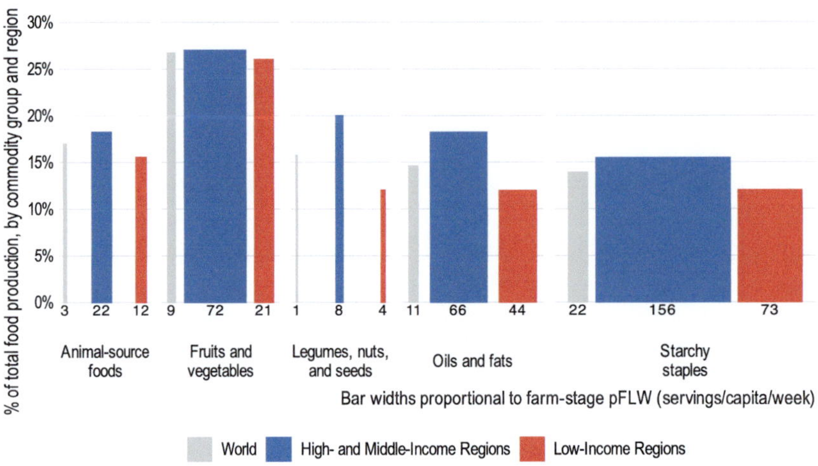

Fig. 5.2 Farmgate physical food loss and waste by food group as a share of total food produced for human consumption, by income region

Bar widths are proportional to farm-stage food losses in terms of number of per capita servings/week. Bar heights represent the percent of total food production intended for human consumption indicated for the income region and commodity. Production data (harvested weight) was calculated from FAOSTAT 2016–2020 (FAOSTAT 2024)

demand for diversified diets increases, but agricultural production has not matched the demand. To facilitate healthy dietary shifts, production systems must diversify to increase perishable food production while simultaneously managing or reducing FLW.

Using the latest loss percentages from Parfitt et al. (2021), we estimate farmgate pFLW for different food categories based on the production data in Fig. 5.1. The analysis of Parfitt et al. (2021) provides the most updated pFLW estimates (including animal losses, unharvested crops, and postharvest losses) for commodities, as a percentage of total food production intended for human consumption in different income regions. Due to data limitations, our analysis is restricted to income regions.

In Fig. 5.2, we compare the absolute total per capita servings lost at the farm level and the percentage of total production lost across commodities and income regions. Contrary to expectations, Fig. 5.2 does not show a clear correlation between farm-level pFLW and regional income level or food perishability. Higher absolute pFLW values are often due to greater overall production (Parfitt et al. 2021). At the same time, the stage of food system transformation influences the structure of specific food commodity value chains, which, in turn, influence both the capacity and incentives to prevent pFLW at the farm level, particularly, in the context of changing climates (Box 5.2).

Regarding FQL, there is a significant knowledge gap at the farm level. Parfitt et al. (2021) did not include FQL, due to insufficient studies measuring it at the farm stage. As discussed in Chap. 4, the difficulties in assessing FQL contribute to the lack of comprehensive FQL data throughout the value chain.

Box 5.2 The Potential for Climate Change Impacts on Farm-Level Food Loss and Waste

Although climate change impacts on food loss and waste (FLW), especially in terms of greenhouse gas emissions, are well recognized (IPCC 2019), the effects of climate change on FLW itself, particularly, at the farm level, are less well explored. Changes in temperature and precipitation can influence FLW by affecting the timing of harvest, the quality of harvested food, and the risk of spoilage. Adapting to these changes will necessitate both technical and institutional interventions to enhance farm resources and market access, thereby influencing harvest decisions and immediate postharvest handling. Discussions on FLW and climate change should not only consider the impact of FLW on emissions but also the strategies needed to sustainably meet nutrient demands through both increased production and reductions in FLW, especially for nutritious, often perishable, foods.

5.3 Perishable Fruits and Vegetables

The relationship between plant life cycles, physiological changes, and quality attributes is a critical factor in preventing FLW of fruits and vegetables. Starting from the time they are ready for harvest, numerous factors influence the freshness, shelf life, and overall quality of perishable produce. The ability of fresh fruits and vegetables to withstand exposure to different farm-level stressors depends on harvesting timing, postharvest handling, and the underlying physiological processes that occur during growth and development.

5.3.1 Plant Life Cycles, Physiological Changes, and Quality Attributes

Plant life cycles and physiological changes reflect their botanical classification, not their culinary definition. Botanically, a fruit is the edible part of a plant that contains the seeds and surrounding tissue. This definition differs from the culinary one for vegetables, which encompasses any edible plant part—including roots, tubers, bulbs, stems, leaves, flower buds, or botanical fruits (Yahia 2019). To avoid confusion, in this section, we will use the term "fruit–vegetable" to refer to foods that are botanically fruits but are typically consumed as vegetables. For example, a tomato is a fruit–vegetable.

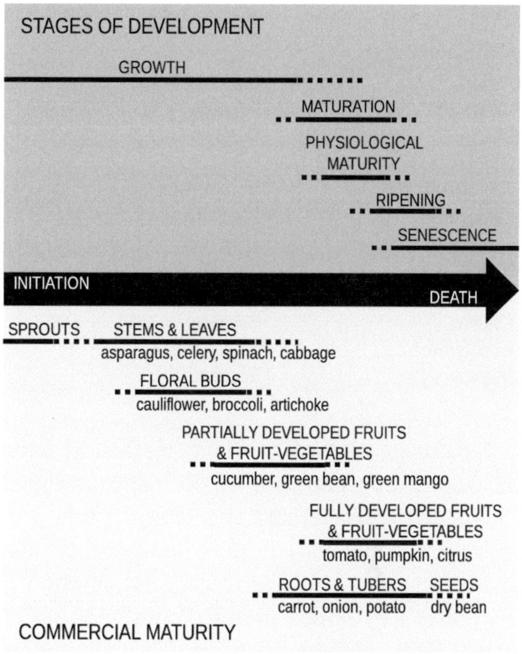

Fig. 5.3 Usage of fruits and vegetables relative to the physiological stages of development
Adapted from Watada et al. (1984)

As illustrated in Fig. 5.3, fruits and vegetables undergo various stages of development, each affecting their quality attributes. The growth phase is characterized by changes in physical properties (for example, size, shape) and ends when the produce reaches its final size on the plant (Watada et al. 1984; Yahia 2019). The maturation stage overlaps with the end of growth and continues until the plant or plant part is physiologically mature. Variations in growing conditions can lead to natural, uneven maturation within a field (Tijskens and Schouten 2014; Johnson et al. 2018b). Changes in color, texture, and sweetness in ripening occurs during physiological maturity and after maturation (Kader 1996; Prasanna et al. 2007; Tijskens and Schouten 2014; Yahia 2019). Even after harvest, fruits and vegetables may continue to develop (for example, rooting, sprouting, seed germination, and fiber development), which can be undesirable and classified as FQL (Kader 1992; Kitinoja and Kader 2015).

Table 5.1 Examples of climacteric and non-climacteric fruit. (From Kitinoja and Kader 2002)

Climacteric fruits		Non-climacteric fruits	
Apples	Muskmelon	Berries	Peas
Avocado	Papaya	Cherries	Peppers
Banana	Passion fruit	Citrus fruits	Pineapple
Blueberries	Pears	Cucumber	Pomegranates
Breadfruit	Persimmon	Dates	Strawberry
Cherimoya	Plantain	Eggplant	Summer squash
Durian	Quince	Grapes	Tamarillo
Feijoa	Sapodilla	Lychee	Watermelon
Fig	Sapote	Okra	
Guava	Soursop		
Kiwifruit	Stone fruits		
Mango	Tomato		

Senescence is the final stage, marked by natural degradation and loss of desirable attributes (Yahia 2019). Fruits and vegetables, while generally perishable, vary in their perishability rates due to respiration (Kader 1992). Respiration is the process by which produce consumes stored energy and releases heat, accelerating perishability (Kader 2002; Dodd and Bouwer 2022). Stress factors, which will be discussed in this section, can influence respiration rates and hasten senescence.

Producers, the initial decision-makers in the food value chain, must evaluate produce quality within the context of the targeted market to determine optimal harvest timing. The producers must be aware of and respond to fluctuating prices and the often conflicting quality demands from traders, retailers, and consumers—all of which can vary due to factors, including seasonality, supply, and intended use (Shewfelt 2014; Bollen and Prussia 2022).

Fruits and vegetables are considered ready for harvest when they reach commercial maturity (Fig. 5.3), which depends on the stage of plant development and desired quality attributes (Watada et al. 1984; Shewfelt 2014). For example, leafy greens may be harvested as "baby" greens (tender and delicate) or as fully grown leaves (tougher, fibrous) (Lucier et al. 2004; Yahia et al. 2019). Postharvest ripening, regulated by ethylene, is an important factor in harvest timing of some fruits and fruit–vegetables. Climacteric fruits and fruit–vegetables can ripen after harvest, while non-climacteric ones cannot (Kader 2002; Fan et al. 2022). Table 5.1 lists examples of both types.

Search attributes, such as size, shape, color, firmness, aroma, and leaf changes, are often used as indicators of commercial maturity (Kader 2002; Barbosa-Cánovas et al. 2003; Shewfelt 2014; Tijskens and Schouten 2014). Producers may also assess compositional factors, including soluble solids, starch content, acidity, and internal ethylene concentration (Barbosa-Cánovas et al. 2003; Kader and Barrett 2005). The selection of limiting quality attributes and assessment methods depends on the target market, producer knowledge, access to information, and available resources (Chap. 4).

5.3.2 Stress Factors and Physiological Changes Drive Food Quality Loss

As food systems transform, and food production diversifies into perishable fruits and vegetables, producers face new challenges and incentives to reduce FLW associated with fresh produce spoilage. Fruits and vegetables are living tissues that undergo biological processes, both before and after harvest, posing challenges for producers and subsequent value chain actors in managing FQL arising from both natural development and stress factors (Yahia 2019; Vázquez-Hernández et al. 2019). Therefore, producers' decisions and resources at every stage of farm operations impact the amount of FLW that occurs on-farm and the quantity and quality of produce that becomes available in post-farmgate value chains.

Categorizing stress factors into environmental (temperature, air composition, moisture, light), biological (pests, pathogens), and physical (surface injuries, bruising) categories helps to assess specific FQL pathways (Kader 1992; Yahia et al. 2019), which are also dependent on initial produce quality and the intensity and duration of stress factor exposure (Lichtenthaler 1996). Different categories of stress factors can interact and increase the susceptibility of produce to other factors, making identification of effective intervention points for reducing FQL more challenging. Therefore, while stress factors may occur at the farm level, observable effects on FQL may not become apparent until later in the value chain.

Table 5.2 summarizes the physiological processes involved in FQL and their relationship with stress factors.

Table 5.2 Physiological processes involved in food quality loss

Physiological process	Natural life cycle development	Response to stress factors	Potential food quality loss
Respiration	Stored energy is broken down and releases heat. The commodity moves toward senescence as energy stores are depleted.	Lower temperatures slow the respiration rate. Higher temperatures increase the respiration rate.	Loss of flavor quality (especially sweetness), loss of salable dry weight; reduced energy value.
Ethylene production	A plant hormone that regulates many aspects of growth, maturation, ripening, and senescence. The extent to which production increases with maturity at harvest depends on the type of commodity.	Rate of ethylene production rises in response to physical injury, disease, elevated temperatures, water stress. Rate of ethylene production decreases in response to low temperature, low oxygen concentration, elevated carbon dioxide concentration.	There is no consistent relationship between the ethylene production capacity of a given commodity and its perishability. Exposure of most commodities to ethylene accelerates their senescence (and the FQL associated with this development stage).
Compositional changes	The rate of development is linked with the rate of compositional changes. Loss of chlorophyll (green color) and development of carotenoids (yellow, orange, red, and pink colors) and anthocyanins (red and blue colors). Conversion of starch-to-sugar and sugar-to-starch; breakdown of pectin/polysaccharides; increased fiber. Changes in flavor quality and production of flavor volatiles over the life cycle.	Oxidation of damaged plant tissue results from physical injury. Elevated temperatures, sunlight, and physical injury linked with micronutrient degradation.	*Pigments*: Loss of green color *Phenolic compounds*: Oxidation of damaged tissue causing browning *Carbohydrates*: Sugar-to-starch ratio; softening from pectin/polysaccharide breakdown; toughening from increased fiber content. *Nutritional quality*: Nutrient degradation (note: These attribute changes are desirable or undesirable, depending on the commodity and desired commercial maturity)
Transpiration and water loss (evaporation of water from plant tissues)	Fruits and vegetables have natural outer protective coverings that regulate water loss. The shape and maturity stage influence the transpiration rate.	Elevated temperatures, lower atmospheric moisture, physical injury accelerate water loss.	*Appearance*: Wilting and shriveling. *Texture*: Softening, flaccidity, loss of crispness and juiciness. *Nutritional quality*: Nutrient leaching.

(continued)

Table 5.2 (continued)

Physiological process	Natural life cycle development	Response to stress factors	Potential food quality loss
Physiological breakdown disorders	Not part of natural development.	*Preharvest nutritional imbalances* result from production-side environmental stress factors, including fertilizer application, soil nutrients, and water quality. *Chilling/freezing injury* results from exposure to low temperatures, above or below freezing point, depending on the commodity. *Heat injury* results from exposure to high temperatures or direct sunlight.	*Preharvest nutritional imbalances*: Tissue breakdown; surface and internal discoloration (browning); pitting; softening. *Chilling/freezing injury:* Surface and internal discoloration (browning); pitting; water-soaked areas; uneven ripening or failure to ripen; off-flavor development; susceptibility to surface molds and decay. *Heat injury*: Bleaching; surface burning or scalding; uneven ripening; excessive softening; moisture loss.
Pathological breakdown disorders	Not part of natural development. Fruits and fruit–vegetables become more susceptible to pathogens with the onset of ripening. Vegetables become more susceptible to pathogens at senescence.	Bacteria and fungi come into contact with the commodity. Physical injury provides a site for pathogen infection. Moisture and warm temperatures provide a suitable growing environment. Physiological breakdown disorders lower fruit and vegetable pathogen resistance.	Deterioration and spoilage; visible discoloration and decay; off-odors and flavor.

Data sources: (Kader 1992; Prusky 2011)
Note: For more information on postharvest fruit and vegetable physiology and biochemistry, refer to Yahia and Carillo-López (2019). Kader and Barrett (2005) offer a detailed analysis of compositional changes after harvest

Harvesting fruits and vegetables with multiple growing seasons requires careful timing (Johnson et al. 2018b; Boiteau and Pingali 2022). Picking too soon or too late in a season can expose produce to stress factors, shorten shelf life, and reduce quality compared to mid-season harvests (Kader and Barrett 2005). Some vegetable crops, like root vegetables and cabbage, are typically harvested at the same time, but variations in field conditions can cause variations in maturity (Johnson et al. 2019). Delays in harvesting can expose mature produce to in-field stress factors, increasing the risk of over-maturing and deterioration (Fig. 5.3 and Table 5.2), which can lead to lower quality harvests or preharvest losses, if producers decide to leave some or all the affected produce unharvested. Therefore, producers must carefully align their production schedules with market demand to ensure that they have commercially mature produce ready to harvest when there is sufficient demand.

Even when producers identify produce as ready for harvest, unforeseen obstacles can arise. Preharvest factors like bad weather, labor shortages, or equipment breakdowns can delay or prevent harvesting. Postharvest challenges, such as inadequate transportation or storage, or market fluctuations, can also hinder the process (Neff et al. 2018; Johnson et al. 2019; Filimonau and Ermolaev 2021; Surucu-Balci and Tuna 2021). Moreover, a lack of knowledge or training among harvest laborers in recognizing maturity can lead to marketable produce being left in the field (Kitinoja and Kader 2015).

Fruits and vegetables are particularly prone to physical damage during harvest. Mechanical harvesters are limited for many perishable commodities, and even when available for specific, high-value crops (grapes, baby greens, blueberries), they may not be accessible or affordable (Huffman 2012; Brondino et al. 2021; Calvin et al. 2022). Therefore, manual harvesting is common across all food system types, but the quality of the process impacts FQL, especially for delicate produce. For example, harvest workers can unintentionally damage plants and produce with tools (knives, clippers, spades), or even with their jewelry (Kitinoja and Kader 2015; Kasso and Bekele 2018). Tender or soft produce is easily damaged by rough handling, especially when placed in containers that are rough or broken (Kitinoja and Kader 2015). Tree fruit, for example, might be thrown from several meters onto the ground (Kasso and Bekele 2018). Implementing food quality standards can incentivize improved harvesting techniques, which can minimize damage and reduce losses (Box 5.3).

For fruits and vegetables that have just been harvested, every step in the postharvest process increases the risk of FQL, with specific factors depending on the value chain context. Immediately after harvest, temperature is a critical factor in preserving the quality of fruits and vegetables. Freshly harvested produce contains residual field heat and continues to release heat through respiration. Therefore, minimizing exposure to farmgate stress factors, such as sunlight, poor ventilation, and improper storage conditions, is crucial. Rapid cooling to remove field heat and limiting exposure to sunlight are essential for slowing down enzyme activity, respiration, the growth of spoilage organisms, water loss, and sprouting (Hardenburg et al. 1986). However, the cooling efficiency depends on the method (simple, nonelectric to advanced, electric), produce density, air circulation, surface area in the container, and relative humidity (Kitinoja and Kader 2015; Dodd and Bouwer 2022). Therefore, producers must have both the knowledge and resources to properly cool their harvested produce. For example, excessively low temperatures can cause chilling or freezing injury, reducing quality (Table 5.2). For a comprehensive list of optimal storage temperatures, refer to Dodd and Bouwer (2022).

Postharvest, produce can also be contaminated by bacteria, fungi, or pests due to inadequate containers, storage practices, or handling environments. Physical damage from harvest or postharvest activities can increase produce susceptibility to spoilage by allowing soil bacteria or other pathogens (for example, from contaminated containers) to enter the produce, leading to acceleration of the decay process (Kasso and Bekele 2018; Yahia et al. 2019). This is why sorting out damaged produce is important to prevent it from contaminating other, better quality produce. At the same time, rough handling during activities, like cleaning, sizing, grading, and sorting, can cause physical damage (Shewfelt 2014). Therefore, effective postharvest incentives are essential to prevent FQL due to poor handling, just as they are for harvesting.

Box 5.3 Quality Standards and Incentives: A Tool for Reducing Food Loss and Waste in Transitional Food Systems

Implementing food quality standards and incentivizing adherence can reduce farm-level food loss and waste (FLW) of perishable fruits and vegetables. In Indonesia, Perdana et al. (2023) compared FLW in traditional and modern vegetable value chains. Although both types of producers faced similar challenges, like rot, pests, diseases, and harvesting damage, contract farming governance led to lower FLW than informal market governance. Modern market producers operate in coordinated value chains with agreed upon quality and ripeness standards, using price-based contracts, thereby incentivizing them to harvest only mature, high quality vegetables and to sort them in the field to minimize losses. In contrast, informal value chains lack such standards, leading to a single price for all quality grades. This disincentivizes producers from being selective in harvesting and packaging, resulting in lower quality produce moving post-farmgate. The presence of low quality or rotten vegetables can contaminate otherwise high quality produce, contributing to overall food quality loss.

5.3.3 Cosmetic Quality Standards Drive Farmgate Physical Food Loss and Waste as Food Systems Modernize

Cosmetic quality standards, which prioritize appearance attributes, can contribute to farm-level pFLW. Although production-side factors, like stress and natural variations, can influence undesirable produce qualities, modern food systems increasingly emphasize cosmetic attributes like uniformity, size, and color. This emphasis can conflict with policies aimed at increasing fruit and vegetable production (Messner et al. 2021). Yet, producers in modern food systems often have limited options for produce that does not meet strict cosmetic standards, unlike those in traditional or transitional systems with greater flexibility of acceptable cosmetic attributes and multiple marketing channels (Gómez and Ricketts 2013; Göbel et al. 2015; Beausang et al. 2017; Neff et al. 2018; Chaboud and Moustier 2021). Therefore, increasing fruit and vegetable production, while adhering to stringent cosmetic standards in modernizing food systems, remains a challenge. Food rescue initiatives have emerged across food value chain stages, including at the farm level, to redirect edible food that is not sold back into human consumption value chains, thereby reducing pFLW. The decision to leave crops in the field versus redirecting them for food rescue is influenced by perceived trade-offs and optimal resource use (see Box 5.4).

Contracts, a key feature of modern food systems, can become the primary marketing channel between producers and post-farmgate value chain actors, including organized retailers. Producers of perishable foods often cite contracts as a cause of farm-level pFLW, especially when they must overproduce to meet contract commitments but also account for fluctuations in harvest quality and quantity (Gillman et al. 2019). However, contracts do not always contribute to pFLW if alternative markets or better information sharing are available. For example, small producers in Scotland had more flexibility in handling surpluses due to multiple sales outlets, while larger producers, often tied to contract farming, benefited from improved communication with retailers regarding surplus and shortage issues (Beausang et al. 2017).

Box 5.4 Balancing Producer Priorities with Farm-Level Food Rescue

Preventing on-farm pFLW involves minimizing the diversion of edible food to non-food uses. Producers aiming to optimize resource use may prioritize strategies that reduce their financial losses, even if these strategies contribute to pFLW. When fruits and vegetables are unsold due to market issues or cosmetic defects, they may still be suitable for human consumption but have lower value. Some producers choose to till such produce into the soil, along with unsalvageable food, as a disposal method. Producers may perceive this approach as cost-effective and environmentally beneficial (Gillman et al. 2019). However, it does not prevent pFLW and may not be the most optimal use of resources for improving soil health (Messner et al. 2021).

Food rescue, including donations to food banks and gleaning operations, offers producers an alternative for reducing farm-level pFLW of edible produce. Although producers often recognize the value of food rescue, financial constraints, logistical challenges, and concerns about liability can hinder their participation (Ceryes et al. 2023). Food rescue organizations also face obstacles, such as transportation costs and capacity limitations, especially when dealing with unpredictable food supplies (Axmann et al. 2024). Addressing these diverse barriers requires a multifaceted approach, involving various interventions and policies to support food rescue initiatives that align with producer financial interests (Neff et al. 2018; Ceryes et al. 2023).

5.4 Animal-Source Foods

Animal-source foods (ASF) are a diverse and perishable group of foods that are sources of quality protein, omega-3 fatty acids, and bioavailable micronutrients (Godfray et al. 2018; Troell et al. 2019). As food systems modernize and diets diversify, ASF are becoming a larger part of many people's diets (Chap. 2). ASF can originate from farmed animals or wild sources and are classified into aquatic and livestock categories, each with distinct production systems, value chains, and quality attributes.

FLW of ASF occur at various farmgate stages, from animal maturity (slaughter, milk production, egg laying, aquaculture harvest, hunting) to postharvest handling (FAO 2014; Pérez Roda et al. 2019). Although the FAO (2011) report included breeding and rearing stages in FLW estimates, we consider these as production-side losses (Chap. 3). However, production-side factors can still cause FLW. Unlike plant-based foods, FLW pathways for ASF are also influenced by animal welfare concerns, alongside food security, environmental, and economic considerations (Klaura et al. 2023).

5.4.1 Aquatic Animal Foods

From wild fisheries to aquaculture farms, aquatic animals including fish, crustaceans, and mollusks are an important part of diets across many food systems (Thilsted et al. 2016). As of 2020, 89% of total fish production was consumed directly by humans (FAO 2022a). These animals offer a variety of nutrients, and their nutritional quality depends on the specific species produced (Hicks et al. 2019). Although the natural physiology of aquatic animals affects FLW similarly across production systems, the specific ways that FLW occurs can vary due to different stress factors.

Since the 1990s, aquaculture has been the primary driver of global growth in fish production (Fig. 5.4). However, this growth is uneven across regions (FAO 2022a). Despite being local operations in coastal waters that use basic technology, small-scale fisheries contribute approximately 40% of the global annual fish catch (WEF 2024). Global fish production is expected to continue expanding, primarily fueled by aquaculture, while wild-caught fish remains relatively stable (Troell et al. 2019; FAO 2022a). However, the most recent analysis of global FLW of aquatic

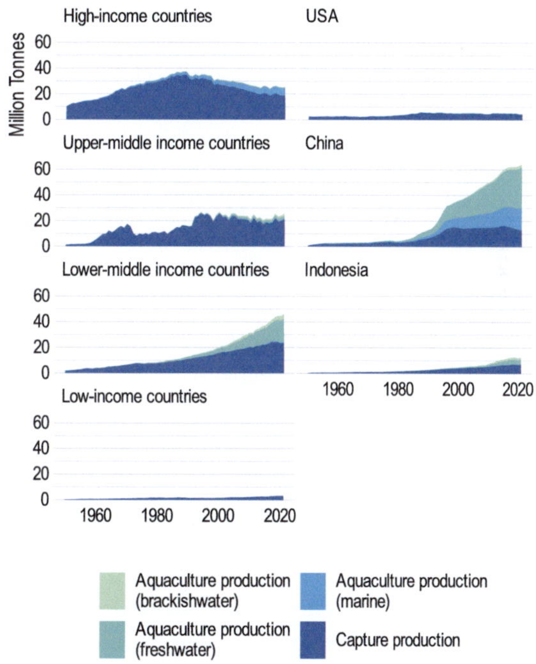

Fig. 5.4 Capture fisheries and aquaculture production by World Bank income group

Data source: FAO: Fisheries and Aquaculture, https://www.fao.org/fishery/en. Accessed August 4, 2023

Given their status as top producers in their respective income groups (high income, upper middle income, and low income), the United States, China, and Indonesia are presented individually in the graphs. Data from FAO (2023a), accessed 4 August 2023

foods by the World Economic Forum (2024) did not include FLW estimates from processing at sea, aquaculture production, or small-scale fisheries, due to lack of reliable data. Addressing these data gaps also requires examining differences in how men and women are involved in fish production systems and value chains (Box 5.5).

Fish enter food value chains in different forms, each with their own unique FLW pathway. Although some live fish are sold, especially in East and Southeast Asia, most fish are processed after being caught (FAO 2022a). Desirable fish qualities include freshness indicators; attributes that affect processing, cooking, and eating quality; and safety assurances against harmful bacteria, parasites, or chemicals (Huss 1995a).

Discards at sea and at-sea processing are two major contributors to FLW of wild-caught fish before the fish reach the landing site (WEF 2024). The decision to discard fish as pFLW depends on their quality attributes, market value, and catch/harvesting method. In addition to catching target species intended for human consumption, capture fisheries often unintentionally catch nontarget species, which may be used for food or feed (Kelleher 2005; Davies et al. 2009; Pérez Roda et al. 2019; Sun et al. 2023). During capture, fish experience stress and injury from environmental factors (for example, temperature, pressure) and mechanical factors (for example, collisions with vessels or fishing gear) (Pérez Roda et al. 2019). Fish mortality, whether it occurs at sea or upon reaching a landing site, is a significant contributor to pFLW in the early stages of the aquatic ASF value chain (Pérez Roda et al. 2019; Papageorgiou and Moutopoulos 2023; WEF 2024).

After being caught, fish stiffen, as their muscles go through a natural process called rigor mortis (Huss 1995b). Properly timing the processing activities around the onset of rigor mortis is important for maintaining fillet quality. The time it takes for rigor mortis to occur depends on the type of fish, how stressed it was during capture, and the temperature at which it is stored (Huss 1995b; Hamada-Sato et al. 2005; Borderías and Sánchez-Alonso 2011). Once rigor mortis passes, the fish muscles soften and relax, as enzymes break down fatty acids. Eventually, chemical changes lead to rancid odors and discoloration (Huss 1995b; Hamada-Sato et al. 2005).

Primary processing can help extend shelf life and reduce FQL from temperature and handling abuses, but it can also contribute to pFLW if not done properly. At-sea fish processing primarily removes inedible parts, which may be discarded at sea or retained for later use as by-products if they have commercial value (WEF 2024). The level of spoilage bacteria affects how long the fish stays fresh; fish with too much bacteria may be rejected (Huss 1995b). Bacteria can contaminate fish through contact with unclean surfaces or unsanitary handling (FAO 2023b). Physical damage, such as abrasions or skin damage, can accelerate tissue breakdown and discoloration, leading to faster spoilage (Huss 1995c; FAO 2023b). While washing and gutting can help reduce bacterial and enzymatic spoilage, especially in larger fish (Borderías and Sánchez-Alonso 2011), proper cooling is crucial to slow down enzyme activity and bacterial growth, thereby preventing spoilage (Huss 1995b).

Aquaculture producers carefully select and breed fish species for specific traits, including rapid growth and disease resistance (Global Panel 2021). These fish are primarily raised for human consumption (FAO 2022a). Unfortunately, some mature farmed fish die before they can be harvested, leading to preharvest pFLW. The main causes of mortality include environmental factors like poor water quality, algal blooms, temperature fluctuations, and extreme weather; biological threats like predators and parasites; and chemical issues like misused veterinary drugs (FAO 2023b).

Box 5.5 A Gender Lens on Fish Production and Early-Stage Food Loss and Waste

Varying gender roles within fish value chains can influence the degree of FLW at different stages, depending on levels of participation and resource access (Kruijssen et al. 2020). Gender roles play a significant part in fish value chains. Although men often dominate fishing activities, women's involvement varies, depending on factors like production systems and cultural norms. For example, Kaminski et al. (2020) demonstrated that, in Zambia, women's household responsibilities, such as farming and childcare, can hinder their participation in traditional fishing. Among women who participated in fishing, there were differences in fishing gear between men and women. Men primarily used nets and handlines from canoes, while women fished with baskets in shallow waters during specific seasons. In contrast, aquaculture, especially backyard pond farming, offers women a more accessible way to engage in fish production (Kawarazuka and Béné 2010). To effectively identify critical loss points in fish value chains, comprehensive gender-sensitive value chain mapping (FAO 2018) must consider variations in production systems and their unique FLW pathways, including at early stages, to address existing data gaps (Kruijssen et al. 2020).

5.4.2 Livestock-Sourced Foods

As populations grow, and incomes rise, the demand for livestock-sourced foods (LSF) increases, leading to shifts in production systems (Chap. 2). Over the next decade, dairy and meat consumption is projected to rise significantly. In many low- and middle-income countries, fresh dairy products (unprocessed, pasteurized, or fermented) are the primary forms of milk consumed, unlike in high income countries where processed dairy products (for example, cheese) are more common (OECD and FAO 2023). Poultry, primarily broiler chickens, currently accounts for the largest share of meat production and consumption globally and is expected to grow further (Parlasca and Qaim 2022; OECD and FAO 2023).

To meet increasing LSF demand in low- and middle-income countries, the number of animals increases before productivity gains, especially for poultry and pigs (Steinfeld et al. 2006; Herrero et al. 2021). The expansion of animals puts pressure on the early stages of the value chain, where the health and well-being of animals, the conditions under which they are raised, the slaughter process, and immediate post-production handling all factor into the extent of FLW at the farm level.

Livestock destined for meat production are exposed to numerous preslaughter stressors that can negatively impact food quality. From excessive heat exposure on farms and in transport vehicles to the chaotic environment of slaughterhouses, these animals face challenges that can compromise their health and well-being, and ultimately, their meat quality (Miranda-de La Lama et al. 2014). Veterinary care, feed, and stocking density all influence animals' vulnerability to stress (Ritz et al. 2005). Improper handling practices, such as overcrowding, harsh environments, prolonged transport, and rough treatment, can further exacerbate stress levels (Nijdam et al. 2004; Delezie et al. 2007; Xing et al. 2019; Bethancourt-Garcia et al. 2019; Pirompud et al. 2023).

Preslaughter mortality of market-ready animals can occur when animals are deemed unfit for human consumption or when animals die from disease, injury, or stress. For example, in poultry, heat stress, physical trauma, and disease are the primary causes of preslaughter mortality (Nijdam et al. 2004; Ritz et al. 2005; Kittelsen et al. 2018). The rate of pFLW can also serve as an initial indicator of animal welfare (Jacobs et al. 2017).

Heat stress, a physiological stressor, creates an ideal environment that facilitates microbial growth and colonization, contributing to animal illness and potential meat quality issues (Zhang et al. 2020). Handling-related stressors, such as overcrowding and unsanitary conditions, further exacerbate the problem by increasing the transmission of pathogens among animals (Ritz et al. 2005). Behavioral responses to stress, including aggression and agitation, can lead to injuries from human handling or animal interactions, further compromising meat quality, such as discoloration (Oppen and Løkketangen 2008). Acute heat stress can also have detrimental effects on meat quality, causing changes in color, texture,

Table 5.3 Livestock production systems

		High income	Upper-middle income	Lower-middle income	Low income
Chicken	*Backyard*	0.03	0.26	0.51	0.08
	Broiler	5.04	5.95	3.76	0.11
	Layer	2.18	5.4	1.36	0.05
Bovine	*Grassland*	3.77	3.27	0.99	0.51
	Mixed	9.71	6.71	9.8	0.65
	Feedlot	1.54	0.83	0.04	–
Small ruminant meat	*Grassland*	0.22	0.35	0.26	0.25
	Mixed	0.42	0.46	0.81	0.23
Pork	*Backyard*	0.03	1.791	0.63	0.11
	Intermediate	0.25	3.3	0.4	0.02
	Industrial	5.59	6.22	0.29	0.02

Meat production by regional income group and production intensity in millions of tons. Production volumes measured in terms of protein. Data from the Global Livestock Environmental Assessment Model (FAO 2022b)

and fluid loss (Mir et al. 2017; Xing et al. 2019; Zhang et al. 2020).

Livestock transport is a particularly stressful experience for animals, influenced by factors such as species mixing, disease risks, transport duration, vehicle size, arrival timing, and slaughterhouse conditions (Oppen and Løkketangen 2008). For example, mammals and birds have diverse physiological responses to heat, making it essential to manage these factors to minimize stress (Silanikove 2000; Gregory 2010; Gonzalez-Rivas et al. 2020). Therefore, the transportation requirements and value chain actors' ability to control these factors are influenced by the scale of their production systems and underlying infrastructure, such as road conditions and networks.

The role of different meat production systems varies across regional income groups (Table 5.3). While intensive and industrial systems often expose animals to stress from high stocking densities, prophylactic antibiotic use, and animal welfare concerns, extensive systems also have their challenges, including heat stress and lower access to feed, water, and veterinary services (Parlasca and Qaim 2022). Climate change can exacerbate stress factors in livestock production systems with varying impacts, depending on the type of system (Nardone et al. 2010). However, more comprehensive data is needed to fully understand the role of production systems in farmgate FLW of livestock foods, as well as the potential benefits and risks of valorizing crop-based pFLW as animal feed to support increasing livestock production (Pinotti et al. 2021; Boumans et al. 2022).

Quality evaluation of both live animals and postmortem meat depends on the type of standards used and how strictly they are enforced. Informal traders, lacking formal grading systems, rely on subjective assessments of live animals, based on visual inspection, estimated live weight, and perceived health. Large processing companies, in contrast, use more detailed and objective grading standards that consider factors like animal type, age, live weight, fat characteristics, and injuries (Alarcon et al. 2017). At slaughter, rejected meat may be completely removed from the value chain (pFLW) or may be rejected from certain markets, potentially leading to diversion to local markets with less stringent quality requirements (Gregory 2010; Miranda-de La Lama et al. 2014; Mir et al. 2017; Forseth et al. 2023). For example, improper slaughter techniques, such as mishandling of organs, can lead to fecal contamination and rejection during inspection (Jaja et al. 2018; Njoga et al. 2023). As slaughtered meat is prepared to move to the post-farmgate value chain, the post-slaughter chilling period is crucial for maintaining food safety and quality attributes (Savell et al. 2005; Mir et al. 2017).

In addition to meat, milk and eggs are also important livestock products vulnerable to FLW due to their perishability. In dairy production, microbial spoilage, even under refrigerated conditions, is a primary contributor to FQL, resulting in off-flavors, odors, and textures (Martin et al. 2021). Physical FLW of milk, typically due to spills or accidents related to infrastructure or management issues, does not necessarily involve quality loss (March et al. 2019). However, antibiotic residues in milk, often from treating diseases like mastitis in dairy cattle, pose a significant food safety concern and lead to pFLW when detected above permissible levels (March et al. 2019). Poor hygiene, sanitation practices, and mechanical milking can increase the risk of mastitis, particularly in systems where subclinical cases go undetected (Motaung et al. 2017; Bari et al. 2022). For eggs, heat stress in egg-laying birds can compromise shell quality, leading to reduced thickness, increased breakage, and a higher risk of foodborne pathogen contamination (Lara and Rostagno 2013; Gautron et al. 2022).

5.5 Conclusions

This chapter underscores the critical role of farm-level FLW in ensuring the availability of nutritious, perishable foods within food value chains. Although diversification and increased production of healthy foods are essential, effective management of FLW at the farm level is equally crucial for delivering healthy diets. Current production of high-value, nutritious foods is not sufficient globally to meet the dietary needs of all people. Physical FLW at farmgate stages, some of which is due to FQL, further reduces the amount of fruits, vegetables, and ASF that enters food value chains to be made accessible to consumers.

For fruits and vegetables, FLW pathways begin once produce reaches a marketable maturity. Understanding the relationship between plant life cycles, physiological changes, and quality attributes is essential for minimizing FLW. Stress factors, including environmental conditions, pests, and physical damage, can significantly impact produce quality and shelf life. Producers must manage FLW by carefully selecting harvest timing, implementing proper handling practices, and ensuring appropriate postharvest storage. However, the challenges of balancing cosmetic quality standards with the need to minimize FLW in modern food systems remain.

FLW pathways of ASF are more diverse than those for fruits and vegetables due to the wide range of products within this category, including aquatic animals, landed animals, and dairy and egg products. From production to postharvest handling, ASF face risks, such as preslaughter stress, disease, and improper handling. Ensuring proper hygiene, sanitation, and veterinary care is crucial for minimizing FLW of mature livestock. In aquatic animals, factors, including capture methods, handling practices, and storage conditions, impact FLW. Addressing these challenges requires a comprehensive approach to improve animal welfare, production practices, market linkages, and infrastructure.

Addressing FLW at the farm level requires a comprehensive approach that involves improved production practices, enhanced infrastructure, strengthened market linkages, policy interventions, and knowledge and capacity building. By addressing these challenges, we can significantly reduce farm-level FLW to increase the quantity of good quality produce made available to food value chains.

References

Alarcon P, Fèvre EM, Murungi MK, Muinde P, Akoko J, Dominguez-Salas P, Kiambi S, Ahmed S, Häsler B, Rushton J (2017) Mapping of beef, sheep and goat food systems in Nairobi—a framework for policy making and the identification of structural vulnerabilities and deficiencies. Agric Syst 152:1–17. https://doi.org/10.1016/j.agsy.2016.12.005. Accessed 24 Jan 2024

Aragie E, Pauw K, Thurlow J (2023) The economywide effects of reducing food loss and waste in developing countries. IFPRI Discussion Paper 2173. International Food Policy Research Institute, Washington, DC

Axmann HB, Soethoudt JM, Dijkink BH, Broeze J, Oostewechel RJA (2024) Saving surplus, nourishing communities: scaling the fight against food loss and hunger in Mexico: case studies on banana and broccoli food loss reduction options. Wageningen Report No. 2574. Wageningen Food & Biobased Research, Wageningen University and Research, Wageningen. https://doi.org/10.18174/662281

Barbosa-Cánovas GV, Fernández-Molina JJ, Alzamora SM, Tapia MS, López-Malo A, Welti CJ (2003) Basic harvest and post-harvest handling considerations for fresh fruits and vegetables (chap. 2). In: Handling and preservation of fruits and vegetables by combined methods for rural areas: technical manual. FAO agricultural services bulletin 149, food and agriculture Organization of the United Nations, Rome

Bari MS, Rahman MM, Persson Y, Derks M, Sayeed MA, Hossain D, Singha S, Hoque MA, Sivaraman S, Fernando P, Ahmad I, Samad A, Koop G (2022) Subclinical mastitis in dairy cows in south-Asian countries: a review of risk factors and etiology to prioritize control measures. Vet Res Commun 46(3):621–640. https://doi.org/10.1007/s11259-022-09948-x. Accessed 22 Dec 2023

Beausang C, Hall C, Toma L (2017) Food waste and losses in primary production: qualitative insights from horticulture. Resour Conserv Recy 126:177–185. https://doi.org/10.1016/j.resconrec.2017.07.042. Accessed 08 Sep 2021

Bethancourt-Garcia JA, Vaz RZ, Vaz FN, Silva WB, Pascoal LL, Mendonça FS, da Vara CC, Nuñez AJC, Restle J (2019) Pre-slaughter factors affecting the incidence of severe bruising in cattle carcasses. Livest Sci 222:41–48. https://doi.org/10.1016/j.livsci.2019.02.009. Accessed 06 Dec 2023

Boiteau JM, Pingali P (2022) Food loss of perishable produce from farm to retail: evidence from tomato supply chains in South India. Am J Clin Nutr 115(6):1535–1548. https://doi.org/10.1093/ajcn/nqac039. Accessed 01 June 2022

Boiteau JM, Pingali P (2023) Can we agree on a food loss and waste definition? An assessment of definitional elements for a globally applicable framework. Glob Food Secur 37:100677. https://doi.org/10.1016/j.gfs.2023.100677. Accessed 28 Feb 2023

Bollen AF, Prussia SE (2022) Sorting for defects. In: Florkowski WJ, Shewfelt RL, Brueckner B, Prussia SE (eds) Postharvest handling: a systems approach, 4th edn. Elsevier Academic Press, Amsterdam, pp 377–398

Borderías AJ, Sánchez-Alonso I (2011) First processing steps and the quality of wild and farmed fish. J Food Sci 76(1):R1–R5. https://doi.org/10.1111/j.1750--3841.2010.01900.x. Accessed 03 Aug 2023

Boumans IJMM, Schop M, Bracke MBM, de Boer IJM, Gerrits WJJ, Bokkers EAM (2022) Feeding food losses and waste to pigs and poultry: implications for feed quality and production. J Clean Prod 378:134623. https://doi.org/10.1016/j.jclepro.2022.134623. Accessed 20 Nov 2023

Brondino L, Borra D, Giuggioli NR, Massaglia S (2021) Mechanized blueberry harvesting: preliminary results in the Italian context. Agriculture 11:1197. https://doi.org/10.3390/agriculture11121197. Accessed 02 Oct 2023

Calvin L, Martin P, Simnitt S (2022) Adjusting to higher labor costs in selected U.S. fresh fruit and vegetable industries. Economic Information Bulletin No. (EIB-235), July. U.S. Department of Agriculture, Economic Research Service

Ceryes C, Heley K, Edwards D, Gao-Rittenberg C, Seifu L, Khan Sohail S, Neff R (2023) "We need a better system": Maryland crop growers' perspectives on reducing food loss through donation. J Agric Food Syst Community Dev 12(4):6784. https://doi.org/10.5304/jafscd.2023.124.002. Accessed 24 Nov 2023

Chaboud G, Moustier P (2021) The role of diverse distribution channels in reducing food loss and waste: the case of the Cali tomato supply chain in Colombia. Food Policy 98:101881. https://doi.org/10.1016/j.foodpol.2020.101881. Accessed 31 Aug 2021

Davies RWD, Cripps SJ, Nickson A, Porter G (2009) Defining and estimating global marine fisheries bycatch. Mar Policy 33(4):661–672. https://doi.org/10.1016/j.marpol.2009.01.003. Accessed 17 April 2023

Delezie E, Swennen Q, Buyse J, Decuypere E (2007) The effect of feed withdrawal and crating density in transit on metabolism and meat quality of broilers at slaughter weight. Poultry Sci 86:1414–1423. https://doi.org/10.1093/ps/86.7.1414. Accessed 30 Nov 2023

Dodd MC, Bouwer J (2022) Cooling fresh produce. In: Florkowski WJ, Shewfelt RL, Brueckner B, Prussia SE (eds) Postharvest handling: a systems approach, 4th edn. Elsevier Academic Press, Amsterdam, pp 435–468

Fabi C, English A (2019) Methodological proposal for monitoring SDG target 12.3.1 sub-indicator 12.3.1.A the food loss index design, data collection methods and challenges. In: FAO statistics working paper ESS / 18–13, food and agriculture Organization of the United Nations, Rome. https://openknowledge.fao.org/server/api/core/bitstreams/6603bc2f-7105-434b-baaa-f440e3505a88/content. Accessed 4 Oct 2024

Fan D, Wang W, Hao Q, Jia W (2022) Do non-climacteric fruits share a common ripening mechanism of hormonal regulation? Front Plant Sci 13:923484. https://doi.org/10.3389/fpls.2022.923484. Accessed 03 Oct 2023

FAO (Food and Agriculture Organization of the United Nations) (2001) Food balance sheets—a handbook. FAO, Rome. https://www.fao.org/4/X9892E/X9892E00.htm. Accessed 03 Sep 2024

FAO (Food and Agriculture Organization of the United Nations) (2011) Global food losses and food waste—extent, causes and prevention. FAO, Rome. https://www.fao.org/4/mb060e/mb060e00.pdf. Accessed 5 Oct 2024

FAO (Food and Agriculture Organization of the United Nations) (2014) Definitional framework of food loss. Working paper, February 27, Global Initiative on Food Loss and Waste Reduction. FAO, Rome. https://openknowledge.fao.org/server/api/core/bitstreams/5931921f-7f68-4169-b731-31d06f5b4835/content. Accessed 8 Oct 2024

FAO (Food and Agriculture Organization of the United Nations) (2018) Gender and food loss in sustainable food value chains: a guiding note. FAO, Rome

FAO (Food and Agriculture Organization of the United Nations) (2022a) The state of world fisheries and aquaculture 2022. Towards blue transformation. FAO, Rome. https://openknowledge.fao.org/handle/20.500.14283/cc0461en. Accessed 8 Oct 2024

FAO (Food and Agriculture Organization of the United Nations) (2022b) Global Livestock Environmental Assessment Model (GLEAM) 3 dashboard. Livestock Emission Data at a Glance. https://foodandagricultureorganization.shinyapps.io/GLEAMV3_Public/. Accessed 05 Sep 2024

FAO (Food and Agriculture Organization of the United Nations) (2023a) Fisheries and Aquaculture: Fishery Statistical Collections fact sheets: Global Production. https://www.fao.org/fishery/en/collection/global_production?lang=en. Accessed 04 Aug 2023

FAO (Food and Agriculture Organization of the United Nations) (2023b) Food loss and waste in fish value chains. https://www.fao.org/flw-in-fish-value-chains/en/. Accessed 2 Aug 2023

FAOSTAT (2024) Food and agriculture data. Food and agriculture Organization of the United Nations (FAO). https://www.fao.org/faostat/en/#home. Accessed 5 Oct 2024

Filimonau V, Ermolaev VA (2021) Mitigation of food loss and waste in primary production of a transition economy via stakeholder collaboration: a perspective of independent farmers in Russia. Sustain Prod Consump 28:359–370. https://doi.org/10.1016/j.spc.2021.06.002. Accessed 25 Oct 2022

Flanagan K, Robertson K, Hanson C (2019) Reducing food loss and waste: setting a global action agenda. World Res Institute Report. https://doi.org/10.46830/wrirpt.18.00130. Accessed 01 Jun 2022

Forseth M, Moe RO, Kittelsen K, Skjerve E, Toftaker I (2023) Comparison of carcass condemnation causes in two broiler hybrids differing in growth rates. Sci Rep 13:4195. https://doi.org/10.1038/s41598-023-31422-0. Accessed 15 Dec 2023

Gautron J, Dombre C, Nau F, Feidt C, Guillier L (2022) Review: production factors affecting the quality of chicken table eggs and egg products in Europe. Animal 16(Suppl. 1):100425. https://doi.org/10.1016/j.animal.2021.100425. Accessed 01 Jun 2023

Gillman A, Campbell DC, Spang ES (2019) Does on-farm food loss prevent waste? Insights from California produce growers. Resour Conserv Recy 150:104408. https://doi.org/10.1016/j.resconrec.2019.104408. Accessed 08 Sep 2021

Global Panel (2020a) Future food systems: for people, our planet, and prosperity. In: Global panel on agriculture and food Systems for Nutrition, London. https://www.glopan.org/wp-content/uploads/2020/09/Foresight-2.0_Future-Food-Systems_For-people-our-planet-and-prosperity.pdf. Accessed 2 Oct 2024

Global Panel (2020b) Rethinking trade policies to support healthier diets. Policy Brief No. 13, February. Global Panel on Agriculture and Food Systems for Nutrition, London. https://www.glopan.org/wp-content/uploads/2020/02/Global-Panel-policy-brief-Rethinking-trade-policies-to-support-healthier-diets.pdf. Accessed 5 Nov 2024

Global Panel (2021) Harnessing aquaculture for healthy diets. Policy brief no. 16. Global panel on agriculture and food Systems for Nutrition, London, UK

Göbel C, Langen N, Blumenthal A, Teitscheid P, Ritter G (2015) Cutting food waste through cooperation along the food supply chain. Sustainability 7(2):1429–1445. https://doi.org/10.3390/su7021429. Accessed 18 Sep 2023

Godfray HCJ, Aveyard P, Garnett T, Hall JW, Key TJ, Lorimer J, Pierrehumbert RT, Scarborough P, Springmann M, Jebb SA (2018) Meat consumption, health, and the environment. Science 361(3699) https://www.science.org/doi/10.1126/science.aam5324. Accessed 02 Sep 2022

Gómez MI, Ricketts KD (2013) Food value chain transformations in developing countries: selected hypotheses on nutritional implications. Food Policy 42:139–150. https://doi.org/10.1016/j.foodpol.2013.06.010. Accessed 26 Aug 2021

Gonzalez-Rivas PA, Chauhan SS, Ha M, Fegan N, Dunshea FR, Warner RD (2020) Effects of heat stress on animal physiology, metabolism, and meat quality: a review. Meat Sci 162:108025. https://doi.org/10.1016/j.meatsci.2019.108025. Accessed 06 Dec 2023

Gregory NG (2010) How climatic changes could affect meat quality. Food Res Int 43(7):1866–1873. https://doi.org/10.1016/j.foodres.2009.05.018. Accessed 06 Dec 2023

Gustavsson J, Cederberg C, Sonesson U, Emanuelsson A (2013) The methodology of the FAO study: "global food losses and food waste—extent, causes and prevention"–FAO, 2011. SIK Report No. 857. Swedish Institute for Food and Biotechnology, Göteborg, Sweden. https://www.diva-portal.org/smash/get/diva2:944159/FULLTEXT01.pdf. Accessed 4 Oct 2024

Hamada-Sato N, Usui K, Kobayashi T, Imada C, Watanabe E (2005) Quality assurance of raw fish based on HACCP concept. Food Control 16(4):301–307. https://doi.org/10.1016/j.foodcont.2004.02.001. Accessed 03 Aug 2023

Hardenburg RE, Watada AE, Wang CY (1986) The commercial storage of fruits, vegetables, and florist and nursery stocks. US Department of Agriculture, Agricultural Research Service, Washington, DC

Herforth A, Venkat A, Bai Y, Costlow L, Holleman C, Masters WA (2022) Methods and options to monitor the cost and affordability of a healthy diet globally. In: Background paper for the state of food security and nutrition in the world 2022. FAO agricultural development economics working paper 22–03. FAO, Rome

Herrero M, Mason-D'Croz D, Thornton PK, Fanzo J, Rushton J, Godde C, Bellows A, de Groot A, Palmer J, Chang J, van Zanten H, Wieland B, DeClerck F, Nordhagen S, Beal T, Gonzalez C, Gill M (2021) Livestock and sustainable food systems: status, trends, and priority actions. In: von

Braun J, Afsana K, Fresco LO, Hassan MHA (eds) Science and innovations for food systems transformation. Springer, Cham, pp 375–399. https://doi.org/10.1007/978-3-031-15703-5_20

Hicks CC, Cohen PJ, Graham NAJ, Nash KL, Allison EH, D'Lima C, Mills DJ, Roscher M, Thilsted SH, Thorne-Lyman AL, MacNeil MA (2019) Harnessing global fisheries to tackle micronutrient deficiencies. Nature 574:95–98. https://doi.org/10.1038/s41586-019-1592-6. Accessed 22 Aug 2023

HLPE (2020) Food security and nutrition: building a global narrative towards 2030. A report by the High Level Panel of Experts on Food Security and Nutrition of the Committee on World Food Security, Rome. https://openknowledge.fao.org/server/api/core/bitstreams/8357b6eb-8010-4254-814a-1493faaf4a93/content. Accessed 3 Oct 2024

Huffman WE (2012) The status of labor-saving mechanization in U.S. fruit and vegetable harvesting. Choices (AAEA). Quarter 2. http://choicesmagazine.org/choices-magazine/theme-articles/immigration-and-agriculture/the-status-of-labor-saving-mechanization-in-us-fruit-and-vegetable-harvesting. Accessed 5 Oct 2024

Huss HH (ed) (1995a) 8. Assessment of fish quality. In: Quality and quality changes in fresh fish. FAO Fisheries Technical Paper 348. Food and Agriculture Organization of the United Nations, Rome. https://www.fao.org/4/v7180e/v7180e00.htm. Accessed 5 Oct 2024

Huss HH (ed) (1995b) 5. Post mortem changes in fish. In: Quality and quality changes in fresh fish. FAO Fisheries Technical Paper 348. Food and Agriculture Organization of the United Nations, Rome. https://www.fao.org/4/v7180e/v7180e00.htm. Accessed 5 Oct 2024

Huss HH (ed) (1995c) 3. Biological aspects. In: Quality and quality changes in fresh fish. FAO Fisheries Technical Paper 348. Food and Agriculture Organization of the United Nations, Rome.. https://www.fao.org/4/v7180e/v7180e00.htm. Accessed 5 Oct 2024

IPCC (Intergovernmental Panel on Climate Change) (2019) Climate Change and Land: An IPCC Special Report on climate change, desertification, land degradation, sustainable land management, food security, and greenhouse gas fluxes in terrestrial ecosystems. https://www.ipcc.ch/srccl/. Accessed 6 Oct 2024

Jacobs L, Delezie E, Duchateau L, Goethals K, Tuyttens FAM (2017) Broiler chickens dead on arrival: associated risk factors and welfare indicators. Poultry Sci 96(2):259–265. https://doi.org/10.3382/ps/pew353. Accessed 28 Nov 2023

Jaja IF, Mushonga B, Green E, Muchenje V (2018) Factors responsible for the post-slaughter loss of carcass and offal's in abattoirs in South Africa. Acta Trop 178:303–310. https://doi.org/10.1016/j.actatropica.2017.12.007. Accessed 13 Dec 2023

Johnson LK, Dara Bloom J, Dunning RD, Gunter CC, Boyette MD, Creamer NG (2019) Farmer harvest decisions and vegetable loss in primary production. Agric Syst 176:102672. https://doi.org/10.1016/j.agsy.2019.102672. Accessed 04 Apr 2023

Johnson LK, Dunning RD, Dara Bloom J, Gunter CC, Boyette MD, Creamer NG (2018a) Estimating on-farm food loss at the field level: a methodology and applied case study on a North Carolina farm. Resour Conserv Recy 137:243–250. https://doi.org/10.1016/j.resconrec.2018.05.017. Accessed 08 Sep 2021

Johnson LK, Dunning RD, Gunter CC, Dara Bloom J, Boyette MD, Creamer NG (2018b) Field measurement in vegetable crops indicates need for reevaluation of on-farm food loss estimates in North America. Agric Syst 167:136–142. https://doi.org/10.1016/j.agsy.2018.09.008. Accessed 16 Mar 2023

Kader AA (1992) Postharvest biology and technology: an overview. In: Kader AA (ed) Postharvest technology of horticultural crops, 2nd edn. University of California, Division of Agriculture and Natural Resources, Davis, pp 15–20

Kader AA (1996) Maturity, ripening, and quality relationships of fruit–vegetables. In Abou-Hadid AF, Jones, RA (eds) International Symposium on Strategies for Market Oriented Greenhouse Production [Alexandria, Egypt, 11–15 March 1995], Acta Hortic 434:249–256. International Society for Horticultural Science (ISHS), Leuven, Belgium

Kader AA (2002) Quality and safety factors: definition and evaluation for fresh horticultural crops. In: Kader AA (ed) Postharvest technology of horticultural crops, 3rd edn. University of California Agriculture and Natural Resources, Davis, pp 279–286

Kader AA, Barrett DM (2005) Classification, composition of fruits, and postharvest maintenance of quality. In: Barrett DM, Somogyi L, Ramaswamy HS (eds) Processing fruits: science and technology, 2nd edn. CRC Press, Boca Raton, pp 3–22

Kaminski AM, Cole SM, Al Haddad RE, Kefi AS, Chilala AD, Chisule G, Mukuka KN, Longley C, Teoh SJ, Ward AR (2020) Fish losses for whom? A gendered assessment of post-harvest losses in the Barotse Floodplain Fishery, Zambia. Sustainability 12(23):10091. https://doi.org/10.3390/su122310091. Accessed 14 Aug 2023

Kasso M, Bekele A (2018) Post-harvest loss and quality deterioration of horticultural crops in Dire Dawa Region, Ethiopia. J Saudi Soc Agric Sci 17(1):88–96. https://doi.org/10.1016/j.jssas.2016.01.005. Accessed 19 Sep 2023

Kawarazuka N, Béné C (2010) Linking small-scale fisheries and aquaculture to household nutritional security: an overview. Food Sec 2:343–357. https://doi.org/10.1007/s12571-010-0079-y. Accessed 14 Aug 2023

KC KB, Dias GM, Veeramani A, Swanton CJ, Fraser D, Steinke D, Lee E, Wittman H, Farber JM, Dunfield K, K MC, Anand M, Campbell M, Rooney N, Raine NE, Acker RV, Hanner R, Pascoal S, Sharif S, Benton TG, EDG F (2018) When too much isn't enough: does current food production meet global nutritional needs?

PLoS One 13(10):e0205683. https://doi.org/10.1371/journal.pone.0205683. Accessed 02 Sep 2021
Kelleher K (2005) Discards in the world's marine fisheries. An update. FAO Fisheries Technical Paper 470, Food and Agriculture Organization of the United Naitons, Rome. https://www.fao.org/4/y5936e/y5936e00.htm. Accessed 6 Oct 2024
Kitinoja L, Kader AA (2002) Small-scale postharvest handling practices: a manual for horticultural crops, 4th edn. University of California, Postharvest Technology Research and Information Center, Davis
Kitinoja L, Kader AA (2015) Small-scale postharvest handling practices: a manual for horticultural crops, 5th edn. University of California, Postharvest Technology Research and Information Center, Davis
Kittelsen K, Granquist E, Aunsmo A, Moe R, Tolo E (2018) An evaluation of two different broiler catching methods. Animals 8(8):141. https://doi.org/10.3390/ani8080141. Accessed 28 Nov 2023
Klaura J, Breeman G, Scherer L (2023) Animal lives embodied in food loss and waste. Sus Prod Consump 43:308–318. https://doi.org/10.1016/j.spc.2023.11.004. Accessed 18 Dec 2023
Kruijssen F, Tedesco I, Ward A, Pincus L, Love D, Thorne-Lyman AL (2020) Loss and waste in fish value chains: a review of the evidence from low and middle-income countries. Glob Food Secur 26:100434. https://doi.org/10.1016/j.gfs.2020.100434. Accessed 20 Apr 2023
Kummu M, de Moel H, Porkka M, Siebert S, Varis O, Ward PJ (2012) Lost food, wasted resources: global food supply chain losses and their impacts on freshwater, cropland, and fertiliser use. Sci Total Environ 438:477–489. https://doi.org/10.1016/j.scitotenv.2012.08.092. Accessed 14 Feb 2022
Lara LJ, Rostagno MH (2013) Impact of heat stress on poultry production. Animals 3(2):356–369. https://doi.org/10.3390/ani3020356. Accessed 07 Dec 2023
Lichtenthaler HK (1996) Vegetation stress: an introduction to the stress concept in plants. J Plant Physiol 148(1–2):4–14. https://doi.org/10.1016/S0176-1617(96)80287-2. Accessed 13 Sep 2023
Lucier G, Allshouse J, Lin B-H (2004) Factors affecting spinach consumption in the United States. Vegetables and Pulses Outlook No. (VGS-300-01). USDA, Economic Research Service
March MD, Toma L, Thompson B, Haskell MJ (2019) Food waste in primary production: Milk loss with mitigation potentials. Front Nutr 6:173. https://doi.org/10.3389/fnut.2019.00173. Accessed 22 Dec 2023
Martin NH, Torres-Frenzel P, Wiedmann M (2021) Invited review: controlling dairy product spoilage to reduce food loss and waste. J Dairy Sci 104(2):1251–1261. https://doi.org/10.3168/jds.2020-19130. Accessed 08 Dec 2023
Messner R, Johnson H, Richards C (2021) From surplus-to-waste: a study of systemic overproduction, surplus and food waste in horticultural supply chains. J Clean Prod 278:123952. https://doi.org/10.1016/j.jclepro.2020.123952. Accessed 13 Nov 2023
Mingione M, Fabi C, Lasinio GJ (2021) Measuring and modeling food losses. J Off Stat 37(1):171–211. https://doi.org/10.2478/jos-2021-0008. Accessed 15 Aug 2024
Mir NA, Rafiq A, Kumar F, Singh V, Shukla V (2017) Determinants of broiler chicken meat quality and factors affecting them: a review. J Food Sci Technol 54:2997–3009. https://doi.org/10.1007/s13197-017-2789-z. Accessed 15 Dec 2023
Miranda-de La Lama GC, Villarroel M, María GA (2014) Livestock transport from the perspective of the pre-slaughter logistic chain: a review. Meat Sci 98(1):9–20. https://doi.org/10.1016/j.meatsci.2014.04.005. Accessed 04 Dec 2023
Motaung TE, Petrovski KR, Petzer I-M, Thekisoe O, Tsilo TJ (2017) importance of bovine mastitis in Africa. Anim Health Res Rev 18(1):58–69. https://doi.org/10.1017/S1466252317000032. Accessed 22 Dec 2023
Nardone A, Ronchi B, Lacetera N, Ranieri MS, Bernabucci U (2010) Effects of climate changes on animal production and sustainability of livestock systems. Livest Sci 130(1–3):57–69. https://doi.org/10.1016/j.livsci.2010.02.011. Accessed 26 Sep 2024
Neff RA, Dean EK, Spiker ML, Snow T (2018) Salvageable food losses from Vermont farms. J JAFSCD 8(2):39–72. https://doi.org/10.5304/jafscd.2018.082.006. Accessed 03 Nov 2021
Nijdam E, Arens P, Lambooij E, Decuypere E, Stegeman JA (2004) Factors influencing bruises and mortality of broilers during catching, transport, and lairage. Poultry Sci 83(9):1610–1615. https://doi.org/10.1093/ps/83.9.1610. Accessed 29 Nov 2023
Njoga EO, Ilo SU, Nwobi OC, Onwumere-Idolor OS, Ajibo FE, Okoli CE, Jaja IF, Oguttu JW (2023) Pre-slaughter, slaughter and post-slaughter practices of slaughterhouse workers in Southeast, Nigeria: animal welfare, meat quality, food safety and public health implications. PLoS One 18(3):e0282418. https://doi.org/10.1371/journal.pone.0282418. Accessed 05 Dec 2023
OECD (Organisation for Economic Co-operation and Development) and FAO (Food and Agriculture Organization of the United Nations) (2023) OECD–FAO Agricultural Outlook 2023–2032. OECD Publishing, Paris. https://www.oecd.org/en/publications/2023/07/oecd-fao-agricultural-outlook-2023-2032_859ba0c2.html. Accessed 10 Oct 2024
Oppen J, Løkketangen A (2008) A tabu search approach for the livestock collection problem. Comput Oper Res 35(10):3213–3229. https://doi.org/10.1016/j.cor.2007.02.021. Accessed 15 Dec 2023
Papageorgiou M, Moutopoulos DK (2023) Small-scale fisheries discards in the eastern Mediterranean Sea: discarding species, quantities, practices and drivers. Fish Res 267:106798. https://doi.org/10.1016/j.fishres.2023.106798. Accessed 04 Aug 2023
Parfitt J, Croker T, Brockhaus A (2021) Global food loss and waste in primary production: a reassessment of its

scale and significance. Sustainability 13(21):12087. https://doi.org/10.3390/su132112087. Accessed 08 Aug 2023

Parlasca MC, Qaim M (2022) Meat consumption and sustainability. Annual Rev Resource Econ 14:17–41. https://doi.org/10.1146/annurev-resource-111820-032340. Accessed 11 Dec 2023

Perdana T, Kusnandar K, Perdana HH, Hermiatin FR (2023) Circular supply chain governance for sustainable fresh agricultural products: minimizing food loss and utilizing agricultural waste. Sus Prod Consump 41:391–403. https://doi.org/10.1016/j.spc.2023.09.001. Accessed 24 Oct 2023

Pérez Roda MA (ed), Gilman E, Huntington T, Kennelly SJ, Suuronen P, Chaloupka M, Medley P (2019) A third assessment of global marine fisheries discards. FAO Fisheries and Aquaculture Technical Paper No. 633, Food and Agriculture Organization of the United Nations, Rome. . https://www.fao.org/responsible-fishing/resources/detail/en/c/1317018/. Accessed 5 Oct 2024

Pinotti L, Luciano A, Ottoboni M, Manoni M, Ferrari L, Marchis D, Tretola M (2021) Recycling food leftovers in feed as opportunity to increase the sustainability of livestock production. J Clean Prod 294:126290. https://doi.org/10.1016/j.jclepro.2021.126290. Accessed 20 Nov 2023

Pirompud P, Sivapirunthep P, Punyapornwithaya V, Chaosap C (2023) Preslaughter handling factors affecting dead on arrival, condemnations, and bruising in broiler chickens raised without an antibiotic program. Poultry Sci 102(8):102828. https://doi.org/10.1016/j.psj.2023.102828. Accessed 28 Nov 2023

Prasanna V, Prabha TN, Tharanathan RN (2007) Fruit ripening phenomena–an overview. Cr Rev Food Sci 47(1):1–19. https://doi.org/10.1080/10408390600976841. Accessed 05 Apr 2023

Prusky D (2011) Reduction of the incidence of postharvest quality losses, and future prospects. Food Secur 3:463–474. https://doi.org/10.1007/s12571-011-0147-y. Accessed 11 Sep 2023

Reardon T, Liverpool-Tasie L, Minten B (2021) Quiet Revolution by SMEs in the midstream of value chains in developing regions: wholesale markets, wholesalers, logistics, and processing. Food Secur 13:1577–1594. https://doi.org/10.1007/s12571-021-01224-1. Accessed 07 Dec 2021

Ritz CW, Webster AB, Czarick M (2005) Evaluation of hot weather thermal environment and incidence of mortality associated with broiler live haul. J Appl Poult Res 14(1):594–602. https://doi.org/10.1093/japr/14.3.594. Accessed 28 Nov 2023

Savell JW, Mueller SL, Baird BE (2005) The chilling of carcasses. Meat Sci 70(3):449–459. https://doi.org/10.1016/j.meatsci.2004.06.027. Accessed 15 Dec 2023

Shewfelt RL (2014) Measuring quality and maturity. In: Florkowski WJ, Shewfelt RL, Brueckner B, Prussia SE (eds) Postharvest handling: a systems approach, 3rd edn. Elsevier Academic Press, Amsterdam, pp 387–410

Silanikove N (2000) Effects of heat stress on the welfare of extensively managed domestic ruminants. Livest Prod Sci 67(1–2):1–18. https://doi.org/10.1016/S0301-6226(00)00162-7. Accessed 27 Sep 2024

Steinfeld H, Wassenaar T, Jutzi S (2006) Livestock production systems in developing countries: status, drivers, trends. Rev Sci Tech 25:505–516

Sun M, Li Y, Chen Y (2023) Unveiling unselective fishing in China: a nationwide meta-analysis of multispecies fisheries. Fish Fish 24(1):142–158. https://doi.org/10.1111/faf.12715. Accessed 02 Aug 2023

Surucu-Balci E, Tuna O (2021) Investigating logistics-related food loss drivers: a study on fresh fruit and vegetable supply chain. J Clean Prod 318:128561. https://doi.org/10.1016/j.jclepro.2021.128561. Accessed 13 Apr 2023

Thilsted SH, Thorne-Lyman A, Webb P, Bogard JR, Subasinghe R, Phillips MJ, Allison EH (2016) Sustaining healthy diets: the role of capture fisheries and aquaculture for improving nutrition in the post-2015 era. Food Policy 61:126–131. https://doi.org/10.1016/j.foodpol.2016.02.005. Accessed 11 Aug 2023

Tijskens LMM, Schouten RE (2014) Modeling quality attributes and quality related product properties. In: Florkowski WJ, Shewfelt RL, Brueckner B, Prussia SE (eds) Postharvest handling: a systems approach, 3rd edn. Elsevier Academic Press, Amsterdam, pp 411–448

Troell M, Jonell M, Crona B (2019) The role of seafood in sustainable and healthy diets. Scoping report: the EAT-*Lancet* Commission report through a blue lens. Stockholm Resilience Centre, Stockholm University. https://eatforum.org/content/uploads/2019/11/Seafood_Scoping_Report_EAT-Lancet.pdf. Accessed 9 Oct 2024

Vázquez-Hernández MC, Parola-Contreras I, Montoya-Gómez LM, Torres-Pacheco I, Schwarz D, Guevara-González RG (2019) Eustressors: chemical and physical stress factors used to enhance vegetables production. Sci Hortic 250:223–229. https://doi.org/10.1016/j.scienta.2019.02.053. Accessed 12 Sep 2023

Watada AE, Herner RC, Kader AA, Romani RJ, Staby GL (1984) Terminology for the description of developmental stages of horticultural crops. HortScience 19(1):20–21. https://doi.org/10.21273/HORTSCI.19.1.20. Accessed 14 Sep 2023

WEF (World Economic Forum) (2024) Investigating global aquatic food loss and waste. White Paper (April). In collaboration with MarFishEco, Friends of Ocean Action and the World Resources Institute. https://www3.weforum.org/docs/WEF_Investigating_Global_Aquatic_Food_Loss_and_Waste_2024.pdf. Accessed 10 Oct 2024

Xing T, Gao F, Tume RK, Zhou G, Xu X (2019) Stress effects on meat quality: a mechanistic perspective.

Comp Rev Food Sci Food Saf 18(2):380–401. https://doi.org/10.1111/1541-4337.12417. Accessed 06 Dec 2023

Yahia EM (2019) Introduction. In: Yahia EM, Carillo-López A (eds) Postharvest physiology and biochemistry of fruits and vegetables. Woodhead Publishing, Duxford, pp 1–17

Yahia EM, Carillo-López A (eds) (2019) Postharvest physiology and biochemistry of fruits and vegetables. Woodhead Publishing, Duxford

Yahia EM, Fonseca JM, Kitinoja L (2019) Postharvest losses and waste. In: Yahia EM (ed) Postharvest technology of perishable horticultural commodities. Woodhead Publishing, Duxford, pp 43–69

Zhang M, Dunshea FR, Warner RD, DiGiacomo K, Osei-Amponsah R, Chauhan SS (2020) Impacts of heat stress on meat quality and strategies for amelioration: a review. Int J Biometeorol 64:1613–1628. https://doi.org/10.1007/s00484-020-01929-6. Accessed 13 Dec 2023

6 Enhancing Value: Post-farmgate Loss Pathways Linked to Food Access

6.1 Introduction

Sustainable food systems must support household food access to diverse foods in order to deliver healthy diets. Food access is a dimension of food security that refers to economic, social, and physical access by which an individual or household has the means to acquire food for an adequate diet that meets their preferences (HLPE 2020). As economies develop and undergo structural transformation, rising consumer incomes lead to shifting food preferences, which are met through diverse value-added approaches. Post-farmgate food value chains involve various operations, such as processing, logistics, and marketing, to bring food to consumers (Fanzo et al. 2017).

Post-farmgate, food moves along the value chain, becoming physically accessible to consumers at increasingly farther distances. Non-staple foods, especially perishable and nutritious ones, require investments in cold chains, packaging, and distribution to meet the growing demand of urban populations. As consumer demands for food quality and safety rise, so do standards, driving product differentiation beyond the basic requirements. The globalization of food trade has intensified this trend, with varying standards between countries influencing FLW pathways and determining which consumers have access to specific foods. Understanding these post-farmgate challenges is important for identifying strategies to enhance food quality, reduce losses, and ensure that nutritious food is accessible and affordable for all.

This chapter explores the intricate relationship between FLW and the diverse post-farmgate value chains that serve various consumer populations. The lack of FLW data is particularly salient in the post-farmgate value chain stages. Therefore, in this chapter we discuss the post-farmgate pathways that we expect based on current evidence. We focus on high-value, perishable food groups, which are typically sold fresh. By examining the distinct characteristics of transitioning food systems, we explore how infrastructure and market dynamics influence the types of foods that are accessible. We also delve into how quality standards for traded food can shape market access for value chain actors and consumers, as well as discuss the destinations of food that does not meet these standards. Finally, we identify the unique features of FLW pathways at the household consumer stage.

6.2 Post-farmgate Value Addition and Shifts in Consumer Demand Patterns

As countries move through structural transformation, typically, on-farm productivity gains reduce farmgate food prices. Rising incomes and

J. Boiteau, P. Pingali, *Wasted Potential*, Sustainable Development Goals Series,
https://doi.org/10.1007/978-3-031-97411-3_6

urbanization shift food demand patterns toward diversified diets, leading to increased demand for products with improved attributes when consumers value different food properties, experience changes in market conditions and social circumstances (Chap. 3), and are willing and able to pay higher prices (Fanzo et al. 2017)

Post-farmgate value addition grows faster than farm value output, as demand increases for improved food packaging, preparation, preservation, quality control, storage, transport, and distribution (Yi et al. 2021). When value addition focuses on improving food safety and reducing spoilage, it can positively impact FLW by ensuring that high quality food reaches consumers. However, value addition strategies, which prioritize product differentiation based on cosmetic attributes or limited supply, can inadvertently increase FLW when edible food is discarded.

Packaging, in some form, is used throughout the entire food system, safeguarding products, streamlining logistics, and serving as a communication channel with consumers (Lindh et al. 2016). Bulk packaging—encompassing bags, boxes, crates, and cartons—facilitates the efficient transportation of large quantities of both packaged and unpackaged food during distribution. Tailored to specific food environments, products may be offered to consumers in smaller formats, such as plastic bags, wraps, cans, or jars, or in a loose state, displayed in crates, baskets, or trays at retail (Verghese et al. 2015). In the past, food packaging materials consisted primarily of glass, metal, paper/cardboard, and plastics. Today, plastics have become the dominant material in food packaging (Fig. 6.1). The balance between minimizing FLW and mitigating environmental impact is particularly acute when considering the widespread use of single-use plastics in consumer packaging (Box 6.1). Therefore, packaging must fit the purpose, with its added value contingent upon the target market and desired product attributes (Lewis 2012). By adhering to these principles, packaging can effectively protect food from FLW while preventing issues stemming from packaging-related factors.

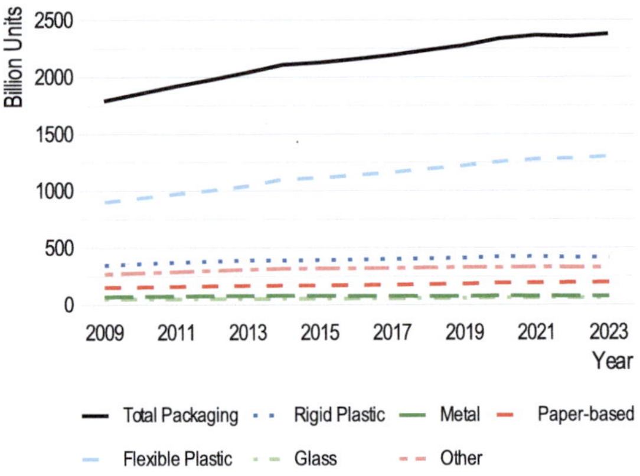

Fig. 6.1 **Trends in global food packaging material, 2009–2023**

Data source: Euromonitor International (2024a)

Box 6.1 The Plastic Paradox in Food Packaging

Consumer food packaging safeguards products, ensuring that they reach consumers in acceptable conditions and preventing FLW at the distribution, retail, and household stages (Grönman et al. 2013). Single-use plastics, as their name implies, are designed for one-time use and are meant to be discarded or recycled after a single use. Only 9% of plastic waste has ever been recycled (UNEP 2018). Food and beverage packaging are among the most common single-use plastics found in the environment (UNEP 2018). Therefore, the tension between minimizing FLW through enhanced packaging and mitigating the environmental impact of plastic food packaging is a pressing concern. More data are required, both on the extent of FLW and data on single-use plastics, to design packaging alternatives that minimize FLW and minimize the use of environmentally problematic plastics (Grönman et al. 2013; UNEP 2018; White and Lockyer 2020).

Closely related to packaging, food processing is critical to adding value by enhancing food quality, safety, and convenience to align with consumer preferences and market demands. Food processing can be divided into two main types: (1) primary processing, which prepares raw ingredients for consumption or use in other foods, and (2) secondary processing, which turns primary-processed foods or ingredients into new products (Parfitt et al. 2010; Park et al. 2014). The extent to which food processing affects perishability depends on the specific processing methods and techniques used. Food processing techniques can be simple or complex, but they usually involve one or more of the following: cleaning, cutting, heating, cooling, fermenting, adding enzymes, and pressing (Park et al. 2014; Monteiro et al. 2017). Processed foods are frequently packaged in consumer-friendly packaging formats to increase convenience.

Increased consumption of processed foods is a feature of dietary transitions that occur with structural transformation of economies (Chap. 2). Efforts to classify food processing consider the extent of change from the state at harvest, nature of change, place of processing, and purpose of processing (van Boekel et al. 2010; Sadler et al. 2021). Although most classification systems focus on how commercial processing affects health, critics argue that the definitions are inconsistent and fail to include important value chain stages, including storage, transportation, or home preparations (Sadler et al. 2021).

FLW can occur throughout the food processing stages, from the preparation of raw materials to after packaging (Darlington et al. 2009; Eičaitė et al. 2023). During the preparation and transformation of ingredients, losses can result from factors such as power outages, equipment malfunctions, and inadequate facilities (Darlington et al. 2009; Raak et al. 2017; Akintola et al. 2022). These inadvertent losses are in contrast to inherent food losses that occur when edible parts of raw materials are considered by-products and not used in the final product (Darlington et al. 2009; Raak et al. 2017). Food processing can inadvertently compromise quality attributes including shape, texture, or color, as well as food safety (Box 6.2). Unacceptable finished products may be removed from the value chain at any point, both before and after packaging.

Box 6.2 Using Solar Drying Approaches for Food Preservation

When food is dried, the removal of moisture helps prevent it from spoiling, making it safer and longer lasting. There are many different drying techniques, each with its own equipment and operational requirements (Cohen and Yang 1995). Some foods can be prepared (for example, cut, salted, fermented, or cooked) before they are dried (Doe and Olley 2020; Natarajan et al. 2022).

Solar drying is a common food preservation method in low- and middle-income countries (LMICs). There are two main types of solar drying: open sun drying, which uses direct sunlight, and convection drying, which uses sunlight to heat air that then dries the food (Bourdoux et al. 2016). Although solar drying can improve food safety and shelf life, it can also lead to losses in food quality. In open sun drying, food is exposed to the elements, making it difficult to control temperature and humidity, as well as insects and pests, which can increase the risk of spoilage and contamination (Bourdoux et al. 2016; Udomkun et al. 2020; Matavel et al. 2022)). Convection dryers protect food from contaminants and dry it more quickly (VijayaVenkataRaman et al. 2012; Bourdoux et al. 2016; Natarajan et al. 2022). However, both methods can cause changes in texture and flavor, and discoloration, depending on the food being dried (Nijhuis et al. 1998; Calín-Sánchez et al. 2020; Natarajan et al. 2022).

Whereas drying is a standard postharvest process to preserve durable foods like legumes, seeds, nuts, and grains (Manandhar et al. 2018), drying perishable foods, like fruits, vegetables, meats, and fish, depends on the context (Kim et al.

2009; Natarajan et al. 2022; Morris 2023). Although drying can extend the shelf life of perishable foods, the higher moisture content of these foods presents unique challenges for preserving quality during solar drying.

As food value chains lengthen, the availability and accessibility of cold chain infrastructure becomes important for maintaining the quality of fresh, perishable foods and reducing post-farmgate FLW. Cold chain infrastructure, encompassing temperature-controlled storage and transportation from farm to consumer, is essential for extending the shelf life and maintaining the quality of fresh, perishable products (Brondy 2019). By slowing down degradation processes, cold chains reduce FQL and enable greater flexibility for value chain actors, including consumers (Heard and Miller 2016). At the same time, cold temperatures can lead to quality losses, such as off-flavors and changes in texture, depending on the specific product, processing methods, packaging, and storage conditions (Bonat Celli et al. 2016).

The implementation and maintenance of cold chains present significant challenges, particularly in transitional food systems (Table 6.1). The availability of clean water, reliable power, well-maintained roads, and appropriate transport vehicles and storage facilities are fundamental requirements for effective cold chains. Despite their benefits, the initial investment and ongoing operational costs can be substantial, particularly, for small-scale producers (Brondy 2019). Moreover, the lack of food safety standards and demand for perishable foods can create barriers to entry and hinder the development of efficient and reliable cold chain systems, reinforcing FLW of nutritious foods. The high costs involved, limited access to finance for small and medium enterprises, inadequate government regulations, and insufficient consumer demand often impede the development and adoption of cold chain infrastructure (Brondy 2019).

Table 6.1 Major constraints affecting food cold chain performance in transitioning food systems

Infrastructure	• Lack of cold storage • Shortage of refrigerated carriers • Poor logistics infrastructure • Unavailability of power and water
Technological	• Lack of awareness about the use of information technology • Lack of modern processing and packing methods • Improper traceability
Operational	• Lack of integration between stakeholders • Improper handling • Lack of information sharing • Inadequate education of farmers • Large number of intermediaries
Regulatory and economic	• Government regulations • Lack of standardization • Lack of customer knowledge • High cost of technology and operations

Adapted from Shashi et al. (2018)

6.3 Fruit and Vegetable Value Chains

Fruits and vegetables reach consumers in both fresh and, to a lesser extent, processed forms. The post-farmgate stages determine the market channels, and value addition occurs through various product differentiation strategies. As food systems transform, the attributes used to differentiate products and limit acceptance shift, influenced by changes in packaging, processing, storage, and distribution.

In early stages of food system transformation, preventing quality losses in fresh fruits and vegetables, including food safety issues, deterioration, and spoilage, enhances product value. Processing into shelf-stable forms adds value by extending shelf life and mitigating food safety hazards. As food systems modernize and FQL prevention related to food safety and spoilage become common, other quality attributes, such as cosmetic attributes for fresh produce and minimal processing for convenience, become attributes for product differentiation.

6.3.1 The Dynamics of Food Quality in Fruits and Vegetables and Adding Value

Fresh fruits and vegetables remain physiologically active throughout storage and distribution. As food value chains lengthen and become more complex, the risk of exposure to environmental, biological, and physical stressors, which can compromise the quality of fresh produce, increases. Although fresh fruits and vegetables require specific conditions to maintain quality, processed products have their own unique requirements. Value chain actors must be cognizant of these distinct needs to effectively manage FQL, according to the target market.

Consumers prioritize distinct quality attributes in fresh and processed fruits and vegetables, often valuing sensory qualities like freshness and appearance in fresh produce, while prioritizing convenience and shelf life in processed products. Similarly, processors may seek raw materials with specific characteristics that are most appropriate for processing to ensure desired product attributes in the final product. These characteristics may differ from than those desired in produce intended for fresh markets (Bonat Celli et al. 2016). Desirable attributes between value chain actors and consumers may conflict, requiring innovative solutions, such as use of storage or packaging (Verghese et al. 2015).

Processors may source lower quality produce from fresh markets based on available supply and market prices, particularly, in traditional and transitional food systems (Subramanian 2016; Boiteau and Pingali 2022). In the tomato processing industry, for example, using tomato varieties intended for fresh markets, which are less suitable for processing, can necessitate the addition of ingredients like sugar to compensate for acceptable quality attributes in the final product (Subramanian 2016; Schreinemachers et al. 2022).

Bulk packaging is a critical component of the value chain for fresh fruits and vegetables, facilitating their movement from farm to market. The selection of appropriate bulk packaging for post-farmgate stages is influenced by packaging options, quantity of produce, and susceptibility to damage during storage, handling, and transportation.

The quality and rate of deterioration of fresh fruits and vegetables, as discussed in Chap. 5, is contingent upon providing adequate protection, ventilation, and temperature control through packaging (Verghese et al. 2015). Even when appropriate packaging is used, improper handling practices, such as failing to sort produce by ripeness, type, or odor sensitivity, can lead to quality deterioration, resulting in produce that is overly ripe, soft, or exhibits unpleasant flavors or odors (Kasso and Bekele 2018; Perdana et al. 2023).

Packing damaged produce with undamaged produce and overfilling containers can lead to cross-contamination, decay, and reduced marketability (Surucu-Balci and Tuna 2021; Perdana et al. 2023). Although bulk packaging practices, such as mixing, mishandling, and overpacking fresh produce, can contribute to FQL and subsequent pFLW, addressing these issues requires a combination of knowledge and incentives to implement improved practices. For example, value chain actors may be more incentivized to invest in improved packaging to reduce FQL when they receive higher prices for higher quality produce in the market.

Limited transport and storage capacity or dedicated facilities can challenge the effective separation of produce, leading to quality variations, premature ripening, softening, off-flavors, and deterioration (Kasso and Bekele 2018; Schudel et al. 2023). For example, fresh produce may be stored in bulk before processing, but lack of storage capacity can lead to FQL (Subramanian 2016). As food value chains lengthen and become more complex, risk of exposure to environmental factors, including sunlight, adverse weather, and unsuitable temperatures and atmospheric conditions, increases. Therefore, bulk packaging requirements need to be considered alongside transport factors, such as transport distance, road conditions, vehicle type, and season, which influence the timing and conditions of transportation between value chain stages and can be challenged by delays, high costs, poor handling, and infrastructure deficiencies (Surucu-Balci and Tuna 2021).

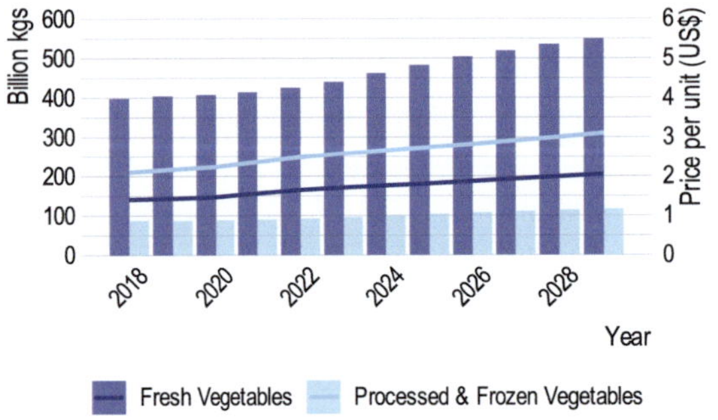

Fig. 6.2 **Global vegetable volumes and revenues by form, estimated from 2018 to 2029**
Source: Statista Market Insights (2024)

Processing fruits and vegetables expands their availability beyond seasonal constraints and market accessibility. Processed fruits and vegetables can be categorized into three broad segments based on storage temperature: ambient (for example, dried products, canned or jarred preserves, and juices); chilled (for example, packaged, peeled, fresh-cut fruits and vegetables); and frozen (Mena et al. 2011; Garrone et al. 2014). As overall vegetable consumption is projected to rise, fresh vegetables are likely to remain the dominant form of supply, compared to processed and frozen varieties (Fig. 6.2). Additionally, fresh vegetables are expected to maintain lower prices than their processed and frozen counterparts. However, in the United States, where processing and cold chain infrastructure are well developed, there is not a consistent pattern in which one format (fresh, canned, or frozen) is more or less expensive—a situation most likely due to factors, such as grower prices, processing costs, handling practices, and spoilage rates (Stewart et al. 2024).

Processing methods for ambient storage can vary widely, from traditional drying and canning techniques to modern technologies, each with their own financial, technical, and operational requirements (Hasan et al. 2019; Schreinemachers et al. 2022; Machala et al. 2022). The quality of the preserved product can differ depending on the chosen method. The chemical and physical transformations from drying or canning can impact the texture, flavor, appearance, and handling of fruits and vegetables, depending on the technique (Nijhuis et al. 1998; Raak et al. 2017; Calín-Sánchez et al. 2020; Natarajan et al. 2022).

Post-processing, dried and canned fruits and vegetables require a dry storage environment to prevent moisture-related spoilage (Surucu-Balci and Tuna 2021). Although warm and dry conditions are acceptable, cold and dry environments are optimal for extending shelf life (Bradford et al. 2018). Damaged packaging can let in moisture, damaging dried products. Additionally, moisture can compromise packaging integrity, as is the case for canned products that are susceptible to rust and seal failure when exposed to moisture, thereby facilitating contamination. Maintaining dry storage conditions can be difficult in hot and humid climates of LMICs in the Global South.

The emergence and later expansion of reliable cold chains facilitate the growth of minimally processed produce markets, responding to the increasing demand for convenient, ready-to-use fruits and vegetables among growing urban, middle-class populations (Mir et al. 2018). The processing methods for ambient and frozen products act on plant physiological activity to slow deterioration and prevent microorganism growth (Rickman et al. 2007). Minimal processing involves techniques, such as cutting, peeling, washing, cooling, and moisture removal (Barrett et al. 2010). However, processing physiologically active plant tissue can induce stress responses, including elevated respiration, ethylene production, browning, and water loss (Brecht 1995; Perera 2020).

Several factors can contribute to FQL during the processing of refrigerated and frozen produce, including pretreatment methods, freezing conditions, packaging, storage, and thawing protocols (Bonat Celli et al. 2016). For frozen products, in particular, the formation and size of ice crystals, as well as the denaturation and inactivation of enzymes, significantly impact the final product's quality attributes, such as texture, color, flavor, and nutrients (Bonat Celli et al. 2016). Balancing consumer demand for convenient, ready-to-use produce with processor and retailer priorities for extended shelf life requires innovations in processing and packaging solutions (Verghese et al. 2015).

Processing inefficiencies can contribute to FLW, even without direct impacts on food quality. Physical FLW occurs during processing when edible food parts, such as skin, seeds, and pulp are removed, according to the desirable attributes for the final product (Secondi et al. 2019; Pop et al. 2021). Material loss during production line changes or cleaning, noncompliance with commercial standards due to production errors, and packaging defects can contribute to additional pFLW (Dora et al. 2020; Eičaitė et al. 2023).

Box 6.3 Controlling Post-farmgate Environments of Perishable Fruits and Vegetables

Fruits and vegetables, whether fresh or frozen, require a reliable cold chain to prevent spoilage. Chilling fresh produce slows down its respiration, extending shelf life. Maintaining freezing temperatures of frozen produce is necessary for preventing quality issues associated with thawing, or with repeated freezing and thawing cycles. In modern food systems, robust cold chains are essential for supporting fresh-cut (refrigerated) and frozen fruits and vegetables, offering consumers time savings in preparation and, particularly, for frozen produce, nutrient retention (Barrett et al. 2010).

Relative humidity is closely linked to temperature. Lower temperatures reduce water loss in produce, while excessive humidity can lead to microbial growth and decay (Kader 1992). However, when relative humidity is too low, produce may wilt, shrivel or shrink, lose weight, and become less firm. Low-tech methods like adding water or using vapor barriers can help adjust humidity, but these methods may also introduce microbial pathogens, depending upon the quality of water and materials (Kasso and Bekele 2018). Although advanced technologies offer more precise control of relative humidity in storage and packaging, they often require additional resources and specialized knowledge.

6.3.2 Fresh Fruit and Vegetable Food Environments

Effective coordination among value chain actors is important for successful value addition strategies, encompassing production planning, market dynamics, packaging, and quality control. Although both formal and informal markets contribute to access to fresh fruits and vegetables, their market shares shift as food systems modernize and urban populations grow. Formal markets prioritize efficiency, marketing, and quality standards, while informal markets often serve geographically dispersed populations, offering convenience and affordability (Cadilhon et al. 2006).

Informal and formal markets often have distinct approaches to quality grading and standards for fresh produce. Retailers and vendors in informal markets frequently sort produce into different quality grades, offering discounted prices for lower quality items (Minten and Reardon 2008; Boiteau and Pingali 2022). By separating lower quality produce that is at risk of spoilage, this approach can minimize FLW and enable consumers to more easily purchase products that have acceptable attributes, including quality and value.

In contrast, formal markets enforce more stringent quality standards, including cosmetic attribute standards, to offer often one quality grade to customers. However, the enforcement of these cosmetic attribute standards can contribute to pFLW of otherwise edible fruits and vegetables. Recent efforts to sell produce with lower grade cosmetic attributes in modern retail have attracted consumers who value environmental sustainability and those seeking more affordable options (Aschemann-Witzel et al. 2018). However, the increasing prevalence of packaged fresh fruits and vegetables can challenge consumers' ability to assess product attributes and retailers' efforts to prevent pFLW of unsold produce (Box 6.4).

Box 6.4 Consumer Shopping Experiences Shape Food Loss and Waste Pathways at the Retail Stage

Consumer shopping experiences, in addition to quality standards, can influence FLW pathways at the retail stage. Loose produce allows consumers to select individual items, enabling them to inspect and choose products that meet their desired quality attributes. However, variations in how retailers grade produce and differences in consumer preferences can result in unsold produce, when consumers have selectively chosen items. The handling of unsold produce, including sorting out spoiled items and regrading edible produce, can vary between informal and formal markets, influencing the amount of food diverted to nonfood uses.

While informal markets often sell fresh produce loose, on display in open-air settings without consumer packaging, formal markets are increasingly offering both loose and prepackaged produce in plastic or cardboard containers, displayed for sale in indoor or shaded areas (White and Lockyer 2020; Downs et al. 2020). Consumer packaging can extend the shelf life of fresh produce but also makes it harder for retailers to identify and discard spoiled items (Verghese et al. 2015). Formal markets often use packaging with additional product information that influences consumer choices. Date labels, visible flaws, and damaged packaging can result in price reductions and pFLW (Lebersorger and Schneider 2014). Formal retailers in transitional food systems face the challenge of balancing quality demands, which can increase prices and lead to slow turnover, with the need to compete with traditional markets, where produce often appears fresher (Cadilhon et al. 2006).

6.4 Animal-Source Food Value Chains

Animal-source foods (ASF) are rich in bioavailable nutrients including protein and micronutrients, but their perishability makes them vulnerable to FQL issues that can compromise their safety and desirability. Relative to plant foods, biological hazards are particularly challenging for ASF that can lead to contamination, spoilage, and foodborne illness (Barbut and Leishman 2022). Therefore, to meet growing demand for safe ASFs, value chain actors must prioritize addressing time, temperature, and handling challenges that arise throughout the stages of processing, storage, and distribution. However, data on FLW in ASF value chains is limited (Karwowska et al. 2021; WEF 2024).

Understanding FLW pathways is crucial for identifying instances where FQL of ASF leads to pFLW, such as when unsafe food is removed, and for determining where to focus efforts on preventing FQL. Processing is included in many of these pathways since most ASF requires processing before it is marketable to consumers. This includes removing parts and cutting carcasses, as well as further processing, like butchering and combining ingredients (Rørå et al. 2001; Mahler et al. 2015; Barbut 2020). Processing choices can affect food hygiene and consumer acceptance in different target markets (Mahler et al. 2015; Barbut 2020).

6.4.1 Animal Meat

After land animals have been slaughtered, including primary processing to remove organs (which may be edible, see Box 6.5) and the carcasses washed, the extent of secondary processing required for consumer acceptability depends on the animal and desirable quality attributes, including tenderness, flavor, and texture (Maddock 2014). Red meat, from larger animals, like beef, pork, lamb, and goat, is typically cut into halves or quarters after slaughter (Maddock 2014). The meats are then further processed into smaller pieces for sale to commercial or household buyers. Poultry, being smaller, are often sold whole, cut into parts, or ground (Smith 2014). Raw meat can be further processed through methods, such as heating, freezing, curing, fermenting, drying, or adding ingredients (Smith 2014).

Box 6.5 Variety Meats Are Not Necessarily Globally Acceptable

The issue of cultural acceptability arises when discussing edible offal, also known as "variety meats," which specifically refers to the edible internal organs of animals. The specific organs considered edible can vary depending on cultural preferences. Variety meats are often sold at lower costs than skeletal muscle meat and can be processed into ground products, used in soups or stews, or served as traditional table meats, especially when they have larger portions of muscle (Schaefer and Arp 2017).

Raw meat is marketed to consumers in three general forms: hot-fresh, chilled-fresh, and frozen (Zhou et al. 2012). Hot-fresh meat, from animals slaughtered and sold on the same day, is prevalent in smaller cities and rural areas, and typically, sold at informal markets (Zhou et al. 2012; Liu et al. 2017). Chilled-fresh meat requires immediate cooling, and a cold chain is more common in larger cities and formal markets (Liu et al. 2017). Frozen meat, processed and stored at freezing temperatures, is also found in formal markets, with cold chains that include freezers (Liu et al. 2017).

The timing of slaughter, purchase, and consumption is critical in preventing FQL, particularly, bacterial growth and spoilage, especially, in regions lacking cold chain infrastructure. In transitional food systems, animals are often slaughtered without prearranged buyers, leading to potential mismatches between slaughter timing and consumer preferences (Alarcon et al. 2017). After slaughter, meat in informal value chains can move through complex channels, including multiple actors, to reach consumers in both rural and urban areas (Alarcon et al. 2017; Wilson 2018; Snel et al. 2021). Without refrigeration or immediate customers, slaughtered meat is exposed to ambient temperatures for extended periods, accelerating quality deterioration and compromising food safety.

In transitional food systems with inadequate food safety standards or enforcement, meat with FQL may remain accessible to consumers. Although this accessibility reduces pFLW, meat with FQL related to food hazards may pose risks to consumer health and safety, challenging the provision of safe and nutritious diets. When adequate standards and enforcement exist, meat portions with FQL can be removed during inspection if they fail to meet quality standards (Mahler et al. 2015; Barbut and Leishman 2022). To minimize pFLW, processors can transform edible trimmings and lower quality cuts into value-added products like sausages or mince, making them available to retailers and consumers in lower income communities (Alarcon et al. 2017).

Packaging meat products helps to prevent both external contamination during storage, transportation, and distribution, as well as cross-contamination with other foods. Packaging also serves as a communication tool, informing consumers about product quality through features, like grades and inspection marks. Inspections become standard practice in processing firms that supply high quality meat to high-income consumers in formal markets. For example, the United States Department of Agriculture has several labels to indicate quality grades, value-addition, and official inspection (USDA 2022). However, consumer preferences for packaged versus unpackaged meat may also be based on cultural context rather than level of economic development (Table 6.2).

Table 6.2 Comparison of packaged meat across selected countries, 2013 and 2023

	Share of packaged meat of total retail volume (%)		
	2013	2023	Change
China	2.5	7.5	+5.00
India	2.0	3.5	+1.50
Japan	80.3	81.3	+1.00
Vietnam	7.2	9.0	+1.80
Russia	63.0	71.2	+8.20
Brazil	56.3	58.5	+2.20
Mexico	43.4	45.7	+2.30
South Africa	21.0	29.0	+8.00
United States	67.4	75.2	+7.80
France	81.2	83.9	+2.70
Italy	39.0	41.1	+2.10
Spain	36.5	39.6	+3.10
United Kingdom	63.0	65.7	+2.70

Meat is either packaged or unpackaged at retail. Packaged meat retail volume is highlighted in gray when it constitutes at least 50% of the total retail volume. Source: Euromonitor (2024b)

6.4.2 Aquatic Animal Foods

The expansion of aquatic food production over the last few decades has been concentrated in aquaculture production in Asia, in contrast to Africa where production has expanded but still remains small (Chap. 5). In Asia, rising incomes have fueled a surge in domestic demand for aquatic foods (Belton et al. 2018; Hernandez et al. 2018). The development and expansion of transportation networks, wholesale markets, and retail channels have facilitated the supply of domestically produced fish—primarily whole fresh fish—to urban areas (Hernandez et al. 2018). In contrast, while the demand for aquatic foods in Africa is increasing due to rising incomes, domestic fish production often falls short. Inadequate infrastructure and logistics hinder the accessibility of both inputs and markets, limiting domestic fish production and making domestically produced fish, particularly fresh fish, an expensive product (Belton et al. 2018; Liverpool-Tasie et al. 2021).

Specific fish species have unique quality attributes and distinct value chains based on their market value. Accessible fish are those typically marketed locally or regionally, while fish classified as a luxury food, often have a wider reach, targeting higher income consumers (Henriksson et al. 2021). In traditional and transitional food systems, low- and medium-value species dominate domestic supplies, while high-value species are often limited (Belton et al. 2018). Fish intended for fresh, international export usually benefit from well-organized value chains, with access to cold chain infrastructure to ensure quality standards (Odoli et al. 2019).

The market value of fish, distance to markets, and availability of transportation impact decisions on whether fish is sold fresh, frozen, or shelf-stable (Kaminski et al. 2020). Traditional processing methods like drying and smoking extend shelf life without requiring cold storage. With the development of cold chain infrastructure, higher value fish are often sold fresh or frozen, while lower value fish are more likely to be processed. In regions with limited infrastructure, processed fish are often more accessible to consumers (Liverpool-Tasie et al. 2021). However, despite growing demand for fresh fish, dried and smoked fish remain popular in Africa, even among higher income consumers (Liverpool-Tasie et al. 2021).

Spoilage is the primary pathway for fish quality loss, especially in regions with inadequate cold chain infrastructure. The time between harvesting aquatic foods and processing or chilling is crucial for minimizing quality deterioration. Even in areas with cold chain infrastructure, improper handling and application of a cold chain can contribute to quality issues in fresh fish (Odoli et al. 2019; Kaminski et al. 2020). Therefore, small-scale fisheries lacking cold storage during fishing or harvesting are limited in their ability to sell fish over long distances, restricting their sales to nearby communities to minimize quality loss (Odoli et al. 2019). Supply fluctuations can further strain existing capacity, leading to challenges in accommodating seasonal availability. Aquaculture, compared to capture, or wild, fisheries, offers greater production control and less variability (Belton et al. 2018).

Open-air drying is a common, traditional method for preserving low-value fish. However, fish size presents unique challenges. Larger, fatty fish require longer processing times for smoking and drying, compared to smaller fish. Although smaller fish dry more quickly, they are more susceptible to breakage (Kaminski et al. 2020). Drying time is a critical factor for quality, because prolonged exposure to weather, insects, and animals increases the risk of FQL, as well as pFLW (Esser et al. 2007; Akintola et al. 2022). Once drying is completed, maintaining a dry chain is essential for preventing quality loss in the final product. However, poor handling, packaging, and transportation practices can expose dried fish to moisture, leading to spoilage (Kaminski et al. 2020). For example, charging transport costs per sack can incentivize overpacking, which can damage dried fish (Akande and Diei-Ouadi 2010).

Seasonal weather changes can impact fish quality. During dry seasons, hot weather can accelerate spoilage in fresh fish, especially during holding and transportation (Akande and Diei-Ouadi 2010). Long distances between value chain stages, bargaining processes, and inadequate infrastructure can worsen quality loss (Akande and Diei-Ouadi 2010). Therefore, prompt chilling of fish is even more essential at this time to reduce quality loss, especially for fish intended for fresh markets. At the same time, too much rain can contribute to incomplete drying of low-value fish, making them vulnerable to insects, mites, and mold, which can contaminate unaffected products when mixing old and new stocks of processed fish (Akande and Diei-Ouadi 2010; Akintola et al. 2022).

Similar to other ASF value chains, fish with FQL may be downgraded and sold at lower prices rather than discarded as pFLW. Downgraded processed fish are often sold at a discount in informal markets with minimal infrastructure, making fish more affordable for lower income consumers (Akande and Diei-Ouadi 2010; Kruijssen et al. 2020).

6.4.3 Dairy

Rising incomes and urbanization have increased the demand for milk, both in terms of quantity and quality, and consumers rely more on retail outlets than household production (Janssen and Swinnen 2019). Value addition in dairy value chains focuses on keeping dairy products free from food hazards and extending the shelf life of raw milk using processing and cold chains. Milk processing includes heating for pasteurization and transforming milk into other products (for example, yogurt, cheese, butter). Pasteurized milk often has longer value chains, involving traders, processing companies, and distributors that connect farmers to urban retailers (Minten et al. 2021).

The location of producers in rural, peri-urban, and urban areas influences their market options (Headey et al. 2024). Small-scale dairy producers, especially those in rural areas without access to modern markets, may rely heavily on informal value chains (Janssen and Swinnen 2019; Kumar et al. 2019). Informal chains with limited chilling or processing are geographically short and involve fewer actors between producers and households due to the perishability of raw milk (Kiambi et al. 2018; Janssen and Swinnen 2019; Kumar et al. 2019; Minten et al. 2021). Efforts to enhance market access have included sourcing milk from urban producers to reduce transportation distances, establishing marketing bodies for milk aggregation, and public–private sector investments in infrastructure, such as chilling units and processing facilities (Headey et al. 2024).

In response to growing demand for high quality dairy products, marketing bodies, such as milk collection centers, cooperatives, and processing centers (domestic and multinational), have emerged, serving as intermediaries between producers and retailers and consumers (Kiambi et al. 2018; Van Campenhout et al. 2021). Milk collection centers, operated by cooperatives or private companies, are important hubs for aggregating milk in bulk, often chilling it before transportation to processors and consumer markets (Van Campenhout et al. 2021). Although milk collection centers typically mark the beginning of the cold chain (Van Campenhout et al. 2021), the availability of chilling facilities can vary depending on their size (Kiambi et al. 2018).

Milk producer groups and cooperatives add value by conducting testing to screen milk for acceptance at collection centers, determining payment systems, and implementing grading systems (Draaiyer et al. 2009). However, establishing collection centers without providing information or incentives for improved safety and quality at the farm or collection center levels can create gaps in FQL prevention (Janssen and Swinnen 2019).

Village-level milk collection centers are strategically located near clusters of rural dairy producers to minimize transportation distances and the time between milking, chilling, and processing (Kiambi et al. 2018; Janssen and Swinnen 2019; Kumar et al. 2019; Van Campenhout et al. 2021). Reducing these distances is important for minimizing FQL, because most milk producers lack their own chilling or processing facilities (Janssen and Swinnen 2019). Therefore, small traders and transporters play an important role in transporting milk from farms to collection centers (Van Campenhout et al. 2021). In contrast, large-scale milk producers often supply milk directly to processing plants, where dairy companies collect milk on-farm (Janssen and Swinnen 2019). Contracts between dairy producers and buyers may be used, depending on the specific context (Kiambi et al. 2018; Janssen and Swinnen 2019).

The timing of milking, chilling, and processing significantly influences FLW in both formal and informal dairy value chains. Maintaining liquid milk in a chilled state until use or consumption is essential to minimize microbial growth (Chap. 5). However, value chain actors may lack appropriate transportation or refrigeration facilities, delaying the movement of milk to subsequent value chain stages. For example, in Kenya, traders in both formal and informal chains have reported transporting milk without refrigeration, either by foot for short distances or in plastic containers placed under passenger seats in public vehicles for longer distances (Kiambi et al. 2018). Seasonal variations in milk production can further create imbalances between supply and demand, leading to excess milk that may not be consumed, stored, or processed into other dairy products (Minten et al. 2021).

The choice of packaging material can also promote or slow down FQL in liquid milk. Plastic containers, especially scratched ones, are difficult to clean and can heat up easily (such as in the sun), promoting microbial growth and accelerating spoilage (Van Campenhout et al. 2021). Therefore, midstream actors in formal value chains typically require improved packaging materials like stainless steel or aluminum cans that prevent FQL (Charad et al. 2019). Yet, even when governments set regulation and licensing requirements, actors in informal dairy value chains often face difficulty complying due to limited infrastructure and resources (Alonso et al. 2018). Advanced processing and packaging innovations for creating shelf-stable liquid milk (Box 6.6) often require substantial resources, limiting their accessibility to modernized dairy processors and higher income consumers.

Milk rejected due to quality issues may be sold to alternative markets, including fresh or processed markets, at discounted prices. However, rejected milk continues to deteriorate over time, especially if not promptly chilled or consumed. Milk that falls below acceptable quality thresholds, such as spoiled milk, may be diverted to nonfood uses like animal feed (Kiambi et al. 2018). In dairy value chains serving large multinational corporations, milk is typically purchased at the farm and received at the factory gate, with strict private practice standards for time limits between milking, chilling, and processing to prevent spoilage (Charad et al. 2019). These companies may not track the final destination of milk lost in transit or rejected due to quality issues at the factory (Charad et al. 2019).

Box 6.6 Trade-Offs of Ultra-High Temperature Milk and Packaging

Ultra-high temperature (UHT) processing rapidly heats milk to high temperatures, virtually eliminating microbial loads. This process enhances food safety and enables ambient storage in aseptic containers. Therefore, this processing and packaging combination allows for liquid milk to be shipped and stored without the need for refrigeration. Despite its advantages, UHT processing and packaging limitations can contribute to FLW and environmental impacts.

One of the challenges associated with UHT processing is the potential for off-flavors and browning that can develop during the high-temperature treatment (Shah et al. 2016). Skilled operators are essential to minimize these undesirable changes. Compared to pasteurized milk, which is heated to a lower temperature and requires refrigeration, changes in UHT milk flavors may not be as well received by consumers. This may influence consumer preferences toward pasteurized milk over UHT milk, particularly, where consumers have year-round availability of pasteurized milk (Shah et al. 2016).

Aseptic packaging, while required for maintaining the shelf stability of UHT milk, can contribute to pFLW and present recycling challenges. Packaging design factors, such as folding at the bottom, corrugations in the internal wall, lid type, container shape, and seal-breaking methods, can influence the amount of milk that remains trapped within the package (Meurer et al. 2017). The coextruded film material commonly used in this packaging is not only difficult to recycle but also lacks value in other industrial processes, leading to landfill disposal (Aparicio-Peralta et al. 2020). Therefore, when implementing FLW reduction strategies, it is important to consider the potential environmental trade-offs associated with packaging innovations.

6.5 Trade

Food trade is a key component of food system transformation, creating new market opportunities for value chain actors and potentially contributing to more stable access to diverse diets. Fresh, nutrient-rich foods are valuable exports, accounting for half of the total value of food and agricultural exports from low-income countries (LICs) (Unnevehr 2000; Maertens et al. 2012). The impact of food trade on livelihoods and food security is influenced by the value chain actors involved, the types of foods traded, and the effects on local food markets. The FLW pathways associated with trade encompass quality loss along different food value chains in the producing country, transportation and distribution challenges during trade, and the food quality, standards, and reporting requirements imposed by importing countries.

6.5.1 Food Quality Standards for Traded Foods

Food trade links countries with varying food safety and quality standards and capacities (HLPE 2017). As consumer preferences for improved food quality increase with rising incomes, food safety and quality standards are established and enforced more rigorously. The level of development of a country often correlates with the effectiveness of institutions responsible for quality control and standards (Swinnen and Vandemoortele 2008). Eventually, countries with industrialized and consolidated food systems have well established public and private food safety and quality standards, primarily focused on promoting and communicating food quality to consumers (Trienekens and Zuurbier 2008; HLPE 2017).

Global, regional, and national regulations may impose specific food quality standards for traded food (Trienekens and Zuurbier 2008; Herzfeld et al. 2011). At the global level, the World Trade Organization's Sanitary and Phytosanitary Measures (SPS) Agreement recognizes the Codex Alimentarius as the international standard-setting body for food safety. The Codex Alimentarius develops international food standards, guidelines, and codes of practice through an international consensus process, aiming to harmonize standards among countries (Unnevehr 2015).

Voluntary standards, often going beyond government regulations, are a powerful tool for enhancing food quality and filling the gaps in public rules (Trienekens and Zuurbier 2008; Herzfeld et al. 2011; Unnevehr 2015). These standards are streamlined by benchmarking organizations. To differentiate their products, individual companies may set even stricter standards (Trienekens and Zuurbier 2008). Multinational retailers can effectively make private standards mandatory, forcing producers and other value chain actors to meet a range of requirements in order to access modern markets (Unnevehr 2015).

High-income countries (HICs) are frequently the destination for high-value, nutritious food products sourced from LMICs (Herzfeld et al. 2011). These imports often necessitate quality certification. Therefore, organized retailers in HICs can exert considerable influence over value chains in LMICs, imposing specific standards on exporters and producers (Dolan and Humphrey 2004; Maertens et al. 2012). However, not all producers and midstream actors can keep pace with the escalating demands of these standards. Those who successfully achieve compliance, certification, and access to high-value export markets are most likely to benefit. This can lead to exclusion among value chain actors who lack the information and resources to adopt quality standards for timely access to new international markets (Henson et al. 2011; Herzfeld et al. 2011).

Higher quality standards in export value chains may not always benefit local and regional value chains. Although there are examples of positive spillover, such as upgraded domestic dairy value chains in Uganda (Van Campenhout et al. 2021), traditional and transitional food systems often lack the resources and consumer demand needed to adopt and achieve these higher standards (Unnevehr 2015).

6.5.2 Foods That Fail to Meet Trade Quality Standards

Importing countries reject food products that do not meet quality standards. Only a small share of firms in LMICs have a competitive advantage by meeting stringent food safety standards of importing countries (Jaffee et al. 2019). As new firms enter the export market, border rejections increase due to their inability to meet quality requirements, allowing firms with competitive advantages to capture the high-value export market and consolidate their position (Jaffee et al. 2019). Compliance issues vary across different food products, often leading to border rejections (Box 6.7).

Reputational spillover effects can also contribute to increased import refusals. Analyzing US import refusals, Jouanjean et al. (2015) found that the likelihood of an exporting country facing at least one import refusal rose by over 100% if a neighboring country experienced a refusal for the same product in the previous year. Further, the odds of refusal increased by 62% if the same country had a previous import refusal for a related product.

Rejected food imports may be destroyed, diverted to markets with lower standards, or repurposed for nonfood uses (Rico-Sole 2012; Fonseca and Njie 2014; Fonseca and Vergara 2014). Import refusals alone do not fully indicate trade compliance capabilities, as exporters that fail to meet food quality standards may not export at all, avoiding refusals (Jouanjean et al. 2015). Unnevehr (2015) found that exporters consistently trade lower quantities to countries with stricter quality standards, often diverting lower quality food to markets with less stringent requirements.

Box 6.7 Compliance Challenges in Agri-Food Exports to the European Union and the United States

Compliance performance data for agri-food exporters to the European Union (EU) and the United States is published in the Rapid Alert System for Food and Feed (RASFF) and the Operational and Administrative System for Import Support (OASIS), respectively. The EU's major compliance issues involve food hazards such as mycotoxins in nuts, pathogens in poultry and fish, pesticide residues in fruits and vegetables, and heavy metals in fish (Henson and Olale 2011; Pigłowski 2020). In the United States, fishery and seafood products, along with vegetables and fruits, are problematic food imports. Sanitary violations are the most common reason for rejecting seafood and fruits at the border, while pesticide residues are the primary concern for vegetables (Henson and Olale 2011; Bovay 2016). The United States has since implemented the Food Safety Modernization Act, which includes a system for inspecting foreign food-producing facilities (Bovay 2016).

6.6 Consumer-Level Food Loss and Waste

FLW pathways at the household stage are unique because consumers are the final stage of the value chain. Therefore, the causes of FLW involve technical aspects as well as consumption decisions, rather than consideration of sending food to another value chain stage. Despite the need for additional data (see Chap. 4), current estimates of consumer-level FLW indicate that household pFLW is similar across country income groups, and there is no apparent relationship between per capita food loss and waste and a country's GDP (UNEP 2024; Hermanussen and Loy 2024).

Household FLW is often influenced by consumer routines and habits, rather than intentional practices (Schanes et al. 2018). Socioeconomic and demographic factors, linked to current or past experiences of scarcity and food insecurity (primarily related to affordability), may influence household FLW (Hermanussen and Loy 2024). Individual factors can influence food-related household behaviors and practices across planning, shopping, storage, cooking, consumption, and disposal stages (Principato et al. 2021).

Household-level FLW pathways differ from other stages of the value chain, as consumers represent the ultimate endpoint. Food quality losses can accumulate throughout the value chain, impacting the quality and perishability of food that reaches households. This is why pFLW occurring at the household stage is connected to FLW pathways that reach across the entire value chain, and are not just isolated to the consumer stage. Local circumstances influence the options available to consumers for unwanted food, such as donation and food rescue organizations. For food diverted to nonfood uses, local factors determine the availability and accessibility of productive applications. Nonetheless, uneaten food is most commonly disposed of through garbage and composting (Giordano and Franco 2021).

Cold chain infrastructure continues to contribute to FQL reduction at the household level. Refrigeration at home allows households to store fresh and perishable foods for longer, thus reducing food waste and safety risks, especially for ASF. Access to refrigeration requires both a refrigerator and reliable electricity. As incomes rise, households can afford to invest in refrigeration, which influences their food purchases, increasing consumption of nutritious foods and potentially replacing less nutritious staples (Heard et al. 2020; Martinez et al. 2021).

Shopping habits and cultural storage practices can impact the connection between household refrigeration, dietary outcomes, and FLW pathways. Although perishable foods are often more expensive than shelf-stable foods, household refrigeration can facilitate economies of scale by enabling less frequent purchases in larger quantities, thereby reducing transportation costs, shopping time, and potentially securing bulk discounts (Martinez et al. 2021; Headey 2023).

6.7 Conclusions

This chapter delves into the complex linkages between FLW and post-farmgate value chains for perishable, nutritious foods. As food systems transform, consumer preferences shift toward diversified diets, driving demand for value-added products. Post-farmgate operations, including processing, packaging, and distribution, play a crucial role in meeting these demands. However, the choice of value addition strategies can impact FLW, with a focus on cosmetic attributes potentially leading to increased losses.

Fruit and vegetable value chains are dominated by fresh products, rather than processed or frozen, and require specific conditions to prevent FQL. Fresh fruits and vegetables can be exposed to stress factors that contribute to spoilage at various stages of the post-farmgate value chain. Capacity limitations in bulk packaging, inadequate storage conditions, insufficient transportation, and processing can lead to quality deterioration. While cold chains can help to slow down degradation, building a comprehensive cold chain requires significant resources and coordination among value chain actors, as well as supportive infrastructure.

The challenges of timing, temperature, and handling in ASF value chains are particularly critical for preventing food safety hazards. As value chains lengthen and processing increases, the risk of FQL rises due to inadequate temperature and hygienic controls, especially the lack of cold chain infrastructure. While stricter quality standards and enforcement can lead to an initial increase in pFLW, investing in capacity improvements such as cold chain infrastructure can help prevent FQL and, therefore, pFLW.

Higher quality standards for exports to HICs may not benefit local value chains in LICs. Although there are positive examples, most value chain actors in transitional food systems often lack the resources to meet these standards. Therefore, lack of capacity to prevent FQL can contribute to FLW in domestic value chains, as well as limited market opportunities for high-value markets. Among exported products, rejected food imports may be destroyed, diverted to lower quality markets, or repurposed.

Household FLW is unique, as consumers are the final stage of the value chain. Losses at this stage are influenced by the quality of food that reaches consumers, consumer habits, socioeconomic factors, and individual behaviors. Despite the need for more data, household FLW appears to be similar across income groups. Cold chain infrastructure reduces FQL at the household level, but access to refrigeration is not universal.

Effective coordination among value chain actors is essential for implementing successful value addition strategies that reduce FLW, in both informal and formal markets. Understanding the complex interplay between FLW and post-farmgate value chains for nutritious, perishable foods is critical for developing effective strategies to reduce FLW and ensure access to healthy diets.

References

Akande G, Diei-Ouadi Y (2010) Post-harvest losses in small-scale fisheries: case studies in five sub-Saharan African countries. FAO, Rome. https://openknowledge.fao.org/items/de4209a3-e521-4587-8252-4f27429702ff. Accessed 25 Nov 2024

Akintola SL, Fakoya KA, Elegbede IO, Odunsi E, Jolaosho T (2022) Postharvest practices in small-scales fisheries. In: Galanakis CM (ed) Sustainable fish production and processing. Academic Press, London, pp 79–110

Alarcon P, Fèvre EM, Murungi MK, Muinde P, Akoko J, Dominguez-Salas P, Kiambi S, Ahmed S, Häsler B, Rushton J (2017) Mapping of beef, sheep and goat food systems in Nairobi—a framework for policy making and the identification of structural vulnerabilities and deficiencies. Agric Syst 152:1–17. https://doi.org/10.1016/j.agsy.2016.12.005. Accessed 24 Jan 2024

Alonso S, Muunda E, Ahlberg S, Blackmore E, Grace D (2018) Beyond food safety: socio-economic effects of training informal dairy vendors in Kenya. Glob Food Secur 18:86–92. https://doi.org/10.1016/j.gfs.2018.08.006. Accessed 17 Feb 2023

Aparicio-Peralta CC, Halabi-Echeverry AX, Puentes-Parodi A (2020) Sustainable requirements and value proposition for milk. Ultra-high temperature (UHT) packaging. Supply Chain Forum 21(1):16–25. https://doi.org/10.1080/16258312.2019.1642138. Accessed 4 Oct 2024

Aschemann-Witzel J, Giménez A, Ares G (2018) Consumer in-store choice of suboptimal food to avoid food waste: the role of food category, communication and perception of quality dimensions. Food Qual Prefer 68:29–39. https://doi.org/10.1016/j.foodqual.2018.01.020. Accessed 18 Jan 2024

Barbut S (2020) Meat industry 4.0: a distant future? Anim Front 10(4):38–47. https://doi.org/10.1093/af/vfaa038. Accessed 24 Jan 2024

Barbut S, Leishman EM (2022) Quality and processability of modern poultry meat. Animals 12(20):2766. https://doi.org/10.3390/ani12202766. Accessed 15 Dec 2023

Barrett DM, Beaulieu JC, Shewfelt R (2010) Color, flavor, texture, and nutritional quality of fresh-cut fruits and vegetables: desirable levels, instrumental and sensory measurement, and the effects of processing. Crit Rev Food Sci Nutr 50(5):369–389. https://doi.org/10.1080/10408391003626322. Accessed 3 Jan 2022

Belton B, Bush SR, Little DC (2018) Not just for the wealthy: rethinking farmed fish consumption in the Global South. Glob Food Secur 16:85–92. https://doi.org/10.1016/j.gfs.2017.10.005. Accessed 22 Aug 2023

Boiteau JM, Pingali P (2022) Food loss of perishable produce from farm to retail: evidence from tomato supply chains in South India. Am J Clin Nutr 115:1535–1548. https://doi.org/10.1093/ajcn/nqac039. Accessed 1 Jun 2022

Bonat Celli G, Ghanem A, Su-Ling Brooks M (2016) Influence of freezing process and frozen storage on the quality of fruits and fruit products. Food Rev Int 32(3):280–304. https://doi.org/10.1080/87559129.2015.1075212. Accessed 18 Feb 2024

Bourdoux S, Li D, Rajkovic A, Devlieghere F, Uyttendaele M (2016) Performance of drying technologies to ensure microbial safety of dried fruits and vegetables: drying technologies and microbial safety. Comp Rev Food Sci Food Saf 15(6):1056–1066. https://doi.org/10.1111/1541-4337.12224. Accessed 12 Apr 2023

Bovay J (2016) FDA refusals of imported food products by country and category, 2005–2013. Economic bulletin number 151. Economic Research Service, United States Department of Agriculture, Washington, DC. https://www.ers.usda.gov/webdocs/publications/44066/57014_eib151.pdf. Accessed 25 Nov 2024

Bradford KJ, Dahal P, Van Asbrouck J, Kunusoth K, Bello P, Thompson J, Wu F (2018) The dry chain: reducing postharvest losses and improving food safety in humid climates. Trends Food Sci Technol 71:84–93. https://doi.org/10.1016/j.tifs.2017.11.002. Accessed 24 Sep 2021

Brecht JK (1995) Physiology of lightly processed fruits and vegetables. HortScience 30(1):18–22. https://doi.org/10.21273/HORTSCI.30.1.18. Accessed 23 Jan 2023

Brondy A (2019) Growing the cold chain in developing and emerging economies. Agric Dev 36:4–9. https://www.cabidigitallibrary.org/doi/pdf/10.5555/20193201960. Accessed 25 Nov 2024

Cadilhon J-J, Moustier P, Poole ND, Tam PTG, Fearne AP (2006) Traditional vs. modern food systems? Insights from vegetable supply chains to Ho Chi Minh City (Vietnam). Dev Policy Rev 24(1):31–49. https://doi.org/10.1111/j.1467-7679.2006.00312.x. Accessed 4 Oct 2023

Calín-Sánchez Á, Lipan L, Cano-Lamadrid M, Kharaghani A, Masztalerz K, Carbonell-Barrachina ÁA, Figiel A (2020) Comparison of traditional and novel drying techniques and its effect on quality of fruits, vegetables and aromatic herbs. Foods 9:1261. https://doi.org/10.3390/foods9091261. Accessed 3 Apr 2023

Charad J, Erhard R, Orias NE (2019) Nestlé's milk losses from farm gate to factory in 30 countries. A FLW Standard case study. https://flwprotocol.org/wp-content/uploads/2019/07/FLW-Standard_Case-Study_July-2019_NestleMilkLosses.pdf. Accessed 25 Nov 2024

Cohen JS, Yang TCS (1995) Progress in food dehydration. Trends Food Sci Technol 6(1):20–25

Darlington R, Staikos T, Rahimifard S (2009) Analytical methods for waste minimisation in the convenience food industry. Waste Manag 29(4):1274–1281. https://doi.org/10.1016/j.wasman.2008.08.027. Accessed 14 Mar 2023

Doe PE, Olley J (2020) Drying and dried fish products. In: Sikorski ZE (ed) Seafood: resources, nutritional composition, and preservation. CRC Press, Boca Raton, pp 125–145

Dolan C, Humphrey J (2004) Changing governance patterns in the trade in fresh vegetables between Africa and the United Kingdom. Environ Plan A 36(3):491–509. https://doi.org/10.1068/a35281. Accessed 20 Feb 2024

Dora M, Wesana J, Gellynck X, Seth N, Dey B, De Steur H (2020) Importance of sustainable operations in food loss: evidence from the Belgian food processing industry. Ann Oper Res 290:47–72. https://doi.org/10.1007/s10479-019-03134-0. Accessed 9 Mar 2023

Downs SM, Ahmed S, Fanzo J, Herforth A (2020) Food environment typology: advancing an expanded definition, framework, and methodological approach for improved characterization of wild, cultivated, and built food environments toward sustainable diets. Foods 9(4):532. https://doi.org/10.3390/foods9040532. Accessed 26 Oct 2021

Draaiyer J, Dugdill B, Bennett A, Mounsey J (2009) Milk testing and payment systems resource book: a practical guide to assist milk producer groups. FAO, Rome

Eičaitė O, Baležentis T, Ribašauskienė E, Morkūnas M, Melnikienė R, Štreimikienė D (2023) Exploring food loss in the food industry: a survey-based approach for Lithuania. J Clean Prod 399:136668. https://doi.org/10.1016/j.jclepro.2023.136668. Accessed 8 Mar 2023

Esser JR, Marriott AT, Salagrama V (2007) Evaluation of a systems-based approach to controlling blowfly infestation of traditionally processed fish at small-scale processing sites. FAO Fish Rep 819:121

Euromonitor International (2024a) Packaging—packaged food: euromonitor from trade sources/national statistics. https://www.euromonitor.com/packaged-food. Accessed 25 Nov 2024

Euromonitor International (2024b) Fresh food: euromonitor from trade sources/national statistics. https://www.euromonitor.com/fresh-food. Accessed 25 Nov 2024

Fanzo JC, Downs S, Marshall QE, de Pee S, Bloem MW (2017) Value chain focus on food and nutrition security. In: de Pee S, Taren D, Bloem MW (eds) Nutrition and health in a developing world. Springer International Publishing, Cham, pp 753–770

Fonseca JM, Njie DN (2014) Addressing losses of fresh produce grown in the Latin America and the Caribbean Region due to non-compliance with quality and safety requirements in export markets. Acta Hortic 1016:179–185. https://www.ishs.org/ishs-article/1016_25. Accessed 25 Nov 2024

Fonseca JM, Vergara N (2014) Logistics systems need to scale up reduction of produce losses in the Latin America and Caribbean Region. Acta Hortic 1047:173–179. https://doi.org/10.17660/ActaHortic.2014.1047.21. Accessed 10 Apr 2022

Garrone P, Melacini M, Perego A (2014) Surplus food recovery and donation in Italy: the upstream process. Br Food J 116(9):1460–1477. https://doi.org/10.1108/BFJ-02-2014-0076. Accessed 9 Mar 2023

Giordano C, Franco S (2021) Household food waste from an international perspective. Sustainability 13(9):5122. https://doi.org/10.3390/su13095122. Accessed 8 Jan 2024

Grönman K, Soukka R, Järvi-Kääriäinen T, Katajajuuri J, Kuisma M, Koivupuro H, Ollila M, Pitkänen M, Miettinen O, Silvenius F, Thun R, Wessman H, Linnanen L (2013) Framework for sustainable food packaging design. Packag Technol Sci 26(4):187–200. https://doi.org/10.1002/pts.1971. Accessed 29 May 2024

Hasan MU, Malik AU, Ali S, Imtiaz A, Munir A, Amjad W, Anwar R (2019) Modern drying techniques in fruits and vegetables to overcome postharvest losses: a review. J Food Process Preserv 43(12). https://doi.org/10.1111/jfpp.14280. Accessed 16 Feb 2023

Headey D (2023) Can dairy help solve the malnutrition crisis in developing countries? An economic analysis. Anim Front 13(1):7–16. https://doi.org/10.1093/af/vfac083. Accessed 21 Feb 2024

Headey DD, Alderman H, Hoddinott J, Narayanan S (2024) The glass of milk half-empty? Dairy development and nutrition in low and middle income countries. Food Policy 122:102585. https://doi.org/10.1016/j.foodpol.2023.102585. Accessed 18 Jan 2024

Heard BR, Miller SA (2016) Critical research needed to examine the environmental impacts of expanded refrigeration on the food system. Environ Sci Technol 50(22):12060–12071. https://doi.org/10.1021/acs.est.6b02740. Accessed 21 Feb 2024

Heard BR, Thi HT, Burra DD, Heller MC, Miller SA, Duong TT, Simioni M, Jones AD (2020) The influence of household refrigerator ownership on diets in Vietnam. Econ Hum Biol 39:100930. https://doi.org/10.1016/j.ehb.2020.100930. Accessed 15 Jan 2024

Henriksson PJG, Troell M, Banks LK, Belton B, Beveridge MCM, Klinger DH, Pelletier N, Phillips MJ, Tran N (2021) Interventions for improving the productivity and environmental performance of global aquaculture for future food security. One Earth 4(9):1220–1232. https://doi.org/10.1016/j.oneear.2021.08.009. Accessed 14 Aug 2024

Henson S, Olale E (2011) What do border rejections tell us about trade standards compliance of developing countries? Analysis of EU and US Data 2002–2008

Henson S, Masakure O, Cranfield J (2011) Do fresh produce exporters in sub-Saharan Africa benefit from global GAP certification? World Dev 39(3):375–386. https://doi.org/10.1016/j.worlddev.2010.06.012. Accessed 2 Feb 2023

Hermanussen H, Loy J-P (2024) Household food waste: a meta-analysis. Environ Chall 14:100809. https://doi.org/10.1016/j.envc.2023.100809. Accessed 8 Jan 2024

Hernandez R, Belton B, Reardon T, Hu C, Zhang X, Ahmed A (2018) The "quiet revolution" in the aquaculture value chain in Bangladesh. Aquaculture 493:456–468. https://doi.org/10.1016/j.aquaculture.2017.06.006. Accessed 17 Aug 2023

Herzfeld T, Drescher LS, Grebitus C (2011) Cross-national adoption of private food quality standards. Food Policy 36(3):401–411. https://doi.org/10.1016/j.foodpol.2011.03.006. Accessed 11 Jan 2024

HLPE (2017) Nutrition and food systems. A report by the high level panel of experts on food security and nutrition of the Committee on World Food Security, Rome. https://openknowledge.fao.org/server/api/core/bitstreams/4ac1286e-eef3-4f1d-b5bd-d92f5d1ce738/content. Accessed 25 Nov 2024

HLPE (2020) Food security and nutrition: building a global narrative towards 2030. A report by the high level panel of experts on food security and nutrition of the Committee on World Food Security, Rome. https://openknowledge.fao.org/server/api/core/bitstreams/8357b6eb-8010-4254-814a-1493faaf4a93/content. Accessed 3 Oct 2024

Jaffee S, Henson S, Unnevehr L, Grace D, Cassou E (2019) The safe food imperative: accelerating progress in low-and middle-income countries. Agriculture and food series. World Bank, Washington, DC. https://openknowledge.worldbank.org/entities/publication/83a34062-c6c8-5edf-857c-62a643c16bb8. Accessed 25 Nov 2024

Janssen E, Swinnen J (2019) Technology adoption and value chains in developing countries: evidence from dairy in India. Food Policy 83:327–336. https://doi.org/10.1016/j.foodpol.2017.08.005. Accessed 29 Jan 2024

Jouanjean M-A, Maur J-C, Shepherd B (2015) Reputation matters: spillover effects for developing countries in the enforcement of US food safety measures. Food Policy 55:81–91. https://doi.org/10.1016/j.foodpol.2015.06.001. Accessed 20 Feb 2024

Kader AA (1992) Postharvest biology and technology: an overview. In: Kader AA (ed) Postharvest technology of horticultural crops, 2nd edn. University of California, Division of Agricultural and Natural Resources, Davis, pp 15–20

Kaminski AM, Cole SM, Al Haddad RE, Kefi AS, Chilala AD, Chisule G, Mukuka KN, Longley C, Teoh SJ, Ward AR (2020) Fish losses for whom? A gendered assessment of post-harvest losses in the Barotse Floodplain Fishery, Zambia. Sustainability 12:10091. https://doi.org/10.3390/su122310091. Accessed 14 Aug 2023

Karwowska M, Łaba S, Szczepański K (2021) Food loss and waste in meat sector—why the consumption stage generates the most losses? Sustainability 13(11):6227. https://doi.org/10.3390/su13116227. Accessed 5 Dec 2023

Kasso M, Bekele A (2018) Post-harvest loss and quality deterioration of horticultural crops in Dire Dawa Region, Ethiopia. J Saudi Soc Agr Sci 17(1):88–96. https://doi.org/10.1016/j.jssas.2016.01.005. Accessed 19 Sep 2023

Kiambi S, Alarcon P, Rushton J, Murungi MK, Muinde P, Akoko J, Aboge G, Gikonyo S, Momanyi K, Kang'ethe EK, Fèvre EM (2018) Mapping Nairobi's dairy food system: an essential analysis for policy, industry and research. Agric Syst 167:47–60. https://doi.org/10.1016/j.agsy.2018.08.007. Accessed 30 Jan 2024

Kim EH-J, Chen XD, Pearce D (2009) Surface composition of industrial spray-dried milk powders. 1. Development of surface composition during manufacture. J Food Eng 94(2):163–168. https://doi.org/10.1016/j.jfoodeng.2008.09.021. Accessed 13 Apr 2023

Kruijssen F, Tedesco I, Ward A, Pincus L, Love D, Thorne-Lyman AL (2020) Loss and waste in fish value chains: a review of the evidence from low and middle-income countries. Glob Food Secur 26:100434. https://doi.org/10.1016/j.gfs.2020.100434. Accessed 20 Apr 2023

Kumar A, Mishra AK, Saroj S, Joshi PK (2019) Impact of traditional versus modern dairy value chains on food security: evidence from India's dairy sector. Food Policy 83:260–270. https://doi.org/10.1016/j.foodpol.2019.01.010. Accessed 29 Jan 2024

Lebersorger S, Schneider F (2014) Food loss rates at the food retail, influencing factors and reasons as a basis for waste prevention measures. Waste Manag 34(11):1911–1919. https://doi.org/10.1016/j.wasman.2014.06.013. Accessed 22 Jan 2024

Lewis H (2012) Designing for sustainability. In: Verghese K, Lewis H, Fitzpatrick L (eds) Packaging for sustainability. Springer, London, pp 41–106

Lindh H, Williams H, Olsson A, Wikström F (2016) Elucidating the indirect contributions of packaging to sustainable development: a terminology of packaging functions and features. Packag Technol Sci 29(4–5):225–246. https://doi.org/10.1002/pts.2197. Accessed 23 Jan 2024

Liu R, Xing L, Zhou G, Zhang W (2017) What is meat in China? Anim Front 7(4):53–56. https://doi.org/10.2527/af.2017.0445. Accessed 16 Feb 2024

Liverpool-Tasie LSO, Sanou A, Reardon T, Belton B (2021) Demand for imported versus domestic fish in Nigeria. J Agric Econ 72(3):782–804. https://doi.org/10.1111/1477-9552.12423. Accessed 11 Aug 2023

Machala ML, Tan FL, Poletayev A, Khan MI, Benson SM (2022) Overcoming barriers to solar dryer adoption and the promise of multi-seasonal use in India. Energy Sustain Dev 68:18–28. https://doi.org/10.1016/j.esd.2022.02.001. Accessed 3 Apr 2023

Maddock R (2014) Meats—beef and pork based. In: Clark S, Jung S, Lamsal B (eds) Food processing, 1st edn. Wiley, Oxford, pp 535–548

Maertens M, Minten B, Swinnen J (2012) Modern food supply chains and development: evidence from horticulture export sectors in sub-Saharan Africa. Dev Policy Rev 30(4):473–497. https://doi.org/10.1111/j.1467-7679.2012.00585.x. Accessed 2 Feb 2023

Mahler X, Coudurier B, Redlingshöfer B (2015) Food losses and waste in the poultry production chain: from farm to retail. In: Conference paper: XXII European symposium on the quality of poultry meal and XVI European symposium on the quality of eggs and egg products, Nantes. https://www.researchgate.net/publication/283302837_Food_losses_and_waste_in_the_poultry_production_chain_from_farm_to_retail. Accessed 25 Nov 2024

Manandhar A, Milindi P, Shah A (2018) An overview of the post-harvest grain storage practices of smallholder farmers in developing countries. Agriculture 8(4):57. https://doi.org/10.3390/agriculture8040057. Accessed 13 Apr 2023

Martinez S, Murguia JM, Rejas B, Winters S (2021) Refrigeration and child growth: what is the connection? Matern Child Nutr 17(2):e13083. https://doi.org/10.1111/mcn.13083. Accessed 15 Jan 2024

Matavel C, Kächele H, Steinke J, Rybak C, Hoffmann H, Salavessa J, Sieber S, Müller K (2022) Effect of passive solar drying on food security in rural Mozambique. Sci Rep 12:17154. https://doi.org/10.1038/s41598-022-22129-9. Accessed 12 Apr 2023

Mena C, Adenso-Diaz B, Yurt O (2011) The causes of food waste in the supplier–retailer interface: evidences from the UK and Spain. Resour Conserv Recycl 55(6):648–658. https://doi.org/10.1016/j.resconrec.2010.09.006. Accessed 18 Jan 2024

Meurer IR, Lange CC, Hungaro HM, Valenzuela Bell MJ, De Carvalho Dos Anjos V, Antonio De Sá Silva C, Aparecida De Oliveira Pinto M (2017) Quantification of whole ultra high temperature UHT milk waste as a function of packages type and design. J Clean Prod 153:483–490. https://doi.org/10.1016/j.jclepro.2016.10.172. Accessed 4 Dec 2024

Minten B, Reardon T (2008) Food prices, quality, and quality's pricing in supermarkets versus traditional markets in developing countries. Rev Agric Econ 30(3):480–490. https://doi.org/10.1111/j.1467--9353.2008.00422.x. Accessed 13 Jul 2022

Minten B, Tamru S, Reardon T (2021) Post-harvest losses in rural-urban value chains: evidence from Ethiopia. Food Policy 98:101860. https://doi.org/10.1016/j.foodpol.2020.101860. Accessed 31 Aug 2021

Mir SA, Shah MA, Mir MM, Dar BN, Greiner R, Roohinejad S (2018) Microbiological contamination of ready-to-eat vegetable salads in developing countries and potential solutions in the supply chain to control microbial pathogens. Food Control 85:235–244. https://doi.org/10.1016/j.foodcont.2017.10.006. Accessed 3 Feb 2023

Monteiro CA, Cannon G, Moubarac J-C, Levy RB, Louzada MLC, Jaime PC (2017) The UN Decade of Nutrition, the NOVA food classification and the trouble with ultra-processing. Public Health Nutr 21:5–17. https://doi.org/10.1017/S1368980017000234. Accessed 20 Sep 2021

Morris S (2023) The case for increased investment in food systems infrastructure in low and middle income countries. GAIN discussion paper series 13. Global Alliance for Improved Nutrition, Geneva. https://www.gainhealth.org/resources/reports-and-publications/gain-discussion-paper-series-13-case-increased-investment-food. Accessed 25 Nov 2024

Natarajan SK, Elangovan E, Elavarasan RM, Balaraman A, Sundaram S (2022) Review on solar dryers for drying fish, fruits, and vegetables. Environ Sci Pollut Res 29:40478–40506. https://doi.org/10.1007/s11356-022-19714-w. Accessed 12 Apr 2023

Nijhuis HH, Torringa HM, Muresan S, Yuksel D, Leguijt C, Kloek W (1998) Approaches to improving the quality of dried fruit and vegetables. Trends Food Sci Technol 9(1):13–20

Odoli CO, Owiti H, Kobingi N, Obiero M, Ogari Z, Mugo J, Nyamweya C, Aura CM (2019) Post-harvest interventions in small-scale fisheries: a boon or bane to food and nutritional security in Kenya? Food Secur 11:855–868. https://doi.org/10.1007/s12571-019-00950-x. Accessed 1 Sep 2023

Parfitt J, Barthel M, Macnaughton S (2010) Food waste within food supply chains: quantification and potential for change to 2050. Philos Trans R Soc B 365(1554):3065–3081. https://doi.org/10.1098/rstb.2010.0126. Accessed 20 Oct 2021

Park SH, Lamsal BP, Balasubramaniam VM (2014) Principles of food processing. In: Clark S, Jung S, Lamsal B (eds) Food processing principles and applications, 2nd edn. John Wiley & Sons, Ltd, Chichester, pp 1–15

Perdana T, Kusnandar K, Perdana HH, Hermiatin FR (2023) Circular supply chain governance for sustainable fresh agricultural products: minimizing food loss and utilizing agricultural waste. Sustain Prod Consump 41:391–403. https://doi.org/10.1016/j.spc.2023.09.001. Accessed 24 Oct 2023

Perera CO (2020) Minimal processing of fruit and vegetables. In: Rahman MS (ed) Handbook of food preservation, 3rd edn. CRC Press, Boca Raton, pp 191–206

Pigłowski M (2020) Food hazards on the European Union market: the data analysis of the Rapid Alert System for food and feed. Food Sci Nutr 8(3):1603–1627. https://doi.org/10.1002/fsn3.1448. Accessed 11 Jan 2024

Pop C, Suharoschi R, Pop OL (2021) Dietary fiber and prebiotic compounds in fruits and vegetables food waste. Sustainability 13(13):7219. https://doi.org/10.3390/su13137219. Accessed 8 Jun 2023

Principato L, Mattia G, Di Leo A, Pratesi CA (2021) The household wasteful behaviour framework: a systematic review of consumer food waste. Ind Mark Manag 93:641–649. https://doi.org/10.1016/j.indmarman.2020.07.010. Accessed 8 Jan 2024

Raak N, Symmank C, Zahn S, Aschemann-Witzel J, Rohm H (2017) Processing- and product-related causes for food waste and implications for the food supply chain. Waste Manag 61:461–472. https://doi.org/10.1016/j.wasman.2016.12.027. Accessed 9 Mar 2023

Rickman JC, Barrett DM, Bruhn CM (2007) Nutritional comparison of fresh, frozen and canned fruits and vegetables. Part 1. Vitamins C and B and phenolic compounds. J Sci Food Agr 87:930–944. https://doi.org/10.1002/jsfa.2825. Accessed 2 Nov 2021

Rico-Sole R (2012) Economic impact of mycotoxins in nuts and dried fruit chain. Acta Hortic 963:155–172. https://doi.org/10.17660/ActaHortic.2012.963.26. Accessed 10 Apr 2022

Rørå AMB, Mørkøre T, Einen R (2001) Primary processing (evisceration and filleting). In: Kestin SC, Warris PD (eds) Farmed fish quality. Blackwell Science, Oxford, pp 249–260

Sadler CR, Grassby T, Hart K, Raats M, Sokolović M, Timotijevic L (2021) Processed food classification: conceptualisation and challenges. Trends Food Sci Technol 112:149–162. https://doi.org/10.1016/j.tifs.2021.02.059. Accessed 27 Feb 2023

Schaefer D, Arp T (2017) Importance of variety meat utilization to the meat industry. Anim Front 7(4):25–28. https://doi.org/10.2527/af.2017.0439. Accessed 25 Jan 2024

Schanes K, Dobernig K, Gözet B (2018) Food waste matters—a systematic review of household food waste practices and their policy implications. J Clean Prod 182:978–991. https://doi.org/10.1016/j.jclepro.2018.02.030. Accessed 8 Jan 2024

Schreinemachers P, Ambali M, Mwambi M, Olanipekun CI, Yegbemey RN, Wopereis MCS (2022) The dynamics of Africa's fruit and vegetable processing sectors. In: Jenane C, Ulimwegu JC, Tadesse G (eds) 2022 ReSAKSS annual trends and outlook report: agrifood processing strategies for successful food systems transformation in Africa. Regional strategic analysis and knowledge support system, AKADEMIYA2063, Kigali, Rwanda, pp 23–32. https://ebrary.ifpri.org/digital/collection/p15738coll2/id/136416. Accessed 25 Nov 2024

Schudel S, Shoji K, Shrivastava C, Onwude D, Defraeye T (2023) Solution roadmap to reduce food loss along your postharvest supply chain from farm to retail. Food Packag Shelf Life 36:101057. https://doi.org/10.1016/j.fpsl.2023.101057. Accessed 8 Mar 2023

Secondi L, Principato L, Ruini L, Guidi M (2019) Reusing food waste in food manufacturing companies: the case of the tomato-sauce supply chain. Sustainability 11:2154. https://doi.org/10.3390/su11072154. Accessed 13 Mar 2023

Shah N, Patel A, Sudhakaran A, Ojha A, Patel PR, Dave JM (2016) Shelf-life study of pouch packed UHT milk—a qualitative approach. Indian J Dairy Sci 69:3

Shashi S, Cerchione R, Singh R, Centobelli P, Shabani A (2018) Food cold chain management: from a structured literature review to a conceptual framework and research agenda. Int J Logist Manag 29:792–821. https://doi.org/10.1108/IJLM-01-2017-0007. Accessed 5 Sep 2024

Smith DP (2014) Poultry processing and products. In: Clark S, Jung S, Lamsal B (eds) Food processing, 1st edn. Wiley, England, pp 549–566

Snel H, Broeze J, Kremer F, Osena E, Muyela J, Erick J, Wanjuu E, van Spronsen A (2021) A food system analysis of Kenya's mango, avocado and poultry sectors: assessing opportunities to reduce food losses. Wageningen Centre for Development Innovation Report WCDI-21-185, Wageningen

Statista Market Insights (2024) Vegetables: market data & analysis. Markets insights report. https://www.statista.com/study/48829/vegetables-market-data-and-analysis/. Accessed 25 Nov 2024

Stewart H, Hyman J, Young S (2024) Satisfying fruit and vegetable recommendations possible for under $3 a day, data analysis shows. USDA Economic Research Service. https://www.ers.usda.gov/amber-waves/2024/september/satisfying-fruit-and-vegetable-recommendations-possible-for-under-3-a-day-data-analysis-shows/#. Accessed 25 Nov 2024

Subramanian R (2016) India processing tomato segment: current status, trends and opportunities for engagement. Prepared by the World Vegetable Center No. 16-806. https://avrdc.org/download/publications/technical-reports/reports/GIZ_India-Processed-Tomato-Study_16Sept2016.pdf. Accessed 25 Nov 2024

Surucu-Balci E, Tuna O (2021) Investigating logistics-related food loss drivers: a study on fresh fruit and vegetable supply chain. J Clean Prod 318:128561. https://doi.org/10.1016/j.jclepro.2021.128561. Accessed 13 Apr 2023

Swinnen JFM, Vandemoortele T (2008) The political economy of nutrition and health standards in food markets. Rev Agric Econ 30(3):460–468. https://www.jstor.org/stable/30225889. Accessed 25 Nov 2024

Trienekens J, Zuurbier P (2008) Quality and safety standards in the food industry, developments and challenges. Int J Prod Econ 113(1):107–122. https://doi.org/10.1016/j.ijpe.2007.02.050. Accessed 11 Jan 2024

Udomkun P, Romuli S, Schock S, Mahayothee B, Sartas M, Wossen T, Njukwe E, Vanlauwe B, Müller J (2020) Review of solar dryers for agricultural products in Asia and Africa: an innovation landscape approach. J Environ Manag 268:110730. https://doi.org/10.1016/j.jenvman.2020.110730. Accessed 11 Apr 2023

UNEP (2018) Single-use plastics, a roadmap for sustainability. United Nations Environment Programme, Nairobi. https://www.unep.org/resources/report/single-use-plastics-roadmap-sustainability. Accessed 25 Nov 2024

UNEP (2024) Food waste index report 2024. United Nations Environment Programme, Nairobi. https://www.unep.org/resources/publication/food-waste-index-report-2024. Accessed 25 Nov 2024

Unnevehr LJ (2000) Food safety issues and fresh food product exports from LDCs. Agric Econ 23(3):231–240. https://www.sciencedirect.com/science/article/abs/pii/S0169515000000955. Accessed 25 Nov 2024

Unnevehr L (2015) Food safety in developing countries: moving beyond exports. Glob Food Secur 4:24–29. https://doi.org/10.1016/j.gfs.2014.12.001. Accessed 12 Jan 2024

USDA (United States Department of Agriculture) (2022) Understanding food quality labels: a guide to AMS shields, value-added lables, and official seals. https://www.ams.usda.gov/publications/content/understanding-food-quality-labels. Accessed 28 Mar 2022

van Boekel M, Fogliano V, Pellegrini N, Stanton C, Scholz G, Lalljie S, Somoza V, Knorr D, Jasti PR, Eisenbrand G (2010) A review on the beneficial aspects of food processing. Mol Nutr Food Res 54(9):1215–1247. https://doi.org/10.1002/mnfr.200900608. Accessed 9 Jun 2022

Van Campenhout B, Minten B, Swinnen JFM (2021) Leading the way—foreign direct investment and dairy value chain upgrading in Uganda. Agric Econ 52(4):607–631. https://doi.org/10.1111/agec.12638. Accessed 30 Jan 2024

Verghese K, Lewis H, Lockrey S, Williams H (2015) Packaging's role in minimizing food loss and waste across the supply chain. Packag Technol Sci 28(7):603–620. https://doi.org/10.1002/pts.2127. Accessed 14 Mar 2023

VijayaVenkataRaman S, Iniyan S, Goic R (2012) A review of solar drying technologies. Renew Sust Energ Rev 16(5):2652–2670. https://doi.org/10.1016/j.rser.2012.01.007. Accessed 12 Apr 2023

WEF (World Economic Forum) (2024) Investigating global aquatic food loss and waste. White paper (April). In collaboration with MarFishEco, Friends of Ocean Action and the World Resources Institute. https://www3.weforum.org/docs/WEF_Investigating_Global_Aquatic_Food_Loss_and_Waste_2024.pdf. Accessed 10 Oct 2024

White A, Lockyer S (2020) Removing plastic packaging from fresh produce—what's the impact? Nutr Bull 45(1):35–50. https://doi.org/10.1111/nbu.12420. Accessed 22 Jan 2024

Wilson RT (2018) The red meat value chain in Tanzania. Anim Husb Dairy Vet Sci 2. https://doi.org/10.15761/AHDVS.1000127. Accessed 9 Feb 2024

Yi J, Meemken E-M, Mazariegos-Anastassiou V, Liu J, Kim E, Gómez MI, Canning P, Barrett CB (2021) Post-farmgate food value chains make up most of consumer food expenditures globally. Nat Food 2:417–425. https://doi.org/10.1038/s43016-021-00279-9. Accessed 14 Apr 2023

Zhou G, Zhang W, Xu X (2012) China's meat industry revolution: challenges and opportunities for the future. Meat Sci 92(3):188–196. https://doi.org/10.1016/j.meatsci.2012.04.016. Accessed 9 Feb 2024

7 Keeping Food Safe and Nutritious: The Role of Food Quality Loss

7.1 Introduction

Safe and nutritious foods are a requisite for achieving food security. Food quality loss (FQL) pathways may directly impact dietary quality when food becomes unsafe or loses specific nutrients but remains within the food value chain. As countries progress through structural transformation, food value chains become more sophisticated and complex. Food systems modernization, including investments in infrastructure and improved practices, can reduce FQL issues associated with transitioning value chains. At the same time, new issues with food safety and nutrient loss emerge, as food value chains become more complex and deal with higher volumes of perishable foods over longer distances.

Food safety risks are multifaceted, stemming from various hazards introduced at any stage of the food value chain. Safeguarding food safety requires understanding market incentives, value chain capacity, and governance. Food hazards can be biological, chemical, or physical, and they can be transmitted through various pathways. These hazards can cause foodborne disease (FBD) and pose risks to human health. Identifying the primary routes and sources of these hazards is critical for developing effective interventions and allocating resources to minimize quality loss and enhance food safety.

Sustainable food systems must also deliver nutritious foods to support healthy diets and prevent nutrient deficiencies. In addition to diversifying diets to increase consumption of nutritious foods, understanding nutrient loss pathways may also contribute to supporting healthy diets. Nutrient losses occur throughout the food value chain, from harvest to consumption. Factors like processing methods, storage conditions, and cooking techniques can affect nutrient retention. While some nutrient losses may be inevitable and have minimal impact on overall dietary quality, elucidating these pathways and identifying critical points of nutrient loss can inform assessments of FLW intervention trade-offs.

In this chapter, we use examples from the literature to describe the pathways through which FQL occurs from farm to consumer while food remains in the supply chain for human consumption. We focus on food safety and nutrient loss pathways that directly affect food security. We first examine the different food hazards and current evidence on FBD. We then consider how food safety issues are linked with diet transitions, market incentives, and food system stakeholder capacities. Shifting the discussion to nutrient loss pathways, we focus on micronutrients, describing how they are lost from farm to consumer, and examine how to contextualize the potential linkage between nutrient losses and diet quality. Finally, we consider the trade-offs between different FQL pathways involving food safety and nutrient loss.

J. Boiteau, P. Pingali, *Wasted Potential*, Sustainable Development Goals Series,
https://doi.org/10.1007/978-3-031-97411-3_7

7.2 Food Safety Loss Pathways

Food safety is the assurance that food will not harm consumers when it is prepared and eaten (WHO 2008). Food quality loss pathways directly impact diet quality when food becomes unsafe but remains in the value chain and is eaten. Ensuring the food safety aspect of food security requires understanding market incentives related to consumer willingness to pay, value chain capacity and incentives for improved food safety, and governance capacity regarding public and private standards and enforcement (Nordhagen et al. 2022; Unnevehr 2022). As food systems transform, changes in market incentives and governance capacity influence FQL pathways linked to food safety.

7.2.1 Food Hazards: Hazard Types, Food Exposure Routes, and Disease Burden

Food safety risks are multifaceted, stemming from various hazards introduced at any stage of the food value chain. Foodborne hazards can be biological, chemical, or physical (Grace 2017). Biological hazards include viruses, parasites, bacteria, and fungi. Chemical hazards can be either natural or artificial. Natural chemicals can include toxic metals and chemicals formed when food is heated, while artificial chemicals often come from industrial sources, such as environmental contaminants and pesticides. Toxins produced by biological compounds, such as mycotoxins from molds, can be classified as either biological or chemical hazards. Physical hazards include visible fragments that are sharp or choking hazards, as well as tiny particles invisible to the human eye (for example, nanomaterials and radionuclides) (Grace 2017).

Food hazards can be transmitted through food alone, or in combination with other pathways, such as environmental (water, soil, air), human-to-human, and animal-to-human contact (Havelaar et al. 2015). The prevalence of these transmission routes varies across different stages of the food supply chain. Identifying the primary routes and sources is necessary for developing effective interventions and allocating resources to minimize FQL and enhance food safety. Figure 7.1 illustrates common food hazard pathways at various stages of the supply chain. Food hazards can originate at the same or different stages of the value chain where they are detected. While FLW pathways start at the time food is ready for harvest or slaughter, food quality can be affected by hazards introduced as early the production stage (Fig. 7.1).

Food hazards can lead to FQL and potentially pose risks to human health (Box 7.1). However, the actual risk depends on both the severity of the hazard and the likelihood of it causing adverse health effects (Roesel et al. 2015; Grace 2023). FBD can cause acute poisoning or chronic illnesses, leading to long-term disability and death. Foodborne illness outbreaks often go unreported due to delayed symptoms, difficulty in identifying the contaminated food, and the lack of significant public health or economic consequences (Grace 2017; Jaffee et al. 2019). Factors like age and immune status influence the severity of FBD, which can have impacts on health and physiology (Nordhagen et al. 2022). Even when food is initially safe, improper consumer storage and handling can introduce hazards and compromise food safety. This can lead to missing or unreliable data on FBD outbreaks (Havelaar et al. 2015; Jaffee et al. 2019).

Box 7.1 Key Food Safety Terms

Food hazard: A biological, chemical, or physical substance, or characteristic in food that has the potential to harm human health, regardless of whether it actually does (Grace 2023).

Foodborne disease: Illness caused by consuming food that is contaminated or contains naturally occurring hazards (Grace 2023).

Food safety risk: A combination of the potential harm of a hazard and the probability of it causing harm to human health (Grace 2023).

Food safety: The assurance that food will not harm consumers when it is prepared and eaten (WHO 2008).

AGRICULTURAL PRODUCTION	HARVEST, SLAUGHTER, OR CATCH	PROCESSING, PACKAGING, AND STORAGE	WHOLESALE AND RETAIL	CONSUMPTION: HOUSEHOLD AND FOOD SERVICE
SOURCES OF CONTAMINATION (not exhaustive)				
Soil, freshwater, salt water	Plant harvesting	Processing and transport conditions, packaging	Food handling	Food handling
Fertilizer, soil amendments, agricultural chemicals	Slaughter	Storage conditions and timing	Storage conditions	Storage conditions
Plants (including feed), livestock, aquatic animals	Aquatic capture	Adulteration with harmful substances	Fomites (passive vectors)	Insufficient heating
Agricultural workers		Peri-domestic pests	Peri-domestic pests	Peri-domestic pests
FOODBORNE HAZARDS (not exhaustive)				
Biological hazards: animal manure, microbial bacteria and viruses, parasites	Biological hazards: water source, infected worker, contamination with gut content or animal skin	Biological hazards: microbial growth, infected worker, equipment, surfaces, insects, rodents, birds	Biological hazards: microbial growth, infected worker, equipment, surfaces, insects, rodents, birds	Biological hazards: microbial growth, infected food preparer, equipment, surfaces, insects, rodents
Chemical hazards: Mycotoxins, heavy metals, veterinary drugs	Chemical hazards: machine oils, cleaning chemicals, mycotoxins	Chemical hazards: unauthorized dyes, packaging migration, mycotoxin accumulation	Chemical hazards: equipment, surfaces, container migration	Chemical hazards: equipment, surfaces, container migration
	Physical hazards: stones, wood splinters	Physical hazards: glass, plastic, metal shards		

Fig. 7.1 Food hazard transmission pathways from production to consumption

Source: Elaborated from Grace (2017)

The stages in dark gray are within the scope of FLW pathways. While food hazards that occur during agricultural production (light gray) fall outside the scope of FLW pathways, contamination with biological or chemical hazards can still lead to FQL from the time food is ready for harvest or slaughter onward

Identifying high-risk foods and hazards in specific regions is essential for controlling food hazards. WHO established the Foodborne Disease Burden Epidemiology Reference Group (FERG) to assess the global burden of FBD. In their initial estimates, published in 2015, FERG identified 12 hazards as exclusively foodborne and 19 as transmitted through food and other pathways (Havelaar et al. 2015; Hald et al. 2016). The vast majority (98%) of the documented health burden from FBD occurred in low- and middle-income countries (LMICs), with biological hazards accounting for the 97% of this burden (Havelaar et al. 2015; Waage et al. 2022). Attention to the connection between nature and food safety hazards in food systems is growing (Box 7.2). A process to update these estimates was announced in 2021, with expected completion in 2025 (WHO 2021).

Most FBD cases result from consuming fresh animal-source foods (ASF) and fresh fruits and vegetables (Grace 2015). Although less likely to contain biological hazards, food safety risks of processed foods are more likely to be chemical hazards (Grace 2015). Modern food systems with intensive production systems have also experienced food safety issues related to the emergence of new human diseases (Grace 2015).

Source attribution for the global burden of FBD, when analyzed by specific food categories, shows that the implicated food sources vary based on the type of food hazard (Hoffmann et al. 2017). For example, vegetables were estimated to contribute to 60–80% of FBD caused by parasites in most regions worldwide, whereas beef was estimated to contribute 40–55% of FBD caused by Shiga toxin-producing *E. coli* in most global regions (Hoffmann et al. 2017). The burden of FBD caused by chemical hazards was highly localized but was not included in estimates of the source attribution by food type (Havelaar et al. 2015; Hoffmann et al. 2017).

Li et al. (2019) analyzed the global burden of FBD linked to ASF, finding that these foods contributed approximately 35% of the total global burden. The lower burden of FBD from in high-income regions, compared to LMICs, combined with the availability of control measures for many ASF hazards, suggests that current control methods may have reached their limits, and that effective ASF safety systems are linked to economic development.

Box 7.2 Climate Change as a Growing Threat to Food Safety

Climate change presents a major challenge to global food safety. Rising temperatures, severe droughts, and unseasonal heavy rainfall can exacerbate the prevalence of biological hazards, particularly, zoonotic diseases transmitted from animals to humans. Addressing food safety in transforming food systems requires a comprehensive approach, which considers both climate change impacts and shifts in livestock, human, and wildlife interactions (Unnevehr 2022). For example, evidence has linked higher temperatures with higher incidences of *Salmonella* spp. and *Campylobacter* spp., two common foodborne pathogens, in various global regions (FAO 2020). Moreover, water scarcity, a consequence of droughts, can prevent the implementation of adequate food safety controls and hygienic practices throughout the food value chain. In contrast, excessive rainfall, coupled with elevated temperatures, can create conditions ideal for pests, fungi, and other biological hazards to grow (FAO 2020).

Market-based approaches can present an opportunity to align zoonotic disease knowledge with effective food safety initiatives by increasing consumer demand and willingness to pay for food safety attributes. These approaches require investments in public infrastructure and training to support food safety and handling practices, consistent enforcement of food safety standards, and consumer education (Aiyar and Pingali 2020). Additionally, integrating market-based strategies with investments in FBD surveillance and data integration requires enhancing local traceability capacity to enable more robust global data sharing and monitoring (Aiyar and Pingali 2020).

7.2.2 Capacity and Food Safety Issues Across Food Systems Transformation

Urbanization and dietary shifts, driven by economic growth, have reshaped food value chains and the challenges associated with FBD. As food systems become geographically longer and more complex, patterns in the burden and sources of FBD emerge (Jaffee et al. 2019), as illustrated in Fig. 7.2. Growing public awareness and private sector investments in food safety, together, drive improvements in food safety standards in domestic markets of LMICs (Jaffee et al. 2019; Unnevehr 2022). However, in early stages of transition, market incentives and governance

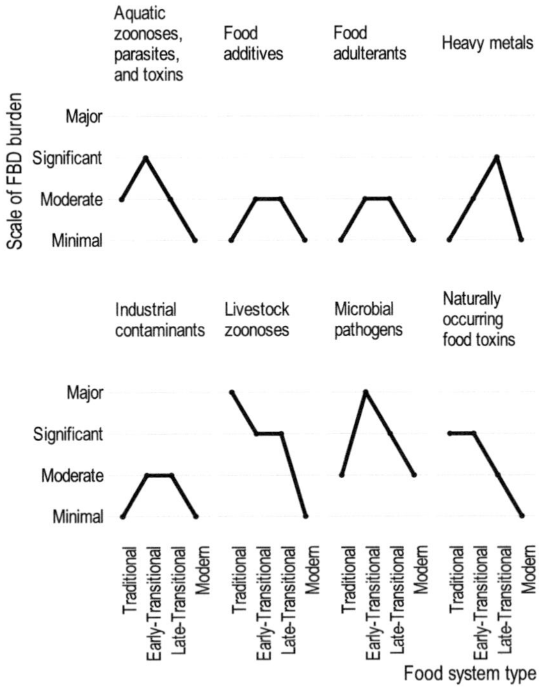

Fig. 7.2 Foodborne hazard sources by food system types

Adapted from Jaffee et al. (2019)

Eight foodborne hazard sources relevant to food safety loss are ranked in terms of the scale of the FBD burden across four food system types. The scale of the FBD burden is intended to be illustrative of the direction of change across each FBS source at different stages of the food safety cycle and does not represent specific quantitative data

capacity may lag, leading to inadequate food safety measures and the presence of unsafe food in the value chain that is eventually eaten. In later stages, the FBD burden tends to decrease as infrastructure for prevention of food hazards and responses to remove unsafe foods improves. Therefore, the burden of FBD often follows an inverted U-shaped curve, increasing initially, and then, decreasing as food systems modernize (Grace 2015; Jaffee et al. 2019). For readers interested in more details of the food safety cycle, we encourage them to look at Jaffee et al. (2019).

In the early stages of structural transformation, staples form a substantial portion of diets. Strategies to enhance staple crop productivity contributed to shifts in staple consumption, as maize, rice, and wheat replaced coarse grains like sorghum, millets, and pulses (Pingali 2012; Khoury et al. 2014; Pingali and Sunder 2017). Aflatoxins, naturally occurring food toxins and type of mycotoxin, are associated with both acute and chronic health problems (Williams et al. 2004). Biological and environmental factors at harvest, storage, handling, and processing stages all affect aflatoxin contamination and accumulation (Neme and Mohammed 2017; Bradford et al. 2018). Despite differences in susceptibility to aflatoxin among staple foods, various staples likely contribute to dietary aflatoxin exposure, with the extent of contribution based on the local food system (Wenndt et al. 2020; Meijer et al. 2021). Credence attributes, such as aflatoxin contamination, can create information asymmetry regarding potential food safety risks (Ricker-Gilbert et al. 2022). However, this information asymmetry may be mitigated when search attributes can serve as proxies for these credence attributes (Hoffmann et al. 2021).

To meet changing consumer demands, food producers adapt by shifting away from staples production toward more perishable, high-value foods. Although this production shift supports dietary diversity, it also increases the risk of FQL due to the perishability of these products (see Chaps. 5 and 6). The specific types of fruits, vegetables, and ASF contributing to the FBD burden vary, based on consumption patterns and production systems (Li et al. 2019; Richardson et al. 2021; Waage et al. 2022). As value chains become more complex and handle larger volumes of perishable food, there are more points at which lapses in food safety management can occur (Unnevehr 2022). Therefore, effective management and coordination within perishable food value chains are important, particularly, for navigating changes in the number of value chain actors that initially increases before consolidating in later, more modern stages (see Chap. 2).

As diets transition, convenience emerges as an important food attribute, particularly, as the opportunity cost of time rises. Although processing can enhance convenience, it may also increase microbial contamination, as demonstrated by a comparison of raw vegetables to value-added purees (Sinayobye and Saalia 2011). However, the risk associated with processed foods can vary depending on factors, including handling, storage, and intended use. Foods requiring further cooking may have lower risks, as long as they are heated to temperatures that kill pathogens. In contrast, ready-to-eat vegetables, despite perceived freshness, can pose significant food safety risks (Box 7.3). These minimally processed products have been linked to numerous foodborne outbreaks worldwide (Stephan et al. 2015; Mir et al. 2018).

Although food safety standards are typically implemented in formal food environments (Downs et al. 2020), discrepancies can remain between the availability of nutritious perishable foods in formal retail outlets, as well as consumer retail outlet preferences. In transitional food systems, small retailers often offer more shelf-stable items and less fresh produce, compared to traditional markets and modern supermarkets (Gómez and Ricketts 2013; Wertheim-Heck et al. 2019). As food systems modernize, informal markets decline, starting with small shops that carry a range of fresh and processed products, followed by wet markets and informal fresh produce outlets (Reardon et al. 2010). Particularly, in transitional food systems, consumers may still prefer to purchase fresh, perishable produce from informal markets over formal markets (Wertheim-Heck et al. 2019).

Box 7.3 Ready-to-Eat Foods Are High in Convenience, But Carry Food Safety Risks

The rise of ready-to-eat foods, both in formal and informal markets, reflects a shift toward convenience foods and food consumed away from home. While formal markets include restaurants, supermarkets, and online platforms, informal markets rely primarily on street vendors and mobile food carts (Downs et al. 2020).

Ready-to-eat foods, particularly those sold in informal markets, face several food safety risks. Unsafe ingredients, improper preparation, and inadequate handling practices can lead to contamination (Alimi 2016). Improper storage and temperature control create ideal conditions for bacterial growth, such as when foods are prepared in batches and kept at ambient temperatures for extended periods during sale (Alimi 2016). Additionally, cross-contamination with contaminated ingredients or handlers can further increase the risk.

Food safety risks can also be present in ready-to-eat foods that are sold in formal markets. In the United States, restaurants accounted for 64% of FBD outbreaks and 44% of associated illnesses in 2017, with sit-down dining establishments being particularly vulnerable (CDC 2019). Norovirus, a virus often transmitted through improper handling, was responsible for 35% of outbreaks and 46% of illnesses (CDC 2019).

Market failures, particularly, information asymmetry and coordination problems, contribute to FQL pathways that may put consumers at risk of food safety hazards (Unnevehr 2022). Lack of information about the presence of food hazards in specific products makes it challenging for consumers to use food quality as a limiting food attribute in determining acceptability. At the same time, value chain actors may not have the capacity or incentives to prioritize food safety. This combination of factors can lead to undetected food safety hazards, allowing unsafe food to remain in the value chain. Addressing food safety in FQL pathways requires incentivizing value chain actors to supply safe, quality food by fostering consumer willingness to pay a premium price and enhancing food safety capacity.

As consumer income rises, food safety emerges as a key quality attribute that can influence product acceptance. Consumer perceptions of food safety risks can significantly influence purchasing decisions. However, consumers' willingness to pay for safe food is limited by the fact that potential hazards are not directly visible. Individuals who perceive a food as unsafe often demonstrate decreased willingness to buy it (Machado Nardi et al. 2020). Perceptions of food safety risk are shaped by various factors, including trust, knowledge, individual characteristics, and sociodemographic attributes (Machado Nardi et al. 2020). Understanding food safety risks and trade-offs is an ongoing process, which evolves as practices and innovations, like plastic packaging, change (Box 7.4).

The information asymmetry along the food value chains limits consumer knowledge about food production, processing, and handling at the point of purchase or consumption (Jaffee et al. 2019). Consumer trust in both government and value chain actors shapes perceptions of food safety. Competent, transparent, ethical, and accountable practices, coupled with clear communication regarding food safety and contamination, are essential for building consumer trust and confidence in food products (WHO 2018; Jaffee et al. 2019; Machado Nardi et al. 2020). Once established, trust is often maintained through repeat purchases from familiar sources (Jaffee et al. 2019; Lee et al. 2022).

Additionally, consumers may be more risk averse toward certain foods, such as ASF, compared to plant-based foods, based on their knowledge of foods that are more or less susceptible to food hazards (Machado Nardi et al. 2020). However, consumer perceptions may not always accurately reflect actual risks, potentially leading to unnecessary avoidance of certain foods or retailers. This can contribute to pFLW of safe and consumable products.

The extent that FQL pathways increase food hazard risks within value chains is determined, in part, by public and private sector governance capacity to establish and enforce food safety standards to build trust among consumers. Food safety standards differentiate the formal and informal markets, where formal markets are distinguished by their licensing and adherence to food safety regulations (Downs et al. 2020). The private sector plays an important role in setting and enforcing food safety standards, including implementing public standards (Reardon et al. 2010). Public food safety and hygiene standards are among the incentives attracting foreign direct investment in modern retail markets that are selling fresh, perishable produce (Pingali and Abraham 2022).

In contrast, informal food markets, which are particularly important for access to nutritious foods for low-income consumers, often lack standardized food safety guidance (Downs et al. 2020). However, informal and formal food supply chains may be interconnected, particularly, in transitional food systems, allowing products to be exchanged between the two (Oguttu et al. 2014; Alarcon et al. 2017). Yet, global food safety guidance for informal food markets remains largely limited to street-vended food and lacks standardization (DeWaal et al. 2022). To effectively implement food safety standards in informal markets at regional and local levels, governments are recommended to comprehensively assess implementation challenges to identify the necessary capacities to promote food safety among food suppliers (DeWaal et al. 2022).

Box 7.4 Food Safety Trade-Offs: The Case of Plastic Packaging

Efforts toward achieving sustainable food systems have focused on the environmental impacts of plastic packaging, creating complex expectations to protect food and the environment (Kasza et al. 2022). Plastic packaging physically protects food from contamination, moisture, and physical damage that can lead to food safety hazards. Therefore, plastic packaging has enabled value chain actors to preserve food safety and quality. However, packaging has become a highly visible sustainability issue. Consumers focus on the negative environmental aspects, potentially overestimating these impacts, while overlooking, and underestimating the role of food packaging in reducing FQL (Kasza et al. 2022; Croad et al. 2024).

At the same time, new evidence is drawing attention to potential food safety hazards that food packaging introduces. Humans are exposed to numerous food contact chemicals through food packaging and other food contact materials. Many of these chemicals have hazard properties of concern or have not been adequately tested for toxicity (WHO 2015). A recent systematic mapping of chemicals from various food contact articles, including food packaging, revealed widespread human exposure to these substances (Geueke et al. 2025). Many of these chemicals pose health risks due to their hazardous properties. Despite existing data gaps, the available evidence supports the need for policies that eliminate the use of hazardous chemicals that migrate from food packaging into foods to protect public health (Geueke et al. 2025).

The local supply of safe foods can be affected by international trade between countries with different food quality standards. International trade of high-value food exports from LMICs to high-income countries (HICs) has led to the implementation of third-party certified process controls to ensure food safety compliance (Unnevehr 2022). In countries with weak public food safety regulations, private and public standards may contribute to improving local food safety practices and preventing the diversion of substandard products to domestic markets (Hawkes et al. 2015).

Although international food safety standards can create trade barriers due to compliance costs and capacity constraints, they can also motivate investments in upgraded food safety management systems to facilitate market entry (Global Panel 2020). In LMICs, food safety monitoring and controls are often prioritized in export supply chains to meet the stringent quality requirements of HIC importers (Jaffee et al. 2019). However, adapting food safety management systems from HICs to LMIC contexts can be challenging and limit market access (Hawkes et al. 2015).

Food value chains directed to domestic food markets often lack robust food safety policies and enforcement measures (Reardon et al. 2010). As a result, food products that fail to meet export standards may be diverted to the domestic market, potentially increasing the risk of food safety hazards (Narrod et al. 2009; Matumba et al. 2015). The public health costs of food safety far exceed the losses in international trade caused by disruptions or stringent compliance requirements, which prevent or redirect the export of high-value food products, such as ASF, fruits, and vegetables, from LMICs (Jaffee et al. 2019; Unnevehr 2022).

High-value food exports from LMICs to other LMICs have grown at a faster pace than those to HICs (Unnevehr 2022). The direct and indirect costs from formal cross-border trade have contributed to the sustained prevalence of small-scale and informal cross-border food trade in LMICs (Lesser and Moisé-Leeman 2009; Afrika and Ajumbo 2012). The problem with informal cross-border trade is that it reduces the governance capacity to enforce food safety policies. At the same time, informal food trade is important to supply informal markets with ASF, fruits, and vegetables to those who rely on the informal sector for both livelihoods and food access (Hara et al. 2017). Therefore, evaluations of FQL pathways involving food safety and other aspects of food security must also consider informal trade.

7.3 Nutrient Loss Pathways

Micronutrient deficiencies, often referred to as "hidden hunger," are a pressing concern in many LMICs where diets commonly lack minerals, such as iron, zinc, iodine, magnesium, and calcium; water-soluble vitamins, such as folate and vitamin B_{12}; and fat-soluble vitamins A and E (Pinstrup-Andersen 2007; Beal et al. 2021; White et al. 2021; Passarelli et al. 2022). To prevent micronutrient deficiencies, food systems must prioritize the production and preservation of a diverse range of nutritious foods (Miller and Welch 2013). We focus the discussion on the natural nutrient losses of individual foods, excluding any added nutrients through fortification or enrichment.

7.3.1 Nutrient Losses from Farm to Consumer

The extent to which nutrient loss, especially, of essential micronutrients, contributes to dietary deficiencies, particularly, among those with limited food diversity, remains uncertain due to the lack of comprehensive data on the nutritive value losses (see Chap. 4). By examining FLW pathways and considering the nutritional value of diverse food sources, we can identify critical points at which nutrients are lost and prioritize areas for data collection and intervention to address these FQL pathways.

When foods are harvested, they are separated from their natural nutrient sources. As they move through processing, storage, transportation, and preparation stages, different nutrients are affected to varying extents based on factors, such as the specific food, the nutrients involved, conditions like time, temperature, handling, and the environment. Figure 7.3 illustrates that the primary causes of nutrient loss between harvest and consumption are

AGRICULTURAL PRODUCTION	HARVEST, SLAUGHTER, OR CATCH	PROCESSING, PACKAGING, AND STORAGE	WHOLESALE AND RETAIL	CONSUMPTION: HOUSEHOLD AND FOOD SERVICE
SOURCES OF NUTRIENT VARIATION (not exhaustive)				
Genetic diversity, crop/animal selection	Plant maturity level	Processing method, initial heat treatment	Product handling	Preparation method
Fertilizer, soil amendments, agricultural chemicals		Type of packaging used, exposure to oxygen, handling	Display conditions	Heating / cooking method
Atmospheric carbon dioxide level		Storage conditions and timing	Storage conditions and timing	Storage conditions and timing
NUTRIENT LOSSES (not exhaustive)				
Reduction in nutrient content in certain crops	Nutrient degradation: susceptibility to damage, rate of degradation	Nutrient degradation: bruising, oxygenation, UV exposure	Nutrient degradation: bruising, oxygenation, UV exposure	Nutrient degradation: Oxygenation, heating
Changes in protein in certain crops	Nutrient leaching: susceptibility to breakage	Nutrient leaching: heat processing in liquid, broken/leaking produce	Nutrient leaching: broken/leaking produce	Nutrient leaching: defrosting, soaking, discarding liquid, cutting
		Specific nutrient-rich, edible food part removed		

Fig. 7.3 Nutrient loss pathways from farm to consumer stages
Lower nutrient content at the agricultural production stage is not considered FQL, because the source of variation occurred throughout production, not just at the time product was ready for harvest or slaughter. Compiled based on Barth and Zhuang (1996), Rickman et al. (2007a, b), Hansen et al. (2012), Smith and Myers (2018) and Sonar et al. (2019)

degradation and leaching. Vitamins are particularly susceptible to degradation, while minerals and water-soluble vitamins may be lost through leaching (Rickman et al. 2007a, b; Barrett et al. 2010).

Certain food preservation, processing, and preparation techniques may be more common for certain food groups, and therefore, have more specific nutrient loss implications. Heating, a common step in canning, freezing, and certain drying processes, can influence the extent of nutrient degradation based on the duration and temperature (Rickman et al. 2007a; Swada et al. 2016). When food is packaged, the packaging also influences exposure to environmental factors, such as sunlight and oxygen, which can contribute to nutrient degradation (Barth and Zhuang 1996; Sonar et al. 2019). Throughout the entire value chain, from farm to consumer, foods are stored and prepared in various ways, including ambient, refrigerated, and frozen storage, as well as cooking, reheating, or serving raw, each of which impact the extent of nutrient degradation and leaching.

In more specific processes, nutrient loss can occur when portions of a food product are physically removed, such as in refining. An example is the loss of nutrients during rice polishing, a process that removes the bran from brown rice to produce white rice (Hansen et al. 2012). The bran contains essential minerals, including iron and zinc, which are lost during polishing, leading to a less nutritious final product (Hansen et al. 2012).

Legumes are valuable sources of protein and essential micronutrients, including folate and non-heme iron. Drying is commonly used to preserve legumes, which then require cooking prior to consumption. Although cooking reduces antinutrients, compounds that interfere with nutrient absorption, and improves the bioavailability of micronutrients (Fabbri and Crosby 2016), water-soluble micronutrients can leach into soaking or cooking liquid, and micronutrients can degrade during heating (Dang et al. 2000; Carvalho Lucia et al. 2012). Although canned or frozen legumes offer convenience, these processing methods can result in greater nutrient leaching, compared to home cooking dried legumes (Stea et al. 2007; Czarnowska and Gujska 2012; Margier et al. 2018). Therefore, FQL pathways for legumes are influenced by factors such as cooking method (for example, boiling, pressure cooking, or steaming), cooking time and temperature, and whether the cooking liquid is consumed alongside the legumes (Dang et al. 2000; Stea et al. 2007; Ferreira et al. 2014; Coe and Spiro 2022).

ASF are essential sources of vitamin B_{12}, a nutrient naturally found only in animal products. Thermal processing, such as pasteurization and cooking, is important for minimizing foodborne hazards in ASF. Although some vitamin B_{12} may be lost during food processing and heat treatment, the concentration per edible portion in cooked meat is often similar or higher than in raw meat due to the loss of moisture and lipids during cooking (Gille and Schmid 2015).

As food systems transform and the distance between producers and consumers grows, the adoption of milk pasteurization and advanced packaging becomes increasingly important to meet growing consumer demand for high quality milk (Grace et al. 2020). Refrigerated milk is typically flash pasteurized and packaged in cardboard cartons, plastic bags, or plastic or glass bottles, ensuring refrigerated stability for 3–21 days, depending on the quality of the raw milk (Melini et al. 2017). Ultra-high temperature (UHT) milk undergoes a more intense heat treatment and is packaged in aseptic containers, making it shelf-stable at ambient temperatures for 3–12 months (Melini et al. 2017). Both pasteurized and UHT milk experience minimal losses of essential micronutrients and amino acids (Claeys et al. 2013; Melini et al. 2017). However, in the production of fermented milk products, the fermenting bacterial strains can consume vitamin B_{12}, resulting in a reduced B_{12} content in the final product (Gille and Schmid 2015).

Fruits and vegetables are essential components of healthy diets, providing fiber, vitamins, minerals, and phytochemicals, like beta-carotene and phenolics. Unlike cereal grains and legumes, fruits and vegetables are consumed either raw or cooked. Due to their high perishability, fresh produce requires efficient value chains when preservation technologies are limited. Canning, drying, refrigeration, and freezing are strategies to extend shelf life and manage seasonal fluctuations in supply.

Nutrient loss in fresh fruits and vegetables is influenced by factors, including acidity, maturity at harvest, time since harvest, postharvest temperature management, and cutting methods (McCarthy and Matthews 1994; Goldman et al. 1999; Lee and Kader 2000; Beaulieu and Lea 2007; Barrett et al. 2010). Water-soluble vitamins and phytochemicals can begin to degrade immediately after harvest, but refrigerated storage slows this process, compared to ambient storage (Rickman et al. 2007a). Fiber and mineral contents generally remain stable during fresh storage, while beta-carotene levels can increase or decrease depending on the specific fruit or vegetable (Rickman et al. 2007b).

Due to its sensitivity to degradation and leaching, vitamin C is often used as an indicator of overall nutrient degradation and freshness (Barrett et al. 2010). However, vitamin C may not accurately represent the loss of all micronutrients, especially those less sensitive to degradation or following different degradation pathways. Changes in micronutrients can be inconsistent and difficult to interpret (Box 7.5), potentially due to factors, including the reporting basis or variations in plant variety (Rickman et al. 2007b; Zhan et al. 2019).

Box 7.5 Measurement Point as a Factor of Micronutrient Changes

Nutrient loss pathways can be complex and nonlinear, and changes in nutrient content may vary depending on the measurement point. Adkison et al. (2018) found that canned or frozen apricots often have similar or higher nutrient content than fresh apricots when measured on a dry weight basis. However, the timing of nutrient measurements reveals underlying trends. Immediately after canning, beta-carotene and total phenol levels increased, while ascorbic acid content decreased. After 3 months, total phenols and ascorbic acid remained relatively stable, but beta-carotene content declined to levels similar to fresh apricots. In contrast, frozen apricots had higher levels of beta-carotene and total phenols than fresh apricots, both immediately after freezing and after 3 months of storage (Adkison et al. 2018).

Nutrient degradation and leaching in fresh produce can occur when the produce is damaged, either intentionally or unintentionally. Rupturing the cells of fresh produce exposes them to oxygen and fluid leakage, contributing to micronutrient losses (Lee and Kader 2000). Unintentional mechanical injuries, such as bruising, abrasions, and cuts, can happen during harvest and postharvest handling (Lee and Kader 2000). The same factors that influence nutrient loss also affect the susceptibility of produce to mechanical injury, including maturity, harvest time, and time since harvest (Hussein et al. 2020). Cutting and trimming produce, while offering convenience to consumers, is also a form of mechanical injury. Refrigerated fresh-cut fruits have been found to lose both water-soluble and fat-soluble vitamins within 6–9 days, with light exposure accelerating some losses (Gil et al. 2006). Relying on search quality attributes may not accurately indicate nutrient loss in fresh-cut produce because visual changes can occur before significant nutrient degradation (Gil et al. 2006).

Nutrient loss in dried, canned, or frozen fruits and vegetables depends on factors, including pretreatment, storage temperature, and storage time (Rickman et al. 2007a, b; Zhan et al. 2019). A comprehensive review by Rickman et al. (2007a) found that initial heat treatment during processing is primarily responsible for the loss of water-soluble and oxygen-sensitive vitamins (for example, vitamins C and B). In canned products, these nutrients remain relatively stable during storage due to the absence of oxygen. In contrast, frozen products experience less nutrient loss during processing, compared to canned products, due to shorter heating times, but can lose nutrients during long-term storage due to oxygen exposure (Rickman et al. 2007a). Although minerals are generally heat-stable, leaching and absorption of minerals in canned and frozen foods can vary, depending on processing conditions (Rickman et al. 2007b). When fruits and vegetables are dried, pretreatment using lower temperatures, or using vacuum or freeze-drying methods, can minimize nutrient losses, but also requires more advanced equipment compared to methods such as sun drying (Sablani 2006; Bhatta et al. 2020).

When fruits and vegetables reach consumers, household storage can further contribute to nutrient loss, particularly, among produce requiring cold chain storage. Freezing fresh produce can cause structural damage and nutrient loss, particularly of water-soluble vitamins, in addition to other quality attributes, including taste and texture (Bulut et al. 2018). Although nutrient leaching can occur during thawing, frozen fruits and vegetables often retain similar nutrient levels to fresh or refrigerated ones, depending on the specific food and micronutrient (Li et al. 2017).

7.3.2 Contextualizing Nutrient Losses

The potential to mitigate nutrient losses and improve dietary intake depends on various factors, including dietary context, extent of nutrient losses, and types of foods in which nutrient losses occur. While nutrient losses can occur throughout the value chain, considering the contribution of individual nutrients to overall dietary requirements is important in determining whether such losses are consequential (Melini et al. 2017). Analyzing nutrient losses in a single nutrient or food type may not provide a comprehensive understanding of overall nutrient losses in a particular food or process (Rickman et al. 2007a). Therefore, examining both important food sources and overall dietary composition are necessary to understand how nutrient loss affects diet quality.

Since nutrient content is a credence attribute, consumers must rely either their own knowledge of general nutritional value of different foods or, when available, on food labeling and nutrient content claims. As food systems modernize, packaged foods increasingly feature labels with nutrition facts that provide details on select micronutrients. However, food packaging, labeling practices, and regulatory requirements vary across different food systems, particularly, for whole and fresh food products. For example, the United States Food and Drug Administration (FDA) mandates food labels indicate information on vitamin D, calcium, iron, and potassium (FDA 2022).

To effectively assess relevant nutrient losses, it is important to identify key sources of specific micronutrients in foods and understand how their content varies with time, temperature, and processing. The Codex Alimentarius Commission (1997) outlines minimum nutrient reference values per serving for foods to be considered "sources of" or "high in" certain nutrients (for example, protein, vitamins, minerals, and dietary fiber). For example, solid foods must contain at least 15% or 30% of the nutrient reference value per 100 g to be classified as a "source of" or "high in" a specific vitamin or mineral, respectively (Codex Alimentarius Commission 1997).[1] These minimum criteria can be useful for evaluating and prioritizing nutrient losses across different foods.

Although losses of several nutrients can occur during food processing, storage, and cooking, analyzing nutrient loss pathways for key micronutrient sources is important for identifying FQL pathways in which nutrient losses result in food that render it less nutritious. For example, while pasteurization of dairy milk degrades vitamin C and folate, these nutrients are naturally present in low quantities in milk, and the loss of these vitamins during processing has a minimal impact on the overall nutritional value of the milk (Melini et al. 2017). Figure 7.4 compares the nutrient content of fresh, fresh-cooked, frozen-cooked, and canned spinach. While raw spinach is "high in" iron, folate, and vitamin C, frozen and canned spinach are downgraded to "sources" of iron and folate due to nutrient losses (Fig. 7.4).

Nutrient losses along the food value chain do not necessarily mean consumers cannot meet their nutritional needs or will experience deficiencies. Measuring nutrient loss at every post-harvest stage is impractical and inefficient. Dietary diversity is a key factor in ensuring adequate nutrient intake (Luckett et al. 2015). Understanding general nutrient loss pathways and potential issues can become part of assessing diet quality and food security, especially in populations with limited dietary diversity. The nutrient losses associated with different food forms (for example, fresh, canned, frozen, stored) can inform dietary assessments that rely on food composition databases, the accuracy of which depends on factors, including when and where food samples are collected (see Chap. 4).

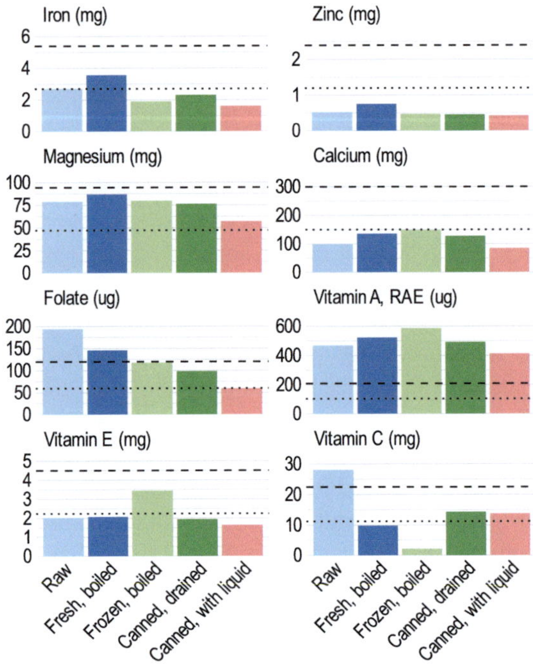

Fig. 7.4 Comparison of selected micronutrients in spinach (per 100 g spinach, fresh weight)

Micronutrient content of different forms of spinach are taken from USDA FoodData Central (USDA 2024). Pointed lines represent thresholds for "source of" the micronutrient; long dashed lines represent the threshold for "high in" the micronutrient (per day). Thresholds are calculated using the Recommended Dietary Allowance for healthy adult nonpregnant, nonlactating females (IOM 2000) and the Codex guidelines for use of nutrition claims (Codex Alimentarius Commission 1997). We applied a moisture correction and converted nutrients, based on boiled or canned weight to fresh weight, following Li et al. (2017)

[1]Alternatively, 5% (10%) of nutrient reference value per 100 kcal or 15% (30%) of nutrient reference value per serving is necessary to be considered a "source of" ("high in") a vitamin or mineral.

7.4 Weighing Food Safety and Nutrient Loss Trade-Offs

The intersection of food safety losses and nutrient losses primarily occurs during food handling, processing, and storage throughout the supply chain. Food safety losses pose a more immediate threat to food security, due to the risk of FBD that can also contribute to undernutrition (Walls et al. 2019).

Inadequate product handling and storage can compromise both food safety and nutritional value. Physical damage, such as broken skin and bruising, can contribute to both food safety issues and nutrient losses as water is leached, nutrients are oxidized, and biological hazards increase. Additionally, storage temperature, humidity, and duration can affect both food safety and nutrient retention.

As food value chains lengthen and the distance between production and consumption grows, commercial food processing becomes more common. Food processing can be classified based on the degree of change from the natural food, the types of changes and added ingredients, the location of processing (for example, commercial or home), and the processing objectives (for example, food safety, taste, appearance) (Sadler et al. 2021). Processed foods play an increasingly important role in providing nutrients in transitional and modern food systems, particularly, as urban and peri-urban populations increase (Herforth et al. 2014; Cumming et al. 2014). For example, processed fruits and vegetables (such as frozen, canned, and dried) are important sources of fiber and micronutrients in US diets (Dwyer et al. 2012).

Food processing is often prioritized to increase shelf life and prevent FQL related to food safety issues in nutritious, perishable foods, but it can also impact food nutrient content. Many processes involve heating, which offers several benefits, including the inactivation of foodborne pathogens, extended shelf life, and improved nutrient digestibility and bioavailability (van Boekel et al. 2010). However, processing can also lead to the formation of chemical hazards, nutrient losses, and undesirable sensory attributes (van Boekel et al. 2010; Forde and Decker 2022). Additionally, packaging materials can pose food safety risks (vom Saal and Hughes 2005; Onyeaka et al. 2022). As noted by Sadler et al. (2021), nutrition messaging or guidelines that emphasize the potential drawbacks of food processing could be misinterpreted as discouraging the consumption of processed foods. This could lead consumers to seek out unprocessed foods or attempt to process foods at home without adequate food safety measures. Therefore, prioritizing specific FQL pathways that directly impact the delivery of safe and nutritious foods must consider the food system context and relevant trade-offs to determine the most effective approaches to reduce FQL and deliver healthy diets.

7.5 Conclusions

In this chapter, we examined food safety and nutrient loss pathways. The specific causes and characteristics of these pathways vary depending on the stage of food system transformation. Perishable ASF, fruits, and vegetables are particularly vulnerable to FQL, especially, food safety. The extent of nutrient losses and impact on diet quality remains less clear and depends on specific nutrients and food sources. For both food safety and nutrient loss, more data is needed to better elucidate the linkages between FQL for these attributes and the immediate impact on diet quality.

Food safety is a crucial aspect of food security, and it involves understanding and addressing various food hazards throughout the food value chain. FBD can have significant health consequences, and the burden is often higher in LMICs. Urbanization and dietary shifts have made food value chains more complex and created FQL challenges associated with food safety. Time, temperature, and handling are the major factors across value chains that contribute to food becoming unsafe. Although food safety standards are important, challenges remain in informal markets and with cross-border trade, in which food safety standards either are not available or are not enforced. Addressing food safety in FQL pathways necessitates enhancing capacity, incentivizing value chain actors, and building consumer trust and demand for safe food.

Although nutrient loss pathways are less immediate threats to food security when compared to food safety losses, they can negatively impact diet quality. As foods move through the value chain, natural nutrient content can change positively or negatively. The extent and timing of nutrient loss depend on various factors, including cultivar, processing, storage, handling, and preparation. More research is needed to fully understand the impact of nutrient losses on nutrient intake, especially considering the changing dietary patterns across different stages of food system transformation. Ultimately, diverse diets rich in nutrient-dense foods are required for meeting nutritional needs. To effectively address FQL and ensure safe, nutritious food, it is essential for governments to consider the specific food system context and weigh the trade-offs between food safety and nutrient loss pathways.

References

Adkison EC, Biasi WB, Bikoba V, Holstege DM, Mitcham EJ (2018) Effect of canning and freezing on the nutritional content of apricots. J Food Sci 83(6):1757–1761. https://doi.org/10.1111/1750-3841.14157. Accessed 6 Feb 2023

Afrika J-GK, Ajumbo G (2012) Informal cross border trade in Africa: implications and policy recommendations. AfDB Africa Econ Brief 3(10). https://www.afdb.org/fileadmin/uploads/afdb/Documents/Publications/Economic%20Brief%20-%20Informal%20Cross%20Border%20Trade%20in%20Africa%20Implications%20and%20Policy%20Recommendations%20-%20Volume%203.pdf. Accessed 5 Dec 2024

Aiyar A, Pingali P (2020) Pandemics and food systems—towards a proactive food safety approach to disease prevention & management. Food Secur 12:749–756. https://doi.org/10.1007/s12571-020-01074-3. Accessed 26 Jan 2023

Alarcon P, Fèvre EM, Murungi MK, Muinde P, Akoko J, Dominguez-Salas P, Kiambi S, Ahmed S, Häsler B, Rushton J (2017) Mapping of beef, sheep and goat food systems in Nairobi—a framework for policy making and the identification of structural vulnerabilities and deficiencies. Agric Syst 152:1–17. https://doi.org/10.1016/j.agsy.2016.12.005. Accessed 24 Jan 2024

Alimi BA (2016) Risk factors in street food practices in developing countries: a review. Food Sci Human Wellness 5(3):141–148. https://doi.org/10.1016/j.fshw.2016.05.001. Accessed 3 Feb 2023

Barrett DM, Beaulieu JC, Shewfelt R (2010) Color, flavor, texture, and nutritional quality of fresh-cut fruits and vegetables: desirable levels, instrumental and sensory measurement, and the effects of processing. Crit Rev Food Sci Nutr 50(5):369–389. https://doi.org/10.1080/10408391003626322. Accessed 3 Feb 2022

Barth MM, Zhuang H (1996) Packaging design affects antioxidant vitamin retention and quality of broccoli florets during postharvest storage. Postharvest Biol Technol 9(2):141–150. https://doi.org/10.1016/S0925-5214(96)00043-9. Accessed 22 Feb 2023

Beal T, White JM, Arsenault JE, Okronipa H, Hinnouho G-M, Murira Z, Torlesse H, Garg A (2021) Micronutrient gaps during the complementary feeding period in South Asia: a comprehensive nutrient gap assessment. Nutr Rev 79(Suppl 1):26–34. https://doi.org/10.1093/nutrit/nuaa144. Accessed 7 Feb 2023

Beaulieu JC, Lea JM (2007) Quality changes in cantaloupe during growth, maturation, and in stored fresh-cut cubes prepared from fruit harvested at various maturities. J Am Soc Hortic Sci 132(5):720–728. https://doi.org/10.21273/JASHS.132.5.720. Accessed 10 Feb 2023

Bhatta S, Stevanovic Janezic TS, Ratti C (2020) Freeze-drying of plant-based foods. Foods 9(1):87. https://doi.org/10.3390/foods9010087. Accessed 22 Feb 2023

Bradford KJ, Dahal P, Van Asbrouck J, Kunusoth K, Bello P, Thompson J, Wu F (2018) The dry chain: reducing postharvest losses and improving food safety in humid climates. Trends Food Sci Technol 71:84–93. https://doi.org/10.1016/j.tifs.2017.11.002. Accessed 24 Sep 2021

Bulut M, Bayer Ö, Kirtil E, Bayindirli A (2018) Effect of freezing rate and storage on the texture and quality parameters of strawberry and green bean frozen in home type freezer. Int J Refrig 88:360–369. https://doi.org/10.1016/j.ijrefrig.2018.02.030. Accessed 6 Feb 2023

Carvalho Lucia MJ, Corrêa MM, Pereira EJ, Nutti MR, Carvalho JLV, Ribeiro EMG, Freitas C (2012) Iron and zinc retention in common beans (*Phaseolus vulgaris* L.) after home cooking. Food. Nutr Res 56:15618. https://doi.org/10.3402/fnr.v56i0.15618. Accessed 6 Feb 2023

CDC (Centers for Disease Control and Prevention) (2019) Surveillance for foodborne disease outbreaks, United States, 2017 annual report. U.S. Department of Health and Human Services, CDC, Atlanta

Claeys WL, Cardoen S, Daube G, De Block J, Dewettinck K, Dierick K, De Zutter L, Huyghebaert A, Imberechts H, Thiange P, Vandenplas Y, Herman L (2013) Raw or heated cow milk consumption: review of risks and benefits. Food Control 31(1):251–262. https://doi.org/10.1016/j.foodcont.2012.09.035. Accessed 10 Feb 2023

Codex Alimentarius Commission (1997) Guidelines for use of nutrition claims. Joint FAO/WHO Food Standards Program, Rome

Coe S, Spiro A (2022) Cooking at home to retain nutritional quality and minimise nutrient losses: a focus on vegetables, potatoes and pulses. Nutr Bull 47(4):538–562. https://doi.org/10.1111/nbu.12584. Accessed 6 Feb 2023

Croad T, Campbell H, Mirosa M (2024) Investigating systemic and social dynamics of food loss and waste: an application of waste regime theory to food production in Aotearoa New Zealand. Clean Waste Syst 7:100125. https://doi.org/10.1016/j.clwas.2023.100125. Accessed 19 Dec 2023

Cumming GS, Buerkert A, Hoffmann EM, Schlecht E, Von Cramon-Taubadel S, Tscharntke T (2014) Implications of agricultural transitions and urbanization for ecosystem services. Nature 515:50–57. https://doi.org/10.1038/nature13945. Accessed 11 Dec 2024

Czarnowska M, Gujska E (2012) Effect of freezing technology and storage conditions on folate content in selected vegetables. Plant Foods Hum Nutr 67:401–406. https://doi.org/10.1007/s11130-012-0312-2. Accessed 9 Feb 2023

Dang J, Arcot J, Shrestha A (2000) Folate retention in selected processed legumes. Food Chem 68(3):295–298. https://doi.org/10.1016/S0308-8146(99)00202-2. Accessed 9 Feb 2023

DeWaal CS, Okoruwa A, Yalch T, McClafferty B (2022) Regional codex guidelines and their potential to impact food safety in traditional food markets. J Food Protect 85(8):1148–1156. https://doi.org/10.4315/JFP-22-052. Accessed 29 Jul 2024

Downs SM, Ahmed S, Fanzo J, Herforth A (2020) Food environment typology: advancing an expanded definition, framework, and methodological approach for improved characterization of wild, cultivated, and built food environments toward sustainable diets. Foods 9(4):532. https://doi.org/10.3390/foods9040532. Accessed 26 Dec 2021

Dwyer JT, Fulgoni VL, Clemens RA, Schmidt DB, Freedman MR (2012) Is "processed" a four-letter word? The role of processed foods in achieving dietary guidelines and nutrient recommendations. Adv Nutr 3(4):536–548. https://doi.org/10.3945/an.111.000901. Accessed 27 Feb 2023

Fabbri ADT, Crosby GA (2016) A review of the impact of preparation and cooking on the nutritional quality of vegetables and legumes. Intl J Gastron Food Sci 3:2–11. https://doi.org/10.1016/j.ijgfs.2015.11.001. Accessed 6 Feb 2023

FAO (Food and Agriculture Organization of the United Nations) (2020) Climate change: unpacking the burden on food safety. Food safety and quality series no. 8. FAO, Rome. https://doi.org/10.4060/ca8185en. Accessed 28 Aug 2024

FDA (United States Food and Drug Administration) (2022) Changes to the nutrition facts label. https://www.fda.gov/food/food-labeling-nutrition/changes-nutrition-facts-label. Accessed 27 Feb 2023

Ferreira AST, Naozuka J, Kelmer GAR, Oliveira PV (2014) Effects of the domestic cooking on elemental chemical composition of beans species (*Phaseolus vulgaris* L.). J Food Process 2014:1–6. https://doi.org/10.1155/2014/972508. Accessed 9 Feb 2023

Forde CG, Decker EA (2022) The importance of food processing and eating behavior in promoting healthy and sustainable diets. Annu Rev Nutr 42:377–399. https://doi.org/10.1146/annurev-nutr-062220-030123. Accessed 27 Feb 2023

Geueke B, Parkinson LV, Groh KJ, Kassotis CD, Maffini MV, Martin OV, Zimmermann L, Scheringer M, Muncke J (2025) Evidence for widespread human exposure to food contact chemicals. J Expo Sci Environ Epidemiol 35(3):330–341. https://doi.org/10.1038/s41370-024-00718-2. Accessed 30 Sep 2024

Gil MI, Aguayo E, Kader AA (2006) Quality changes and nutrient retention in fresh-cut versus whole fruits during storage. J Agric Food Chem 54(12):4284–4296. https://doi.org/10.1021/jf060303y. Accessed 16 Feb 2023

Gille D, Schmid A (2015) Vitamin B_{12} in meat and dairy products. Nutr Rev 73(2):106–115. https://doi.org/10.1093/nutrit/nuu011. Accessed 22 Feb 2023

Global Panel (2020) Rethinking trade policies to support healthier diets. Policy brief no. 13. Global Panel on Agriculture and Food Systems for Nutrition, London. https://www.glopan.org/wp-content/uploads/2020/02/Global-Panel-policy-brief-Rethinking-trade-policies-to-support-healthier-diets.pdf. Accessed 5 Dec 2024

Goldman IL, Kader AA, Heintz C (1999) Influence of production, handling, and storage on phytonutrient content of foods. Nutr Rev 57(9):46–52. https://doi.org/10.1111/j.1753-4887.1999.tb01807.x. Accessed 3 Jan 2022

Gómez MI, Ricketts KD (2013) Food value chain transformations in developing countries: selected hypotheses on nutritional implications. Food Policy 42:139–150. https://doi.org/10.1016/j.foodpol.2013.06.010. Accessed 26 Aug 2021

Grace D (2015) Food safety in low and middle income countries. Int J Environ Res Public Health 12(9):10490–10507. https://doi.org/10.3390/ijerph120910490. Accessed 30 Mar 2022

Grace D (2017) White paper: food safety in developing countries: research gaps and opportunities. Prepared for United States Agency for International Development (USAID), Washington, DC

Grace D (2023) Burden of foodborne disease in low-income and middle-income countries and opportunities for scaling food safety interventions. Food Secur 15:1475–1488. https://doi.org/10.1007/s12571-023-01391-3. Accessed 27 Aug 2024

Grace D, Wu F, Havelaar AH (2020) MILK symposium review: foodborne diseases from milk and milk products in developing countries—review of causes and health and economic implications. J Dairy Sci 103(11):9715–9729. https://doi.org/10.3168/jds.2020-18323. Accessed 14 Feb 2023

Hald T, Aspinall W, Devleesschauwer B, Cooke R, Corrigan T, Havelaar AH, Gibb HJ, Torgerson PR, Kirk MD, Angulo FJ, Lake RJ, Speybroeck N, Hoffmann S (2016) World Health Organization estimates of the relative contributions of food to the burden of disease due to selected foodborne hazards: a structured expert elicitation. PLoS One 11(1):e0145839. https://doi.org/10.1371/journal.pone.0145839. Accessed 21 Jul 2022

Hansen TH, Lombi E, Fitzgerald M, Laursen KH, Frydenvang J, Husted S, Boualaphanh C, Resurreccion A, Howard DL, de Jonge MD, Paterson D, Schjoerring JK (2012) Losses of essential mineral nutrients by polishing of rice differ among genotypes due to contrasting grain hardness and mineral distribution. J Cereal Sci 56(2):307–315. https://doi.org/10.1016/j.jcs.2012.07.002. Accessed 29 Mar 2022

Hara M, Greenberg S, Thow AM, Chimarito S, du Toit A (2017) Trade and investment in fish and fish products between South Africa and the rest of SADC: implications for food and nutrition security. Working paper 47. Institute for Poverty, Land and Agrarian Studies (PLAAS), University of Western Cape (UWC), Cape Town. https://uwcscholar.uwc.ac.za:8443/server/api/core/bitstreams/1b603071-714c-41fb-b73f-49f95ff9970d/content. Accessed 5 Dec 2024

Havelaar AH, Kirk MD, Torgerson PR, Gibb HJ, Hald T, Lake RJ, Praet N, Bellinger DC, de Silva NR, Gargouri N, Speybroeck N, Cawthorne A, Mathers C, Stein C, Angulo FJ, Devleesschauwer B, On behalf of World Health Organization Foodborne Disease Burden Epidemiology Reference Group (2015) World Health Organization global estimates and regional comparisons of the burden of foodborne disease in 2010. PLoS Med 12(2):e1001923. https://doi.org/10.1371/journal.pmed.1001923. Accessed 21 Jul 2022

Hawkes C, Grace D, Throw AM (2015) Trade liberalization, food, nutrition and health. In: Smith R, Blouin C, Mirza Z, Beyer P, Drager N (eds) Trade and health: building a national strategy. World Health Organization, Geneva, pp 92–116

Herforth A, Frongillo EA, Sassi F, Mclean MS, Arabi M, Tirado C, Remans R, Mantilla G, Thomson M, Pingali P (2014) Toward an integrated approach to nutritional quality, environmental sustainability, and economic viability: research and measurement gaps. Ann N Y Acad Sci 1332:1–21. https://doi.org/10.1111/nyas.12552. Accessed 11 Dec 2024

Hoffmann S, Devleesschauwer B, Aspinall W, Cooke R, Corrigan T, Havelaar A, Angulo F, Gibb H, Kirk M, Lake R, Speybroeck N, Torgerson P, Hald T (2017) Attribution of global foodborne disease to specific foods: findings from a World Health Organization structured expert elicitation. PLoS One 12(9):e0183641. https://doi.org/10.1371/journal.pone.0183641. Accessed 20 Jul 2022

Hoffmann V, Mutiga SK, Harvey JW, Nelson RJ, Milgroom MG (2021) Observability of food safety losses in maize: evidence from Kenya. Food Policy 98:101895. https://doi.org/10.1016/j.foodpol.2020.101895. Accessed 31 Aug 2021

Hussein Z, Fawole OA, Opara UL (2020) Harvest and postharvest factors affecting bruise damage of fresh fruits. Hortic Plant J 6(1):1–13. https://doi.org/10.1016/j.hpj.2019.07.006. Accessed 16 Feb 2023

IOM (Institute of Medicine) (2000) DRI dietary reference intakes: applications in dietary assessment—micronutrients. https://www.ncbi.nlm.nih.gov/books/NBK222881. Accessed 23 Feb 2023

Jaffee S, Henson S, Unnevehr L, Grace D, Cassou E (2019) The safe food imperative: accelerating progress in low- and middle-income countries. Agriculture and food series. World Bank, Washington, DC. http://hdl.handle.net/10986/30568. Accessed 5 Dec 2025

Kasza G, Veflen N, Scholderer J, Münter L, Fekete L, Csenki EZ, Dorkó A, Szakos D, Izsó T (2022) Conflicting issues of sustainable consumption and food safety: risky consumer behaviors in reducing food waste and plastic packaging. Foods 11(21):3520. https://doi.org/10.3390/foods11213520. Accessed 27 Aug 2024

Khoury CK, Bjorkman AD, Dempewolf H, Ramirez-Villegas J, Guarino L, Jarvis A, Rieseberg LH, Struik PC (2014) Increasing homogeneity in global food supplies and the implications for food security. Proc Natl Acad Sci U S A 111(11):4001–4006. https://doi.org/10.1073/pnas.1313490111. Accessed 1 Feb 2023

Lee SK, Kader AA (2000) Preharvest and postharvest factors influencing vitamin C content of horticultural crops. Postharvest Biol Technol 20(3):207–220. https://doi.org/10.1016/S0925-5214(00)00133-2. Accessed 14 Feb 2023

Lee J, Pelto GH, Nordhagen S (2022) Beliefs, values, and sociocultural patterns related to food safety in low- and middle-income countries: a synthesis of the descriptive ethnographic literature. Appetite 178:106265. https://doi.org/10.1016/j.appet.2022.106265. Accessed 20 Feb 2023

Lesser C, Moisé-Leeman E (2009) Informal cross-border trade and trade facilitation reform in Sub-Saharan Africa. OECD trade policy papers no. 86. OECD Publishing, Paris. https://doi.org/10.1787/225770164564. Accessed 30 Aug 2024

Li L, Pegg RB, Eitenmiller RR, Chun J-Y, Kerrihard AL (2017) Selected nutrient analyses of fresh, fresh-stored, and frozen fruits and vegetables. J Food Compos Anal 59:8–17. https://doi.org/10.1016/j.jfca.2017.02.002. Accessed 5 Dec 2024

Li M, Havelaar AH, Hoffmann S, Hald T, Kirk MD, Torgerson PR, Devleesschauwer B (2019) Global disease burden of pathogens in animal source foods, 2010. PLoS One 14(6):e0216545. https://doi.org/10.1371/journal.pone.0216545. Accessed 21 Jul 2022

Luckett BG, DeClerck FAJ, Fanzo J, Mundorf AR, Rose D (2015) Application of the nutrition functional diversity indicator to assess food system contributions to dietary diversity and sustainable diets of Malawian households. Public Health Nutr 18(13):2479–2487.

https://doi.org/10.1017/S136898001500169X. Accessed 27 Feb 2023

Machado Nardi VA, Teixeira R, Ladeira WJ, De Oliveira Santini F (2020) A meta-analytic review of food safety risk perception. Food Control 112:107089. https://doi.org/10.1016/j.foodcont.2020.107089. Accessed 16 Oct 2024

Margier M, Georgé S, Hafnaoui N, Remond D, Nowicki M, Du Chaffaut L, Amiot M-J, Reboul E (2018) Nutritional composition and bioactive content of legumes: characterization of pulses frequently consumed in France and effect of the cooking method. Nutrients 10(11):1668. https://doi.org/10.3390/nu10111668. Accessed 6 Feb 2023

Matumba L, Van Poucke C, Monjerezi M, Njumbe Ediage E, De Saeger S (2015) Concentrating aflatoxins on the domestic market through groundnut export: a focus on Malawian groundnut value and supply chain. Food Control 51:236–239. https://doi.org/10.1016/j.foodcont.2014.11.035. Accessed 21 Feb 2023

McCarthy MA, Matthews RH (1994) Nutritional quality of fruits and vegetables subject to minimal processes. In: Wiley RC (ed) Minimally processed refrigerated fruits & vegetables. Springer US, Boston, pp 313–326. https://doi.org/10.1007/978-1-4615-2393-2_8. Accessed 16 Feb 2023

Meijer N, Kleter G, de Nijs M, Rau M-L, Derkx R, van der Fels-Klerx HJ (2021) The aflatoxin situation in Africa: systematic literature review. Compr Rev Food Sci Food Saf 20(3):2286–2304. https://doi.org/10.1111/1541-4337.12731. Accessed 1 Feb 2023

Melini F, Melini V, Luziatelli F, Ruzzi M (2017) Raw and heat-treated milk: from public health risks to nutritional quality. Beverages 3(4):54. https://doi.org/10.3390/beverages3040054. Accessed 8 Feb 2023

Miller DD, Welch RM (2013) Food system strategies for preventing micronutrient malnutrition. Food Policy 42:115–128. https://doi.org/10.1016/j.foodpol.2013.06.008. Accessed 10 Mar 2022

Mir SA, Shah MA, Mir MM, Dar BN, Greiner R, Roohinejad S (2018) Microbiological contamination of ready-to-eat vegetable salads in developing countries and potential solutions in the supply chain to control microbial pathogens. Food Control 85:235–244. https://doi.org/10.1016/j.foodcont.2017.10.006. Accessed 3 Feb 2023

Narrod C, Roy D, Okello J, Avendaño B, Rich K, Thorat A (2009) Public–private partnerships and collective action in high value fruit and vegetable supply chains. Food Policy 34(1):8–15. https://doi.org/10.1016/j.foodpol.2008.10.005. Accessed 21 Feb 2023

Neme K, Mohammed A (2017) Mycotoxin occurrence in grains and the role of postharvest management as a mitigation strategies. A review. Food Control 78:412–425. https://doi.org/10.1016/j.foodcont.2017.03.012. Accessed 1 Feb 2023

Nordhagen S, Lambertini E, DeWaal CS, McClafferty B, Neufeld LM (2022) Integrating nutrition and food safety in food systems policy and programming. Glob Food Secur 32:100593. https://doi.org/10.1016/j.gfs.2021.100593. Accessed 27 Aug 2024

Oguttu JW, McCrindle CME, Makita K, Grace D (2014) Investigation of the food value chain of ready-to-eat chicken and the associated risk for staphylococcal food poisoning in Tshwane Metropole, South Africa. Food Control 45:87–94. https://doi.org/10.1016/j.foodcont.2014.04.026. Accessed 17 Feb 2023

Onyeaka H, Passaretti P, Miri T, Al-Sharify ZT (2022) The safety of nanomaterials in food production and packaging. Curr Res Food Sci 5:763–774. https://doi.org/10.1016/j.crfs.2022.04.005. Accessed 27 Feb 2023

Passarelli S, Free CM, Allen LH, Batis C, Beal T, Biltoft-Jensen AP, Bromage S, Cao L, Castellanos-Gutiérrez A, Christensen T, Crispim SP, Dekkers A, De Ridder K, Kronsteiner-Gicevic S, Lee C, Li Y, Moursi M, Moyersoen I, Schmidhuber J, Shepon A, Viana DF, Golden CD (2022) Estimating national and subnational nutrient intake distributions of global diets. Am J Clin Nutr 116(2):551–560. https://doi.org/10.1093/ajcn/nqac108. Accessed 22 Feb 2023

Pingali P (2012) Green revolution: impacts, limits, and the path ahead. Proc Natl Acad Sci U S A 109(31):12302–12308. https://doi.org/10.1073/pnas.0912953109. Accessed 25 Aug 2021

Pingali P, Abraham M (2022) Food systems transformation in Asia—a brief economic history. Agric Econ 53(6):895–910. https://doi.org/10.1111/agec.12734. Accessed 1 Aug 2022

Pingali P, Sunder N (2017) Transitioning toward nutrition-sensitive food systems in developing countries. Annu Rev Resour Econ 9:439–459. https://doi.org/10.1146/annurev-resource-100516-053552. Accessed 24 Aug 2021

Pinstrup-Andersen P (2007) Agricultural research and policy for better health and nutrition in developing countries: a food systems approach: agricultural research and policy for better health and nutrition in developing countries. Agric Econ 37(s1):187–198. https://doi.org/10.1111/j.1574-0862.2007.00244.x. Accessed 7 Feb 2023

Reardon T, Henson S, Gulati A (2010) Links between supermarkets and food prices, diet diversity and food safety in developing countries. In: Hawkes C, Blouin C, Henson S, Drager N, Dubé L (eds) Trade, food, diet and health: perspectives and policy options. Blackwell Publishing Ltd, Chichester, pp 111–130

Richardson LC, Cole D, Hoekstra RM, Rajasingham A, Johnson SD, Bruce BB (2021) Foods implicated in U.S. outbreaks differ from the types most commonly consumed. J Food Protect 84(5):869–875. https://doi.org/10.4315/JFP-20-293. Accessed 22 Jul 2022

Ricker-Gilbert J, Omotilewa O, Kadjo D (2022) The economics of postharvest loss and loss-preventing technologies in developing countries. Annu Rev Resour Econ 14:243–265. https://doi.org/10.1146/annurev-resource-111820-020601. Accessed 10 May 2023

Rickman JC, Barrett DM, Bruhn CM (2007a) Nutritional comparison of fresh, frozen and canned fruits and vegetables. Part 1. Vitamins C and B and phenolic compounds. J Sci Food Agric 87(6):930–944. https://doi.org/10.1002/jsfa.2825. Accessed 2 Nov 2021

Rickman JC, Bruhn CM, Barrett DM (2007b) Nutritional comparison of fresh, frozen, and canned fruits and vegetables II. Vitamin A and carotenoids, vitamin E, minerals and fiber. J Sci Food Agric 87(7):1185–1196. https://doi.org/10.1002/jsfa.2824. Accessed 2 Nov 2021

Roesel K, Grace D, Makita K, Bonfoh B, Kang'ethe E, Kurwijila L, Hendrickx S, McCrindle C, Tano-Debrah K, Zewde G, Matusse H (2015) Hazards do not always translate to risks. In: Roesel K, Grace D (eds) Food safety and informal markets: animal products in Sub-Saharan Africa. Routledge, Oxford, pp 33–44

Sablani SS (2006) Drying of fruits and vegetables: retention of nutritional/functional quality. Dry Technol 24(2):123–135. https://doi.org/10.1080/07373930600558904. Accessed 22 Feb 2023

Sadler CR, Grassby T, Hart K, Raats M, Sokolović M, Timotijevic L (2021) Processed food classification: conceptualisation and challenges. Trends Food Sci Technol 112:149–162. https://doi.org/10.1016/j.tifs.2021.02.059. Accessed 27 Feb 2023

Sinayobye E, Saalia FK (2011) Size reduction of food commodities and food safety in some Accra markets. Afr J Food Agric Nutr Dev 11(6):5282–5299. https://doi.org/10.18697/ajfand.30.11035. Accessed 26 Jan 2023

Smith MR, Myers SS (2018) Impact of anthropogenic CO_2 emissions on global human nutrition. Nat Clim Change 8:834–839. https://doi.org/10.1038/s41558-018-0253-3. Accessed 6 Feb 2023

Sonar CR, Paccola CS, Al-Ghamdi S, Rasco B, Tang J, Sablani SS (2019) Stability of color, β-carotene, and ascorbic acid in thermally pasteurized carrot puree to the storage temperature and gas barrier properties of selected packaging films. J Food Process Eng 42(4):e13074. https://doi.org/10.1111/jfpe.13074. Accessed 22 Feb 2023

Stea TH, Johansson M, Jägerstad M, Frølich W (2007) Retention of folates in cooked, stored and reheated peas, broccoli and potatoes for use in modern large-scale service systems. Food Chem 101(3):1095–1107. https://doi.org/10.1016/j.foodchem.2006.03.009. Accessed 9 Feb 2023

Stephan R, Althaus D, Kiefer S, Lehner A, Hatz C, Schmutz C, Jost M, Gerber N, Baumgartner A, Hächler H, Mäusezahl-Feuz M (2015) Foodborne transmission of *Listeria monocytogenes* via ready-to-eat salad: a nationwide outbreak in Switzerland, 2013–2014. Food Control 57:14–17. https://doi.org/10.1016/j.foodcont.2015.03.034. Accessed 3 Feb 2023

Swada JG, Keeley CJ, Ghane MA, Engeseth NJ (2016) Synergistic potential of papaya and strawberry nectar blends focused on specific nutrients and antioxidants using alternative thermal and non-thermal processing techniques. Food Chem 199:87–95. https://doi.org/10.1016/j.foodchem.2015.11.087. Accessed 8 Feb 2023

Unnevehr LJ (2022) Addressing food safety challenges in rapidly developing food systems. Agric Econ 53(4):529–539. https://doi.org/10.1111/agec.12724. Accessed 27 Aug 2024

USDA (United States Department of Agriculture) (2024) USDA FoodData Central. https://fdc.nal.usda.gov. Accessed 5 Dec 2024

van Boekel M, Fogliano V, Pellegrini N, Stanton C, Scholz G, Lalljie S, Somoza V, Knorr D, Jasti PR, Eisenbrand G (2010) A review on the beneficial aspects of food processing. Mol Nutr Food Res 54(9):1215–1247. https://doi.org/10.1002/mnfr.200900608. Accessed 9 Jun 2022

vom Saal FS, Hughes C (2005) An extensive new literature concerning low-dose effects of bisphenol A shows the need for a new risk assessment. Environ Health Perspect 113(8):926–933. https://doi.org/10.1289/ehp.7713. Accessed 27 Feb 2023

Waage J, Grace D, Fèvre EM, McDermott J, Lines J, Wieland B, Naylor NR, Hassell JM, Chan K (2022) Changing food systems and infectious disease risks in low-income and middle-income countries. Lancet Planet Health 6(9):e760–e768. https://doi.org/10.1016/S2542-5196(22)00116-4. Accessed 25 Jan 2024

Walls H, Baker P, Chirwa E, Hawkins B (2019) Food security, food safety & healthy nutrition: are they compatible? Glob Food Secur 21:69–71. https://doi.org/10.1016/j.gfs.2019.05.005. Accessed 20 Jul 2022

Wenndt A, Sudini HK, Pingali P, Nelson R (2020) Exploring aflatoxin contamination and household-level exposure risk in diverse Indian food systems. PLoS One 15(10):e0240565. https://doi.org/10.1371/journal.pone.0240565. Accessed 31 Jan 2023

Wertheim-Heck S, Raneri JE, Oosterveer P (2019) Food safety and nutrition for low-income urbanites: exploring a social justice dilemma in consumption policy. Environ Urban 31(2):397–420. https://doi.org/10.1177/0956247819858019. Accessed 17 Feb 2023

White JM, Beal T, Arsenault JE, Okronipa H, Hinnouho G-M, Chimanya K, Matji J, Garg A (2021) Micronutrient gaps during the complementary feeding period in 6 countries in Eastern and Southern Africa: a comprehensive nutrient gap assessment. Nutr Rev 79(Suppl 1):16–25. https://doi.org/10.1093/nutrit/nuaa142. Accessed 7 Feb 2023

WHO (World Health Organization) (2008) Foodborne disease outbreaks: guidelines for investigation and control. World Health Organization, Geneva. https://iris.who.int/bitstream/handle/10665/43771/9789241547222_eng.pdf. Accessed 5 Dec 2024

WHO (World Health Organization) (2015) Human biomonitoring: facts and figures. WHO Regional Office for Europe, Copenhagen. https://iris.who.int/bitstream/handle/10665/164588/WHO-EURO-2015-3209-42967-60040-eng.pdf. Accessed 5 Dec 2024

WHO (World Health Organization) (2018) Regional framework for action on food safety in the Western Pacific. World Health Organization, Regional Office for the Western Pacific, Manila. https://iris.who.int/bitstream/handle/10665/272681/9789290618478_eng.pdf. Accessed 5 Dec 2024

WHO (World Health Organization) (2021) WHO steps up action to improve food safety and protect people from disease: new handbook helps countries assess causes, magnitude and distribution of foodborne diseases, June 7. https://www.who.int/news/item/07-06-2021-who-steps-up-action-to-improve-food-safety-and-protect-people-from-disease. Accessed 15 Oct 2024

Williams JH, Phillips TD, Jolly PE, Stiles JK, Jolly CM, Aggarwal D (2004) Human aflatoxicosis in developing countries: a review of toxicology, exposure, potential health consequences, and interventions. Am J Clin Nutr 80(5):1106–1122. https://doi.org/10.1093/ajcn/80.5.1106. Accessed 28 Aug 2024

Zhan X, Zhu Z, Sun D-W (2019) Effects of pretreatments on quality attributes of long-term deep frozen storage of vegetables: a review. Crit Rev Food Sci Nutr 59(5):743–757. https://doi.org/10.1080/10408398.2018.1496900. Accessed 6 Feb 2023

8 Unlocking Potential: Identifying Critical Loss Points and Driving Change

8.1 Introduction

To address high-level objectives (for example, food security, economic efficiency, environmental impacts), interventions to reduce and manage food loss and waste (FLW) must fit the food system context and investment priorities. All food systems have challenges and limitations that may block potential pathways to support healthy, sustainable diets. Examining the critical loss points, the causes of FLW, and food system investment priorities are essential for setting appropriate and specific FLW reduction targets and interventions for action to achieve higher-level objectives.

To effectively address FLW, it is to identify critical loss points where food quality is affected and food is removed from the value chain. While previous chapters explored methods for measuring physical food loss and waste (pFLW) and assessing food quality loss (FQL), this chapter focuses on identifying critical loss points for pFLW across different value chain stages and food commodities, and their implications for achieving broader FLW reduction objectives. The lack of FLW data remains a challenge, and attempts to identify critical loss points highlight the ongoing need for definitional and measurement frameworks that align FLW reduction activities to broader sustainable food systems objectives. For example, FLW was not included as an indicator in the Food Systems Countdown, a collaborative initiative to monitor global food systems (see Chap. 2) due to lack of sufficient country-level data to track and inform food system transformation (Schneider et al. 2023).

As described in Chaps. 5–7, the FLW pathways are complex, and the causes of FLW are often interrelated. Therefore, the causes of FLW along different pathways require consideration of the different levels (HLPE 2014):

- *Micro-level*: causes at the same stage involving individual entities in a particular food value chain.
- *Meso-level*: structural causes involving larger groups of actors in the food value chain.
- *Macro-level:* systemic causes involving the broader food system, institutional or policy conditions, and enabling environment.

Different stakeholders within food systems have varying perspectives on the causes and consequences of FLW, which can influence measurement and prevention strategies. While previous chapters focused on stage-specific causes of FLW, this section delves deeper into the meso- and macro-level factors that drive FLW and create feedback loops within food value chains. By understanding these broader technical, infrastructural, regulatory, and environmental drivers, we can identify potential entry points for effective interventions.

Experts on food security and nutrition have identified key investment priorities for transition-

J. Boiteau, P. Pingali, *Wasted Potential*, Sustainable Development Goals Series,
https://doi.org/10.1007/978-3-031-97411-3_8

ing food systems toward sustainability and health (HLPE 2017). For traditional food systems, early-stage investments in production, storage, and distribution are crucial for increasing food quality, safety, and food affordability. In contrast, transitional and modern systems require significant investments in mid-to-downstream stages, including processing, packaging, and marketing. As food systems modernize, investment priorities are required for promoting demand for healthy diets in addition to improving access to nutritious foods. This chapter explores how addressing FLW can align with these investment priorities to achieve food security and nutrition goals.

Identifying incentives for FLW reduction across public and private sectors is necessary for feasibly addressing FLW. These points may not always align with the stages where losses occur. For instance, addressing upstream issues like overproduction or inefficient harvesting practices can prevent downstream losses. Additionally, trade-offs may exist between different sustainability objectives, such as food security and environmental impact.

To effectively address FLW, it is crucial to accurately identify critical loss points and understand the complex interplay of factors driving FLW. This knowledge, combined with a clear understanding of stakeholder incentives and public–private partnerships, can inform the development of targeted interventions that support sustainable food systems and promote healthy diets.

8.2 Critical Loss Points Across Foods and Food Systems

Pinpointing critical loss points or "hotspots" within food systems is essential to prioritize policies and investments aimed at effectively addressing FLW. Understanding the extent and distribution of FLW across regions, food value chain stages, and commodities enables deeper examination of the underlying causes and development of targeted strategies. However, identifying critical loss points is not an end in itself; it must be evaluated within the broader context of FLW reduction objectives, including improved nutrition and food security, environmental impacts, and economic efficiency. This section will focus on the critical loss points of pFLW.

In the FLW literature, hotspot analyses have primarily focused on pFLW, using different units depending on the specific objective and scope of the analysis (Guo et al. 2020; Gatto and Chepeliev 2024). Physical FLW is quantified in terms of mass or volume (as detailed in Chap. 4). These physical units can be converted into other metrics, such as nutritional values (for example, calories, macronutrients, micronutrients) or environmental indicators (for example, greenhouse gas emissions, embedded land use, water use), to examine linkages between the amount of pFLW and specific impacts.

Converting pFLW from physical units to nutritional value metrics can introduce bias and obscure important details related to healthy diets. Recent studies have reported nutritional value metrics to analyze FLW in the context of nutrition and food security (Spiker et al. 2017; van den Bos Verma et al. 2020; Guo et al. 2020; Gatto and Chepeliev 2024). Although calorie-based metrics are most commonly used for analyses, followed by those based on macronutrients, micronutrient-based assessments are less frequent. However, focusing solely on macronutrients can overlook micronutrient-rich and perishable foods, such as fruits and vegetables that are not macronutrient-dense. To address this limitation, a food-based approach is more informative for comparing the impact of pFLW on healthy diets. Food groups are also more practical for cross-context comparisons, due to varying core food baskets worldwide (Fabi and English 2019).

Examining pFLW through several environmental units provides insights into how reducing pFLW can contribute to different environmental sustainability objectives. For example, Cattaneo et al. (2021) highlighted the varying environmental impacts and resource use efficiency implications of reducing pFLW at different value chain stages. Similarly, Gatto and Chepeliev (2024) demonstrated how identifying the top 10 country pFLW hotspots changes, depending on the unit of comparison used (for example, million tonnes, grams/capita/day; embedded land use; water use). This research underscores the importance of considering differences across food groups, value chain stages, and environmental indicators when identifying critical loss points that are most relevant in a specific context.

To effectively identify critical loss points in the food system, detailed estimates of pFLW are needed for each value chain stage, beyond simple aggregation into postharvest and retail/consumer stages. Previous studies (FAO 2011; Gatto and Chepeliev 2024) have categorized pFLW across five value chain stages: agricultural production; postharvest handling and storage; processing and manufacturing; distribution and retail; and consumption. These more specific value chain stages have different characteristics based on the food system type and commodity. Therefore, disaggregating pFLW estimates by value chain stage is important for gaining valuable insights into the specific activities and actors involved in the FLW pathways (Fabi and English 2019; Parfitt et al. 2021).

As diets shift with transforming food systems, economic and environmental factors can influence production patterns and value chain operations, potentially shifting the location of critical loss points over time. The percentage of pFLW for each commodity group is a useful indicator for identifying critical loss points across different geographies and value chain stages (Fabi and English 2019). While the absolute value of pFLW may fluctuate with changes in production levels, the percentage indicator remains useful for isolating critical loss points. Therefore, the pFLW percentage indicator helps identify changes in how different food value chains are organized and function. Meanwhile, the absolute values of pFLW are essential for assessing the potential impact on diets and environmental indicators, ultimately determining acceptable pFLW percentages for each commodity group in a given context.

Contextualizing critical loss points is essential for identifying which value chain actors, both at the loss point and at upstream and downstream stages, have incentives to reduce pFLW or maintain current practices. Although one stage may have the highest share of pFLW, it is important to consider the relative contributions of other stages, the underlying causes of pFLW leading to the critical loss point, and the potential consequences of intervening at different points in the value chain. As explored in previous chapters, FLW pathways for the same commodity can vary significantly, depending on the specific food system context.

A significant limitation in identifying critical loss points is the ongoing lack of primary data for pFLW across various commodity groups and value chain stages. As discussed in Chap. 4, numerous data gaps exist, particularly, for perishable foods. While the Sustainable Development Goal (SDG) 12.3 Indicators, the Food Loss Index and Food Waste Index, aim to address data gaps, limitations persist in terms of coverage across value chain stages and comparability issues arising from different definitions and estimation approaches (see Chaps. 3 and 4). Harmonizing a single FLW definition and establishing a FLW data value chain are crucial steps toward accurately identifying critical loss points at various scales.

Gatto and Chepeliev (2024) recently compiled a country-level database to address data gaps in the FLW database (2019) of the Food and Agriculture Organization of the United Nations (FAO) and to assess the magnitude, composition, and geographic distribution of pFLW across global value chains. The authors contextualize critical loss points within the framework of food trade, highlighting that approximately 15.7% of global FLW in 2014 was associated with traded food (Gatto and Chepeliev 2024). Consequently, as food trade between exporting low- and middle-income countries (LMICs) and importing high-income countries (HICs) increases, trade plays an important role in FLW pathways, exacerbating pFLW in LMICs, particularly, at the farm level.

Figure 8.1 summarizes the estimates of pFLW by global region and perishable food group, along with the corresponding data sources. Most data for fruits and vegetables is sourced from the FAO database, while estimates for animal-source foods

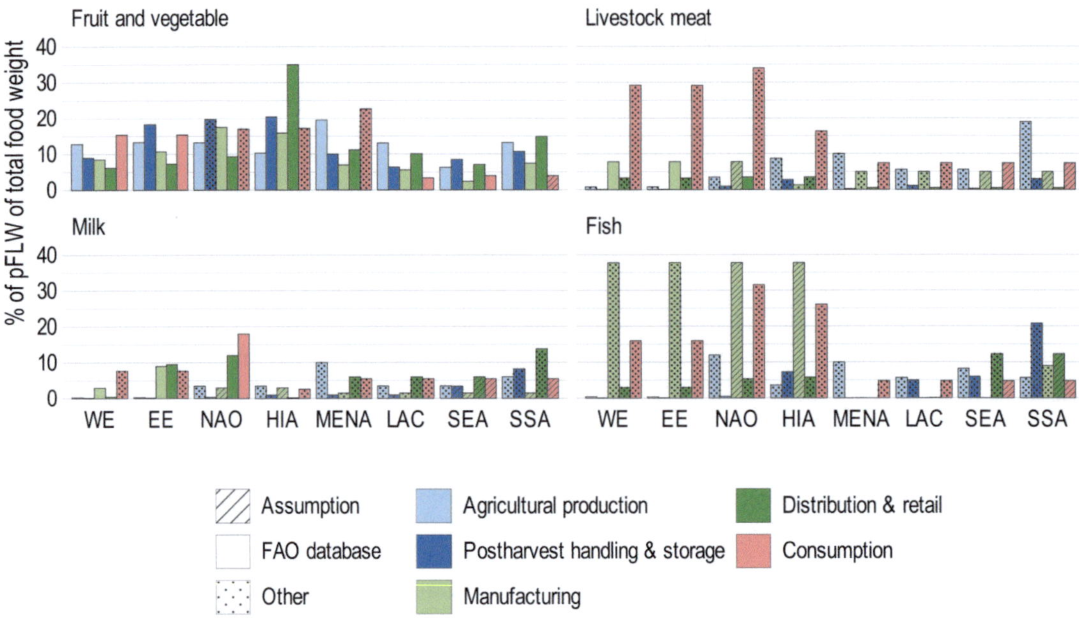

Fig. 8.1 Share of physical food loss and waste of total food weight by food group and global region

Note: Data from Gatto and Chepeliev (2024). The striped pattern on the bars indicates unavailable data, the lack of which was replaced with a consistent gap-filling methodology based on pFLW in comparable regions, commodities, and stages of the food value chain (hence, labeled "assumption"). An absence of pattern in the color bars (that is, solid-colored) indicates the data came from the FAO FLW database (2019). The dotted pattern in the bars indicates other sources were used, based on the authors' literature review

Abbreviations: WE (Western Europe); EE (Eastern Europe); NAO (North America and Oceania); HIA (High-Income Asia); MENA (Middle East and North Africa); LAC (Latin American and Caribbean); SEA (Southeast Asia); SSA (Sub-Saharan Africa)

(ASF) and consumer stages, especially in LMICs, rely on literature review and assumptions. Given the limitations of the FAO database, additional literature, and the overall scarcity of data in the literature, significant data gaps persist across perishable commodities, stages, and regions.

8.3 Meso- and Macro-level Food and Loss Waste Drivers

Higher-level drivers of FLW, including environmental factors, market conditions, policies, and social norms are distinct from more proximate, micro-level pressures that directly lead to FLW (Spang et al. 2019). These meso- and macro-level drivers involve nonhuman factors (for example, weather, pests, natural variations, and life cycle), economic, social, and structural factors based in institutional and organizational arrangements (Gille 2012). Meso- and macro-level causes of FLW affect value chain actors and other food systems stakeholders differently, leading to possibly conflicting incentives and interests, particularly, in terms of public versus private sectors. Challenges, such as limited access to technology and knowledge, inadequate infrastructure, insufficient resources, and a complex regulatory environment, can hinder the implementation of FLW reduction strategies and the ability of value chain actors, especially, small-scale producers, to adopt effective practices and technologies (HLPE 2014; Xue et al. 2017; Soethoudt et al. 2021). Therefore, FLW is not solely caused by lack of sufficient technology or inefficiency, but also from risk avoidance, value generation, and uncertainty along food value chains (Gille 2012; Croad et al. 2024). As food systems modernize, these higher level drivers can shift FLW pathways. Instead of losing food with clear market value due to FQL, food may be discarded due to stricter quality standards that are intended to create value (Croad et al. 2024).

8.3.1 Technologies and Practices

Technology and practices are often connected with FLW pathways at the micro-level, as discussed in Chaps. 5–7 of this book. Therefore, approaches to address FLW often look toward technology innovations. For example, a review of 276 innovations, focused on nutrient-dense foods, FLW, and food safety, identified 154 (56%) as products or technologies, while 55 (20%) were practices (Webb and Shrestha 2023). Meso- and macro-level factors influence the availability and access to technologies, practices, information, and resources required for decision-making, as well as the incentives and enabling environments to use technologies or implement best practices. The entry points for intervention prioritize the potential economic gains that can be achieved if the causes at different levels are addressed.

Technological advancements within food value chains encompass tools and equipment designed to optimize specific stages, such as harvesting, storage, and processing. These technologies primarily aim to minimize FQL through methods like cold chain storage, packaging, and enhanced coordination across value chain stages. The availability of certain technologies in a given context is influenced by factors, including innovation and distribution channels (Soethoudt et al. 2021). The design, level of advancement, and quality of these technologies can also impact the extent to which pFLW is embedded in their routine operation. For example, food manufacturing and processing equipment may generate pFLW during regular use and maintenance (Raak et al. 2017; Rösler et al. 2021).

Among processed products, technical and human error can lead to incorrect product specifications. Products may be discarded when they fall outside of the attribute specifications, such as when there is physical damage of smaller produce rolling around in packaging (Gillman et al. 2019), or when packaging breaks from the product exceeding its package capacity (Surucu-Balci and Tuna 2021). Additionally, technology errors or user errors can lead to FQL and pFLW, such as errors in packaging that lead to breakage or inadequate protection during subsequent stages of the value chain (Eičaitė et al. 2023).

Technology accessibility is influenced by factors, such as affordability and ease of use. The ability of value chain actors to invest in available technology depends on entity characteristics (for example, size, activities), labor costs, and access to financial resources. The lack of financial services, including funding and credit at the meso-level, can obstruct investments in technology that could help minimize FLW (Soethoudt et al. 2021). Without adequate training and knowledge in proper technology use and maintenance, value chain actors may not fully realize the potential for technology to address existing inefficiencies that contribute to FLW.

Similar to technology, best practices for minimizing FLW often require access to information through training and knowledge sharing. While some best practices require specific technical skills for equipment operation, processing, and communication, others focus on behavior changes, handling procedures, and strategic approaches to management and marketing (Kitinoja and Kader 2015; Soethoudt et al. 2021; Vos and Cattaneo 2021). The lack of meso-level support and investment in training materials, programs, and knowledge-sharing platforms can prevent the dissemination of information on best practices, limiting their impact on reducing FLW.

As food systems modernize, changes in the meso- and macro-level drivers, related to the use of technology and improved practices, can have feedback loops that exacerbate existing FLW pathways or create new pathways. At the same time, using technology and best practices that minimize FLW at one stage can inadvertently lead to cascading effects that push FLW downstream. This often occurs with the introduction of technology at midstream packaging and processing, as well as downstream changes in consumer income and preferences for convenience.

Bulk and consumer food packaging becomes an increasingly important technology as food systems modernize, particularly, for high-value, perishable products. Beyond protecting food and providing information, packaging can also influence upstream food quality specifications, such as produce shape and size—leading to the downgrading of edible food that no longer meets the functional requirements for the packaging (Gillman et al. 2019). Depending on market channels and processing requirements, food that does not meet fresh value chain standards may be diverted to processing value chains. Therefore, while packaging is meant to reduce FLW, it can also become a reason for food value loss or complete diversion to nonfood destinations.

Consumer packaging can specifically influence the extent of pFLW at the consumer stage. For example, packaged foods have predetermined units that can lead consumers to purchase greater quantities than they require (Aschemann-Witzel et al. 2017; Wilson et al. 2017). Food packaging can also prevent consumers from using sensorial cues to determine food quality, leading to purchase of food with qualities that do not meet their preferences (Barbut and Leishman 2022).

Processing can absorb lower quality products initially intended for fresh food value chains. However, as quality standards for both fresh and processed foods increase, the raw material requirements for the processing sector may limit its capacity to absorb lower quality produce, effectively restricting an alternative value chain and potentially increasing pFLW of unmarketable products (Redlingshöfer et al. 2017). More advanced processing and packaging technologies often have narrower tolerances for food product variations. For example, mechanized meat processing requires specific size, shape, orientation, and density standards for animal tissues (Barbut 2020). When considering trade-offs between acceptable levels of FLW and other factors, such as occupational hazards in food processing, a comprehensive approach is necessary.

8.3.2 Infrastructure Organization

Value chain actors operate within the constraints of their environment. Organizational drivers of FLW affect the enabling environment, or lack thereof, related to distribution and service networks and infrastructure that are essential for efficient food value chains. As the distances between rural producers and urban consumers increases, so does the importance of adequate storage, transport, and communication infrastructure. Organizational issues that act as meso- and macro-level drivers of FLW include both private and public infrastructure.

Food value chains depend on the availability and access to rural infrastructure, including quality roads, reliable electricity, storage facilities, and telecommunications. Socioeconomic factors that drive FLW in Africa, Latin America, and developing Asia are not only related to country wealth, but also to agricultural machinery, transportation, and telecommunications (Krishna Bahadur et al. 2016). Depending on the FLW pathway, there may be one or more infrastructure constraints, such as rural and feeder road access, electricity supply, temperature-controlled storage access, and wholesale market development (Vos and Cattaneo 2021).

The limiting infrastructure may depend on the food product, value chain stage, and activities carried out. In the case of fresh perishable products, particularly, fruits and vegetables, poor roads and lack of adequate transport vehicles at the farm level, together, create inadequate transport conditions, leading to FQL, particularly, in value chains without cold chain storage and transport (Kader and Rolle 2004). In contrast, the level and variability of energy costs related to inadequate energy infrastructure may be particularly salient for food processors and distributors whose key functions rely on consistent and affordable access to electricity (Vos and Cattaneo 2021). For example, the decisions on processing techniques of perishable foods are dictated by the availability, access, and profitability of each technique. Cooling preservation methods, such as refrigeration and freezing, require consistent electrification, which becomes a major limitation in contexts with limited infrastructure and high energy costs (Hasan et al. 2019). Although drying is more energy intensive than cooling preservation methods on an absolute basis, energy from the sun is a practical energy source if electricity supply is limited or unaffordable (Machala et al. 2022).

Nonhuman factors combine with human-factors in FLW pathways when fluctuations in food supply exacerbate the lack of coordination between actors. The seasonal production of perishable foods contributes to the technical, financial, and operational challenges of processing, what Machala et al. (2022) refer to as "seasonally stranded" processing facilities. The unstable supply of fresh perishable foods suitable for processing can either lead to an inadequate supply of low-cost raw material or an oversupply of raw material that exceeds storage and processing capacity (Subramanian 2016; Akintola et al. 2022; Schreinemachers et al. 2022). Seasonal production also becomes an issue when perishables with overlapping harvest seasons compete for limited available processing capacity (Beausang et al. 2017). The lack of coordination and communication between value chain actors can also contribute to capacity issues, for example, when production sizes exceed trader packing facilities, leading to improper storage conditions (Surucu-Balci and Tuna 2021; Perdana et al. 2023).

In addition to changes in storage and transport infrastructure requirements, longer physical distances between value chain stages make it more difficult for sharing information through in-person interactions. As a result, there is an increasing reliance on information and communication technologies, which require access to the Internet and mobile phone services (Vos and Cattaneo 2021). For rural actors who are more isolated, including smallholder producers, access to telecommunications and knowledge of its use can provide important information about market requirements and prices, enabling the smallholders to make informed decisions that can impact FLW pathways.

Private and public sectors have different objectives for investing in infrastructure. The private sector may invest in roads, electrification, telecommunication, or marketing infrastructure, because these specific investments are required to support their operation or as part of other aspects of their business strategy (for example, corporate social responsibility) (Feyaerts et al. 2020). In contrast, public investment in similar infrastructure may be specific to an area to attract private investment in a particular food industry and export or, more broadly, to improve socioeconomic development and resilience (Feyaerts et al. 2020; Vos and Cattaneo 2021). Improved infrastructure, particularly, for high-value perishable foods, can incentivize additional investment with spillover benefits to support "spontaneous clusters" of downstream small- and medium-sized enterprises and local value chains (Feyaerts et al. 2020; Vos and Cattaneo 2021). Public investments in rural infrastructure also support objectives to support inclusive market development.

8.3.3 Public and Private Regulations

FLW pathways are influenced by public and private regulations, particularly, those governing food quality attributes. Regulations set standards and requirements for different food attributes, which then become limiting attributes for product acceptance (see Chap. 4). FQL may be implicated in limiting attributes, but pFLW can occur with or without FQL of a limiting attribute. These regulations can improve food quality for consumers and enhance value for value chain actors but may also contribute to pFLW of food suitable for human consumption. Therefore, a closer examination of the role of regulations in FLW pathways is important for interpreting critical loss points and understanding how to minimize FQL of limiting attributes or keep food fit for human consumption in food value chains.

As discussed in Chap. 7, FQL can directly affect diet quality, particularly, when unsafe food reaches consumers. FQL can also be considered a meso-level cause of pFLW when FQL that occurs at one or multiple stages eventually leads to pFLW at another stage. Moreover, FQL is also closely linked to market value loss. When FQL results in noncompliance with quality standards, marketable food items may be diverted to lower value channels or rejected entirely, ultimately exiting the food value chain and diverted to nonfood uses or landfill. Quality standards may include different limiting attributes for several marketable grades or may take a binary approach in which foods must meet all limiting attributes for a specific market, each affecting FLW pathways differently (Gillman et al. 2019).

At the macro level, government institutions set quality standards for different commodity groups. Government standards for food safety attributes typically emerge first and are treated separately from other food quality attributes. Public food safety standards are legal requirements that must be met to enter a market, with entry denied for goods that to not comply with these standards (Unnevehr and Ronchi 2014). In traditional and transitional food systems, informal food sectors are prevalent and lie outside of food safety regulatory control (Hoffmann et al. 2019). When foods fail to meet food safety standards set in modern retail outlets, food may be resold through informal outlets (Roesel and Grace 2015; Hoffmann et al. 2019).

Over the last few decades, there has been a noticeable rise in food and nutrition regulations, especially, those concerning food safety and quality (Fig. 8.2). While food safety standards prioritize public health, quality standards based on search attributes primarily aim to enhance market competitiveness in trade (De Hooge et al. 2018; Porter et al. 2018; Walsh 2022). This results in a misalignment between the objective of the standards criteria and the intended outcome. However, macro-level cosmetic regulations may inadvertently imply that food is unsuitable for human consumption when natural variations fall outside of specified parameters (Porter et al. 2018). Consequently, public regulatory objectives may be misaligned when nutritious foods are removed from the food value chain, worsening the nutrient supply gap.

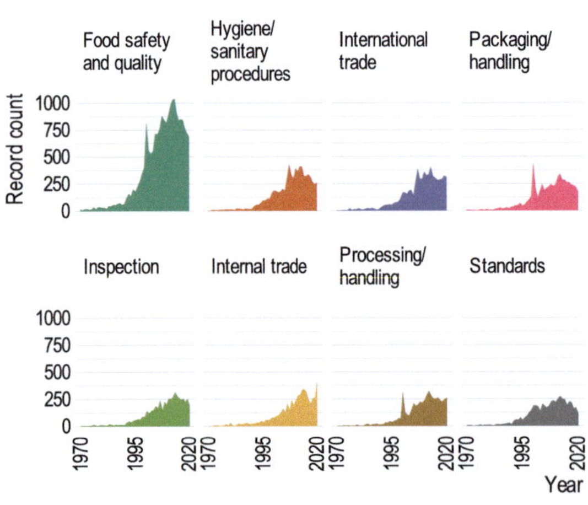

Fig. 8.2 National legislation, policies, and agreements on food and nutrition by specific topic areas from 1970 to 2023

The x-axis is the year of original text or, if amended, the year last amended. (Source: FAOLEX database (FAO 2024))

Quality standards established by private sector actors, including processors and retailers, aim to differentiate and add value to their products. These standards frequently exceed government regulations, if they exist, and are linked with shifts in power dynamics as value chains consolidate (Box 8.1). Retailers in formal food environments, particularly, supermarkets, prioritize search attributes like size, shape, color, and ripeness when setting and enforcing quality standards, rather than experience or credence attributes (Beausang et al. 2017; Gillman et al. 2019; Messner et al. 2021).

Value chain actors may implement their own quality requirements to prevent FQL downstream (Box 8.2) or to present themselves as high quality suppliers in the market (De Hooge et al. 2018). For example, when buyers reject entire loads based on a small percentage of products being of borderline quality, producers respond with increased caution and effectively apply a more stringent quality threshold than is specified (Gillman et al. 2019).

Although informal market wholesalers and retailers lack formalized quality standards, they may rely on search attributes to distinguish food quality, based on the potential returns from quality improvements in specific products and markets (Vandeplas and Minten 2015). However, when farmers and local traders receive the same price for their product, regardless of the quality level, they are disincentivized to invest in grading, sorting, or other efforts to reduce FQL

(Cadilhon et al. 2006; Perdana et al. 2023). The incentives to minimize FQL depends on the types of quality attributes that are deemed important and derive market value. In their study of perishable food markets in India, Fafchamps et al. (2008) found that price premiums for some quality attributes stop midstream, implying that attributes important for wholesalers are not important for processors and retailers, who are more consumer-facing. Food ownership also influences these incentives and financial risks that can change, depending on the value chain actors who are responsible for moving food along the value chain without assuming ownership of the food. The lack of policy and regulation or enforcement of important FQL factors through licenses and monitoring can contribute to value chain power imbalances that perpetuate FQL.

Producers often serve as the initial gatekeepers in adhering to quality standards, as many of these attributes are established or initiated preharvest (for example, size, shape, color, and ripeness). Private standards can influence producers' production, harvest, and postharvest practices, even among factors that are outside of producers' control. Nonhuman factors, including weather and pests, can contribute to FQL of search attributes that do not affect the edibility of the food itself (Beausang et al. 2017; Gillman et al. 2019). To meet these standards, producers may choose to leave fresh produce unharvested, sort it out in the field, or divert edible produce to lower value channels, such as processing (Pietrangeli and Cicatiello 2024). Failure to meet these standards can also lead to post-farmgate rejection. Retailers and processors may reject product even after producers have invested in packing and transporting the product (Soma et al. 2021). The value of the fresh produce continues to drop and the cost to the value chain actor who has ownership increases as the need arises to transport, store, and find a new buyer, or cover the cost of disposal (or diversion to nonfood use) (Gillman et al. 2019; Soma et al. 2021; Boiteau and Pingali 2022).

Box 8.1 Value Chain Consolidation and Shift in Power Dynamics

Power dynamics between value chain actors change as food systems transform and the midstream value chain actors consolidate to create market structures with few processors and retailers that connect many suppliers and consumers. This consolidation strengthens the influence that post-farmgate value chain actors have to set and enforce private standards and their power to shift the risks of supply and demand fluctuations to producers (Gille 2012; Rindt and Mouzas 2015; Gillman et al. 2019). Additionally, consolidation narrows market access for producers, leaving few, if any, alternative mass distribution channels (Messner et al. 2021).

Private sector actors, such as retailers and processors, often influence quality standards according to their business objectives. These standards, particularly for search attributes, like size and color, can be subject to fluctuations based on supply and demand. When supply surpasses demand, quality specifications are narrowed to the point that there may be no market for low-grade produce, leading to pFLW of perfectly edible food. This happens, particularly, during peak season for lower quality products that would otherwise have economic value in nonpeak seasons (Messner et al. 2021). Supermarkets may create alternative markets through discounts or packaging strategies, but this often requires producers to adapt their practices to align with shifting market demands (Messner et al. 2021).

Since producers depend on securing and retaining buyers, processors and retailers protect themselves from market uncertainties using requirements for consistent volumes that meet specific quality standards (Gille 2012). Producers in modern food systems, in particular, often view farmgate pFLW as acceptable compared to post-farmgate pFLW, based on perceptions of resource use (Beausang et al. 2017).

Box 8.2 Balancing Functional Quality Attributes with Consumer Preferences

Important food quality attributes encompass those that align with consumer preferences and those that serve specific functional purposes. Functional quality attributes and their acceptable thresholds are determined by the value chain structure and access to infrastructure that support minimizing FQL. Attributes can often overlap, meeting both value chain actors' needs and consumer preferences. For example, consumer preferences for fruit firmness align with firmness as a functional quality attribute to withstand transportation and handling. However, as fruits naturally soften during ripening, they may become overly soft, increasing susceptibility to physical damage, decay, and spoilage of other produce. To avoid this, value chain actors may remove still-edible fruit that does not meet their firmness criteria, based on anticipated quality changes during distribution (Gillman et al. 2019).

Although cosmetic specifications are often perceived as consumer preferences, they may actually reflect a buyer's potentially limited, interpretation of consumer acceptance (Porter et al. 2018; Gillman et al. 2019; Messner et al. 2021). When retailers offer products with specific quality attributes, consumer experiences can create feedback loops, as they associate search and experience attributes of foods based on a narrow range of search attributes. As a result, limited exposure to "abnormal" produce in the market can hinder the broadening of consumer preferences and acceptance (Porter et al. 2018).

8.3.4 Environmental Changes

Although climate and environmental issues are frequently part of the FLW conversation, it is often with regard to the impacts of FLW on the environment. How climate change can be a driver of FLW is less discussed. Higher temperatures, changes in rainfall, and increased frequency and intensity of extreme weather events can all impact each stage of food value chains, contributing to FQL and pFLW. Regions in the Global South, which are particularly vulnerable to climate change also have transitional food systems that are susceptible to various meso- and macro-level FLW drivers, discussed throughout this chapter (FAO 2018b). For example, higher temperatures will increase the need for adequate cold chain transport, storage, and distribution to prevent FLW of perishable foods. Additionally, extreme weather events, such as floods, can disrupt timely food transportation, particularly, in regions with inadequate road infrastructure and vehicles. Further research is needed to fully understand the role of climate change as a driver of FLW within the context of other macro- and meso-level factors and to identify effective interventions.

8.4 Leverage Points and Investment Priorities

FLW and its linkage to food security has been a globally recognized issue since the 1970s (Shen et al. 2023). Interest in FLW was renewed after the FAO (2011) report estimated global FLW at 33% of total production, highlighting its impact on nutrition, food security, the environment, and economic efficiency. Currently, the 2030 Agenda for Sustainable Development, adopted in 2015, has set the current global targets to reduce food waste by 50% by 2030. While it has set a target for reducing postharvest food loss, a quantitative reduction target has not been specified (UN 2015). Reaching these targets will require both public and private investments (Lipinski et al. 2017).

8.4.1 Sustainable Development Goal 12.3 Drives Public Institution Reduction Targets

While commitment to the UN SDGs are legally nonbinding, governments are encouraged to establish national frameworks for policies, plans, and programs, including progress monitoring. Most of the countries that have prioritized FLW reduction as a national strategy are HICs (see Box 8.3). As of 2021, 13 countries and regional blocs,[1] covering about 55% of the global population, have committed to specific targets in alignment with SDG 12.3 (Lipinski 2022). Additionally, United Nations regional commissions in Latin America and the Caribbean and the Near East and North Africa Region have outlined regional strategic frameworks for reducing FLW (FAO 2015a, b).

Countries have adopted various approaches to institutionalize national FLW frameworks and actions. The chosen indicators and actions for FLW reduction often reflect institutional priorities. For example, India's SDG national indicator framework uses per capita food availability as an indicator for food loss before and up to the retail stage, and postharvest losses in central and state government wheat and rice stockpiles as an indicator for food waste at the retail and consumer level (Government of India 2023). The African Union's Malabo Declaration commits to reducing postharvest losses by 50% by 2025 (African Union 2014). However, global coverage of country-specific FLW reduction targets aligned with SDG 12.3 remains insufficient to achieve the 2030 goal (Lipinski 2022).

Box 8.3 National-Level Food Loss and Waste Reduction Policies

FLW legislation at the national level is quite limited. In a systematic review of national-level FLW reduction policies, Shen et al. (2023) found only eight countries (Germany, Norway, the United Kingdom, United States, Argentina, South Korea, Japan, and China) that have enacted anti-FLW laws. With regard to specific laws directly targeting FLW, the authors found a total of five laws across three European countries (France, Italy, and Spain) and China, the majority of which apply to the entire food value chain and use penalties (versus encouragement) as the main policy approach (Shen et al. 2023). Most recently, legally binding targets have been proposed for EU member states for FLW reduction by 10% at processing and manufacturing stages and 30% (per capita) at retail and consumption stages by 2030 (Lipinski 2023). Some national governments (United States, Italy, Argentina, Czech Republic, and France) have laws to encourage or require food donation, either through tax incentives or liability disclaimers (Shen et al. 2023). In addition to national-level approaches, some major cities have anti-FLW initiatives, including London, Milan, Seoul, Cape Town, and several cities in the United States (Lipinski 2023).

To leverage the private sector's role in achieving national FLW targets, public–private coalitions have formed to mobilize action, build capacity, and publicize commitments. In the United States, the US Food Loss and Waste 2030 Champions, a group of 50 food businesses across retail, foodservice, processing, and man-

[1] African Union, Argentina, Australia, China, European Union, Indonesia, Japan, Malaysia, South Africa, United Arab Emirates, United Kingdom, United States, and Vietnam.

ufacturing stages, have committed to reducing FLW in their operations by 50% by 2030, with periodic public reporting (USDA 2023). Although this initiative encourages the use of the FLW Protocol and Reporting Standard to track progress, it lacks a standardized definition of FLW, leaving individual companies to define it for themselves (Hanson et al. 2016). At a more global scale, Champions 12.3 is a coalition of executives from public and private institutions that are actively involved in FLW reduction (Champions 12.3 2023). As part of their initiatives, over 10 of the world's largest food retailers and food service providers collectively operate in over 80 countries, engaged with over 200 of their suppliers to reduce FLW by half by 2030 (WRI 2020). To fully leverage the potential of public–private partnerships, a shared definition of the FLW problem is necessary.

8.4.2 The Business Case for Food Loss and Waste Reduction

While the private sector is involved in all stages of the food value chain, commitment to FLW reduction varies. As of 2021, 39 of the 50 largest global food companies had set FLW reduction targets (Lipinski 2022). Although most manufacturers and retailers have established targets, the remaining companies, primarily involved in food production and distribution, have not. Greater coordination among value chain actors is necessary to support FLW reduction goals.

A strong business case for FLW reduction is essential for influencing public and private sector decision-makers involved in food value chains. Although the potential positive impacts on food security, environmental sustainability, and economic conditions are generally recognized, the specific benefits and costs vary across contexts. For example, Hanson and Mitchell (2017) analyzed 1,000 business sites across 17 countries and found that 99% of sites in the processing and food service/retail sectors could achieve a positive return on investment from FLW reduction efforts, with a median benefit–cost ratio of 14:1. The companies with higher benefit–cost ratios were often those that had not previous implemented significant FLW reduction measures, or required only minor adjustments to best practices or minimal capital investment (Hanson and Mitchell 2017). To gain deeper insights into the costs and benefits across various FLW pathways and levels of loss, more comprehensive data on FLW across different value chain stages are needed.

The cascading effects and feedback loops of FLW can create complex financial incentives for private and public sector entities. For example, public–private interventions in the United Kingdom to reduce FLW have shown that the financial costs were shared between the private sector (manufacturers and retailers) and the public sector (national and local governments) (Britton et al. 2014; Hanson and Mitchell 2017). While both sectors benefited from these interventions, the primary financial benefits were realized by individual households (Britton et al. 2014; Hanson and Mitchell 2017).

In addition to financial benefits, nonfinancial motivations can drive FLW reduction efforts for both public and private sector entities. These motivations, often related to food security, environmental sustainability, stakeholder relationships, and social responsibility, can be particularly important for consumer-facing food companies that are seeking to differentiate their brands in a competitive market.

The underlying motivations and food system context influence the prioritization of FLW management approaches. Messner et al. (2020) described the "Prevention Paradox", in which governments and food companies publicly support FLW prevention but their actions are more focused on FLW management. For example, common strategies like diverting surplus food to donation or rescue organizations, educating consumers to reduce waste, and aiming for zero waste to landfill often fail to address the root causes of FLW (Messner et al. 2020). Diverting surplus food may not address overproduction, educating consumers overemphasizes individual responsibility, and zero waste to landfill may still result in nonfood uses.

Power dynamics between value chain actors can influence the distribution of responsibility for FLW reduction. For example, retailers may implement corporate social responsibility (CSR) initiatives to reduce FLW at the retail level, such as redirecting rejected produce to food rescue organizations. However, these initiatives may not address underlying causes of FLW and can shift the burden onto upstream actors. In their study of Australian food retailers, Devin and Richards (2018) found that retailers often redirected rejected produce to food rescue organizations before taking ownership of the food. This meant that producers, not the retailers, bore the financial burden for the rejected produce, highlighting the importance of considering the entire value chain when developing FLW reduction strategies. By leveraging efforts to build inclusive and resilient food systems, we can better understand differences in critical loss points, identify gaps in access to and adoption of technologies and information, and assess the trade-offs associated with FLW reduction strategies (Box 8.4).

Addressing FLW across the entire value chain requires sufficient data across stages, including data from private companies. As of 2021, only 19 countries,[2] representing 12% of the global population, had measured pFLW across entire food value chains (Lipinski 2022). Similarly, while 28 of the world's 50 largest food companies were measuring pFLW, only 19 publicly reported their data (Lipinski 2022). Gille (2012) suggests that risk-avoidance strategies and incentives to keep FLW hidden have contributed to this data gap. While upstream drivers of FLW are well-understood, downstream factors remain less explored. Further research is needed to identify effective intervention points and to consider the potential trade-offs between FLW reduction and inclusivity (Spang et al. 2019; Gaupp et al. 2021).

Box 8.4 Leveraging Efforts for Inclusion for More Targeted FLW Reduction

Inclusive and resilient food systems are essential for ensuring stable and equitable livelihoods among value chain actors. Food value chains involve diverse actors, offering opportunities for both inclusion and exclusion. Truly inclusive food systems empower marginalized groups to participate in value chain activities and earn fair wages that support livelihoods, and enable food security for all people (Vos and Cattaneo 2021; Gaupp et al. 2021). Addressing FLW presents a significant opportunity to enhance food system inclusivity. However, it is crucial to consider how different groups are affected by the drivers of FLW.

Vulnerable groups often face systemic and institutional constraints that contribute to FLW. In sub-Saharan Africa and lower income Asia, 25% of rural employment is tied to midstream food value chains (Vos and Cattaneo 2021). Organizational challenges, such as inadequate infrastructure for transportation, electricity, and telecommunications, can limit the ability of rural value chain actors to connect with urban markets. As demand for higher value, perishable foods grows, inclusive FLW reduction approaches can support economic development and provide additional income opportunities for rural value chain actors.

In addition to rural and smallholder value chain actors, women's status, in particular, requires closer examination as to how gender and gender equity issues are linked with FLW pathways. Women play a significant role in food value chains, contributing to both farm-level and postharvest activities, processing, and informal retail (Nordhagen 2021). However, they often face barriers related to access to resources and information, which are also drivers of FLW. Although the extent to which FLW

[2]Argentina, Australia, Canada, Colombia, Denmark, Israel, Italy, Japan, Finland, Mexico, the Netherlands, New Zealand, Norway, Saudi Arabia, Slovenia, Spain, Sweden, the United Kingdom, and the United States.

impacts women differently from men remains unclear due to data limitations, gender-sensitive approaches are essential to address these disparities. Such approaches involve mapping value chains to identify gender-specific loss points, understanding gender-based constraints, and developing targeted interventions (FAO 2018a). Addressing the root causes of gender inequality is important to ensure that FLW reduction strategies are effective, inclusive, and sustained (Farhall and Rickards 2021).

8.4.3 Limiting Environmental Impact Trade-Offs

Minimizing environmental impacts is a key objective of FLW reduction. As discussed in Chap. 3, edible and inedible food parts have distinct pathways from farm to consumer stages. Preventing FLW ensures that edible food remains suitable for human consumption and is eventually eaten. However, when edible food is diverted to nonfood uses, the specific environmental impacts vary depending on the end use. Therefore, there are trade-offs between the environmental impacts of preventing FLW and managing it. These trade-offs differ from those associated with inedible food parts, which are never intended for human consumption and follow separate pathways to nonfood uses.

Priorities for FLW reduction vary, depending on which FLW reduction outcomes are prioritized. In their study on FLW-associated greenhouse gas (GHG) emissions (accounting for edible food parts only), Guo et al. (2020) found that while vegetables and fruits accounted for nearly half of total FLW in 2017, they contributed only 16.8% to total FLW-associated GHG emissions. In contrast, bovine meat, though representing a smaller portion of total FLW (0.7%), contributed 16.3% of FLW-associated GHG emissions. The emission intensity (emissions per unit of food produced) is a useful metric to provide a standardized measure (Nash et al. 2017; Gromko and Abdurasulova 2019).

Changes in pFLW and resulting shifts in food supply and demand can have significant environmental implications. Evaluating the environmental impacts of FLW reduction interventions requires understanding adjustments to food supply and demand, identifying where along the value chain environmental impacts occur and the extent of impact, where along the value chain FLW occurs and the extent of FLW, and the potential for FLW reduction (Cattaneo et al. 2021). Additionally, the geographic scope of environmental impacts can vary, ranging from localized effects (for example, land and water use) to global impacts (for example, GHG emissions). When leveraging environmental objectives to promote FLW reduction, it is essential to consider current and future food needs to ensure that efforts to reduce FLW do not compromise food security. This can be assessed by measuring resource use efficiency to compare the food consumed or lost to different environmental impacts of food produced (Gatto and Chepeliev 2024).

8.5 Conclusions

To effectively reduce FLW, tailored interventions must address critical loss points within specific food system contexts. Understanding the complex linkages of economic, social, and environmental factors driving FLW is essential. This chapter explores approaches for identifying critical loss points, analyzing underlying causes, and leveraging opportunities to improve food security, environmental sustainability, and economic efficiency.

Identifying critical loss points is essential for developing effective FLW reduction strategies. Careful consideration of FLW indicators, including how pFLW is measured and converted into nutritional or environmental units, is important for understanding the relationship between FLW and nutrition and environmental outcomes. Disaggregating FLW by food commodity group can help identify these linkages and improve cross-regional comparisons. Both absolute values and percentages of pFLW are necessary to monitor trends and assess impacts. However, significant data gaps persist, particularly, across regions, commodities, and value chain stages, hindering our ability to accurately identify critical loss points.

Developing FLW reduction strategies also requires comprehensive understanding of FLW drivers that act on food value chains at different levels. Although technologies and best practices can play a role in reducing FLW, their effectiveness is influenced by several factors, including access, affordability, and appropriate implementation. Inadequate transportation, storage, and communication infrastructure, particularly, in rural areas, can hinder the efficient movement and storage of food products or the adoption and implementation of technologies and best practices. Public and private regulations on food quality standards can affect FLW pathways differently, depending on which quality attributes serve as limiting attributes. Regulations can also alter power dynamics within food value chains. Finally, affecting all food systems is the role of climate change and how increased temperatures, altered rainfall patterns, and extreme weather events contribute to FQL and pFLW.

While the SDG 12.3 target has prompted national governments to commit to some FLW targets, the private sector will also need to be leveraged for achieving global FLW reduction goals. Therefore, leveraging the business case to incentivize public and private sector investments will require comprehensive assessments of potential cascading effects and feedback loops associated with different FLW reduction approaches and intended food security, environmental, and economic outcomes.

References

African Union (2014) Malabo declaration on accelerated agricultural growth and transformation for shared prosperity and improved livelihoods. https://www.resakss.org/sites/default/files/Malabo%20Declaration%20on%20Agriculture_2014_11%2026-.pdf. Accessed 5 Dec 2024

Akintola SL, Fakoya KA, Elegbede IO, Odunsi Odunayo E, Jolaosho Lekan T (2022) Postharvest practices in small-scale fisheries. In: Galanakis CM (ed) Sustainable fish production and processing. Academic Press, London, pp 79–110. https://doi.org/10.1016/B978-0-12-824296-4.00008-6. Accessed 30 Aug 2023

Aschemann-Witzel J, Jensen JH, Jensen MH, Kulikovskaja V (2017) Consumer behaviour towards price-reduced suboptimal foods in the supermarket and the relation to food waste in households. Appetite 116:246–258. https://doi.org/10.1016/j.appet.2017.05.013. Accessed 26 Feb 2024

Barbut S (2020) Meat industry 4.0: a distant future? Anim Front 10(4):38–47. https://doi.org/10.1093/af/vfaa038. Accessed 24 Jan 2024

Barbut S, Leishman EM (2022) Quality and processability of modern poultry meat. Animals 12(20):2766. https://doi.org/10.3390/ani12202766. Accessed 15 Dec 2023

Beausang C, Hall C, Toma L (2017) Food waste and losses in primary production: qualitative insights from horticulture. Resour Conserv Recycl 126:177–185. https://doi.org/10.1016/j.resconrec.2017.07.042. Accessed 8 Sep 2021

Boiteau JM, Pingali P (2022) Food loss of perishable produce from farm to retail: evidence from tomato supply chains in South India. Am J Clin Nutr 115(6):1535–1548. https://doi.org/10.1093/ajcn/nqac039. Accessed 8 Sep 2021

Britton E, Brigdon A, Parry A, LeRoux S (2014) Econometric modelling and household food waste. Waste & Resources Action Programme (WRAP), Oxford. https://www.wrap.ngo/sites/default/files/2020-12/Econometric-modelling-and-household-food-waste-2014.pdf. Accessed 5 Dec 2024

Cadilhon J-J, Moustier P, Poole ND, Tam PTG, Fearne AP (2006) Traditional vs. modern food systems? Insights from vegetable supply chains to Ho Chi Minh City (Vietnam). Dev Policy Rev 24(1):31–49. https://doi.org/10.1111/j.1467-7679.2006.00312.x. Accessed 4 Oct 2023

Cattaneo A, Federighi G, Vaz S (2021) The environmental impact of reducing food loss and waste: a critical assessment. Food Policy 98:101890. https://doi.org/10.1016/j.foodpol.2020.101890. Accessed 24 Aug 2021

Champions 12.3 (2023) Champions 12.3. https://champions123.org/. Accessed 4 Mar 2023

Croad T, Campbell H, Mirosa M (2024) Investigating systemic and social dynamics of food loss and waste: an application of waste regime theory to food production in Aotearoa New Zealand. Clean Waste Syst 7:100125. https://doi.org/10.1016/j.clwas.2023.100125. Accessed 19 Dec 2023

De Hooge IE, Van Dulm E, Van Trijp HCM (2018) Cosmetic specifications in the food waste issue: supply chain considerations and practices concerning suboptimal food products. J Clean Prod 183:698–709. https://doi.org/10.1016/j.jclepro.2018.02.132. Accessed 8 Mar 2024

Devin B, Richards C (2018) Food waste, power, and corporate social responsibility in the Australian food supply chain. J Bus Ethics 150:199–210. https://doi.org/10.1007/s10551-016-3181-z. Accessed 5 Mar 2024

Eičaitė O, Baležentis T, Ribašauskienė E, Morkūnas M, Melnikienė R, Štreimikienė D (2023) Exploring food loss in the food industry: a survey-based approach for Lithuania. J Clean Prod 399:136668. https://doi.org/10.1016/j.jclepro.2023.136668. Accessed 8 Mar 2023

Fabi C, English A (2019) Methodological proposal for monitoring SDG target 12.3. Sub-indicator 12.3.1.a the food loss index design, data collection methods and challenges. Working paper ESS/18-13. FAO, Rome. https://openknowledge.fao.org/server/api/core/bitstreams/6603bc2f-7105-434b-baaa-f440e3505a88/content. Accessed 5 Dec 2024

Fafchamps M, Hill RV, Minten B (2008) Quality control in nonstaple food markets: evidence from India. Agric Econ 38(3):251–266. https://doi.org/10.1111/j.1574--0862.2008.00297.x. Accessed 8 Sep 2021

FAO (Food and Agriculture Organization of the United Nations) (2011) Global food losses and food waste—extent, causes and prevention, Rome. https://www.fao.org/4/mb060e/mb060e.pdf. Accessed 5 Dec 2024

FAO (Food and Agriculture Organization of the United Nations) (2015a) Food losses and waste in Latin America and the Caribbean. Bulletin no. 2, April, FAO Regional Office for Latin American and the Caribbean. https://openknowledge.fao.org/server/api/core/bitstreams/daa2a3ee-28b0-4b42-8a14-d9a993c26312/content. Accessed 4 Dec 2024

FAO (Food and Agriculture Organization of the United Nations) (2015b) Regional strategic framework reducing food losses and waste in the near East & North Africa Region, Cairo. https://openknowledge.fao.org/server/api/core/bitstreams/4c311e4a-e1df-4be8-8da6-e3ea4d36a085/content. Accessed 5 Dec 2024

FAO (Food and Agriculture Organization of the United Nations) (2018a) Gender and food loss in sustainable food value chains: a guiding note, Rome. https://openknowledge.fao.org/server/api/core/bitstreams/0e730b95-9ead-4ced-9fd9-23cdcf01a189/content. Accessed 5 Dec 2024

FAO (Food and Agriculture Organization of the United Nations) (2018b) The state of agricultural commodity markets 2018. Agricultural trade, climate change and food security. FAO, Rome. https://openknowledge.fao.org/handle/20.500.14283/i9542en. Accessed 4 Mar 2022

FAO (Food and Agriculture Organization of the United Nations) (2024) FAOLEX database. https://www.fao.org/faolex/en/. Accessed 9 Mar 2024

Farhall K, Rickards L (2021) The "Gender Agenda" in agriculture for development and its (lack of) alignment with feminist scholarship. Front Sustain Food Syst 5:573424. https://www.frontiersin.org/journals/sustainable-food-systems/articles/10.3389/fsufs.2021.573424/full. Accessed 5 Dec 2024

Feyaerts H, Van Den Broeck G, Maertens M (2020) Global and local food value chains in Africa: a review. Agric Econ 51(1):143–157. https://doi.org/10.1111/agec.12546. Accessed 24 Oct 2024

Gatto A, Chepeliev M (2024) Global food loss and waste estimates show increasing nutritional and environmental pressures. Nat Food 5:136–147. https://doi.org/10.1038/s43016-023-00915-6. Accessed 5 Feb 2024

Gaupp F, Ruggeri Laderchi C, Lotze-Campen H, DeClerck F, Bodirsky BL, Lowder S, Popp A, Kanbur R, Edenhofer O, Nugent R, Fanzo J, Dietz S, Nordhagen S, Fan S (2021) Food system development pathways for healthy, nature-positive and inclusive food systems. Nat Food 2:928–934. https://doi.org/10.1038/s43016-021-00421-7. Accessed 21 Mar 2022

Gille Z (2012) From risk to waste: global food waste regimes. Sociol Rev 60(2 Suppl):27–46. https://doi.org/10.1111/1467-954X.12036. Accessed 19 Dec 2023

Gillman A, Campbell DC, Spang ES (2019) Does on-farm food loss prevent waste? Insights from California produce growers. Resour Conserv Recycl 150:104408. https://doi.org/10.1016/j.resconrec.2019.104408. Accessed 8 Sep 2021

Government of India (2023) Sustainable development goals national indicator framework 2023. Ministry of Statistics and Programme Implementation. https://www.mospi.gov.in/sites/default/files/publication_reports/SDGs_NIF_Progress_Report_2023N_0.pdf. Accessed 4 Dec 2024

Gromko D, Abdurasulova G (2019) Climate change mitigation and food loss and waste reduction: exploring the business case. CCAFS report no. 18. CGIAR Research Program on Climate Change, Agriculture and Food Security (CCAFS), Wageningen. https://cgspace.cgiar.org/items/a937cb13-4d7f-4e95-af72-208a7560a04c. Accessed 5 Dec 2024

Guo X, Broeze J, Groot JJ, Axmann H, Vollebregt M (2020) A worldwide hotspot analysis on food loss and waste, associated greenhouse gas emissions, and protein losses. Sustainability 12(18):7488. https://doi.org/10.3390/su12187488. Accessed 18 Dec 2023

Hanson C, Mitchell P (2017) The business case for reducing food loss and waste. Champions 12.3. https://champions123.org/sites/default/files/2020-08/business-case-for-reducing-food-loss-and-waste.pdf. Accessed 5 Dec 2024

Hanson C, Lipinski B, Robertson K, Dias D, Gavilan I, Gréverath P, Ritter S, Fonseca J, van Otterdijk R, Timmermans T, Lomax J, O'Connor C, Dawe A, Swannell R, Berger V, Reddy M, Somogyi D, Tran B, Leach B, Quested T (2016) Food loss and waste accounting and reporting standard. World Resources Institute, Washington, DC. https://flwprotocol.org/wp-content/uploads/2017/05/FLW_Standard_final_2016.pdf. Accessed 5 Dec 2024

Hasan MU, Malik AU, Ali S, Imtiaz A, Munir A, Amjad W, Anwar R (2019) Modern drying techniques in fruits and vegetables to overcome postharvest losses:

a review. J Food Process Preserv 43. https://doi.org/10.1111/jfpp.14280. Accessed 16 Feb 2023

HLPE (2014) Food losses and waste in the context of sustainable food systems. A report by the high level panel of experts on food security and nutrition of the committee on world food security, Rome. https://openknowledge.fao.org/server/api/core/bitstreams/b1949fae-23d4-473c-8b87-8c4359b74d6c/content. Accessed 5 Dec 2024

HLPE (2017) Nutrition and food systems. A report by the high level panel of experts on food security and nutrition of the committee on world food security, HLPE report 12. Rome. http://www.fao.org/3/a-i7846e.pdf. Accessed 5 Dec 2024

Hoffmann V, Moser C, Saak A (2019) Food safety in low and middle-income countries: the evidence through an economic lens. World Dev 123:104611. https://doi.org/10.1016/j.worlddev.2019.104611. Accessed 29 Mar 2022

Kader AA, Rolle RS (2004) The role of post-harvest management in assuring the quality and safety of horticultural produce, FAO Agricultural Services Bulletin 152. Food and Agriculture Organization of the United Nations, Rome. https://openknowledge.fao.org/server/api/core/bitstreams/2a68415b-9a83-418a-8c72-d40815477cf2/content. Accessed 5 Dec 2024

Kitinoja L, Kader AA (2015) Small-scale postharvest handling practices: a manual for horticultural crops, 5th edn. Postharvest horticulture series no. 8E. Davis Postharvest Technology Research and Information Center, University of California. https://ucanr.edu/sites/Postharvest_Technology_Center_/files/231952.pdf. Accessed 5 Dec 2024

Krishna Bahadur KC, Haque I, Legwegoh AF, Fraser EDG (2016) Strategies to reduce food loss in the global south. Sustainability 8(7):595. https://doi.org/10.3390/su8070595. Accessed 1 Sep 2021

Lipinski B (2022) SDG target 12.3 on food loss and waste: 2022 progress report. An annual update on behalf of Champions 12.3. https://champions123.org/sites/default/files/2022-09/22_WP_SDG%20Target%2012.3_2022%20Progress%20Report_v3_0.pdf. Accessed 5 Dec 2024

Lipinski B (2023) SDG target 12.3 on food loss and waste: 2023 progress report. An annual update on behalf of Champions 12.3. https://champions123.org/sites/default/files/2023-10/2023%20Champions%20Progress%20Report.pdf. Accessed 5 Dec 2024

Lipinski B, Clowes A, Goodwin L, Hanson C, Swannell R, Mitchell P (2017) SDG target 12.3 on food loss and waste: 2017 progress report. An annual update on behalf of Champions 12.3. https://www.rockefellerfoundation.org/wp-content/uploads/SDG-Target-12.3-on-Food-Loss-and-Waste_2017-Progress-Report.pdf. Accessed 5 Dec 2024

Machala ML, Tan FL, Poletayev A, Khan MI, Benson SM (2022) Overcoming barriers to solar dryer adoption and the promise of multi-seasonal use in India. Energy Sustain Dev 68:18–28. https://doi.org/10.1016/j.esd.2022.02.001. Accessed 3 Apr 2023

Messner R, Richards C, Johnson H (2020) The "prevention paradox": food waste prevention and the quandary of systemic surplus production. Agric Hum Values 37:805–817. https://doi.org/10.1007/s10460-019-10014-7. Accessed 18 Oct 2022

Messner R, Johnson H, Richards C (2021) From surplus-to-waste: a study of systemic overproduction, surplus and food waste in horticultural supply chains. J Clean Prod 278:123952. https://doi.org/10.1016/j.jclepro.2020.123952. Accessed 12 Nov 2023

Nash J, Peña O, Galford G, Gurwick N, Pirolli G, White J, Wollenberg E (2017) Reducing food loss in agricultural development projects through value chain efficiency. CCAFS working paper no. 204. CGIAR Research Program on Climate Change, Agriculture and Food Security (CCAFS), Wageningen. https://cgspace.cgiar.org/items/4801a3d2-8799-4901-b62a-1084b7a4bb54. Accessed 5 Dec 2024

Nordhagen S (2021) Gender equity and reduction of post-harvest losses in agricultural value chains. Global Alliance for improved nutrition (GAIN) working paper #20, Geneva. https://doi.org/10.36072/wp.20. Accessed 18 Aug 2023

Parfitt J, Croker T, Brockhaus A (2021) Global food loss and waste in primary production: a reassessment of its scale and significance. Sustainability 13(21):12087. https://doi.org/10.3390/su132112087. Accessed 8 Aug 2023

Perdana T, Kusnandar K, Perdana HH, Hermiatin FR (2023) Circular supply chain governance for sustainable fresh agricultural products: minimizing food loss and utilizing agricultural waste. Sustain Prod Consum 41:391–403. https://doi.org/10.1016/j.spc.2023.09.001. Accessed 24 Oct 2023

Pietrangeli R, Cicatiello C (2024) Lost vegetables, lost value: assessment of carrot downgrading and losses at a large producer organisation. J Clean Prod 478:143873. https://doi.org/10.1016/j.jclepro.2024.143873. Accessed 24 Oct 2024

Porter SD, Reay DS, Bomberg E, Higgins P (2018) Avoidable food losses and associated production-phase greenhouse gas emissions arising from application of cosmetic standards to fresh fruit and vegetables in Europe and the UK. J Clean Prod 201:869–878. https://doi.org/10.1016/j.jclepro.2018.08.079. Accessed 11 Nov 2021

Raak N, Symmank C, Zahn S, Aschemann-Witzel J, Rohm H (2017) Processing- and product-related causes for food waste and implications for the food supply chain. Waste Manag 61:461–472. https://doi.org/10.1016/j.wasman.2016.12.027. Accessed 9 Mar 2023

Redlingshöfer B, Coudurier B, Georget M (2017) Quantifying food loss during primary production and processing in France. J Clean Prod 164:703–714. https://doi.org/10.1016/j.jclepro.2017.06.173. Accessed 15 Jan 2024

Rindt J, Mouzas S (2015) Exercising power in asymmetric relationships: the use of private rules. Ind Mark Manag 48:202–213. https://doi.org/10.1016/j.indmarman.2015.03.018. Accessed 13 Feb 2024

Roesel K, Grace D (eds) (2015) Food safety and informal markets: animal products in sub-Saharan Africa. Routledge, London

Rösler F, Kreyenschmidt J, Ritter G (2021) Recommendation of good practice in the food-processing industry for preventing and handling food loss and waste. Sustainability 13(17):9569. https://doi.org/10.3390/su13179569. Accessed 9 Mar 2023

Schneider KR, Fanzo J, Haddad L, Herrero M, Moncayo JR, Herforth A, Remans R, Guarin A, Resnick D, Covic N, Béné C, Cattaneo A, Aburto N, Ambikapathi R, Aytekin D, Barquera S, Battersby J, Beal T, Molina PB, Cafiero C, Campeau C, Caron P, Conforti P, Damerau K, Di Girolamo M, DeClerck F, Dewi D, Elouafi I, Fabi C, Foley P, Frazier TJ, Gephart J, Golden C, Fischer CG, Hendriks S, Honorati M, Huang J, Kennedy G, Laar A, Lal R, Lidder P, Loken B, Marshall Q, Masuda YJ, McLaren R, Miachon L, Muñoz H, Nordhagen S, Qayyum N, Saisana M, Suhardiman D, Sumaila UR, Cullen MT, Tubiello FN, Vivero-Pol J-L, Webb P, Wiebe K (2023) The state of food systems worldwide in the countdown to 2030. Nat Food 4:1090–1110. https://doi.org/10.1038/s43016-023-00885-9. Accessed 16 Jan 2024

Schreinemachers P, Ambali M, Mwambi M, Olanipekun CI, Yegbemey RN, Wopereis MCS (2022) The dynamics of Africa's fruit and vegetable processing sectors. In: Jenane C, Ulimwegu JC, Tadesse G (eds) 2022 ReSAKSS annual trends and outlook report: agrifood processing strategies for successful food systems transformation in Africa. Regional strategic analysis and knowledge support system, AKADEMIYA2063, Kigali, Rwanda, pp 23–32. https://ebrary.ifpri.org/digital/collection/p15738coll2/id/136416. Accessed 25 Nov 2024

Shen G, Li Z, Hong T, Ru X, Wang K, Gu Y, Han J, Guo Y (2023) The status of the global food waste mitigation policies: experience and inspiration for China. Environ Dev Sustain 26:8329–8357. https://doi.org/10.1007/s10668-023-03132-0. Accessed 19 Feb 2024

Soethoudt JM, Pedrotti M, Bos-Brouwers HEJ, Castelein RB (2021) Adoption of food loss and waste-reducing interventions in low- and middle-income countries. Wageningen Food & Biobased Research, Wageningen

Soma T, Kozhikode R, Krishnan R (2021) Tilling food under: barriers and opportunities to address the loss of edible food at the farm-level in British Columbia. Can Resour Conserv Recycl 170:105571. https://doi.org/10.1016/j.resconrec.2021.105571. Accessed 8 Feb 2024

Spang ES, Moreno LC, Pace SA, Achmon Y, Donis-Gonzalez I, Gosliner WA, Jablonski-Sheffield MP, Momin MA, Quested TE, Winans KS, Tomich TP (2019) Food loss and waste: measurement, drivers, and solutions. Annu Rev Environ Resour 44:117–156. https://doi.org/10.1146/annurev-environ-101718-033228. Accessed 9 Jan 2023

Spiker ML, Hiza HAB, Siddiqi SM, Neff RA (2017) Wasted food, wasted nutrients: nutrient loss from wasted food in the United States and comparison to gaps in dietary intake. J Acad Nutr Diet 117(7):1031–1040.e22

Subramanian R (2016) India processing tomato segment: current status, trends and opportunities for engagement. World Vegetable Center, no. 16-806. https://avrdc.org/download/publications/technical-reports/reports/GIZ_India-Processed-Tomato-Study_16Sept2016.pdf. Accessed 5 Dec 2024

Surucu-Balci E, Tuna O (2021) Investigating logistics-related food loss drivers: a study on fresh fruit and vegetable supply chain. J Clean Prod 318:128561. https://doi.org/10.1016/j.jclepro.2021.128561. Accessed 13 Apr 2023

UN (United Nations) (2015) A/RES/70/1 transforming our world: the 2030 agenda for sustainable development. https://www.un.org/en/development/desa/population/migration/generalassembly/docs/globalcompact/A_RES_70_1_E.pdf. Accessed 5 Dec 2024

Unnevehr L, Ronchi L (2014) Food safety and developing markets: research findings and research gaps. IFPRI discussion paper 1376. https://cgspace.cgiar.org/server/api/core/bitstreams/40539cd0-79b9-4ef9-885c-d395ad2964b9/content. Accessed 30 Mar 2022

USDA (United States Department of Agriculture) (2023) U.S. food loss and waste 2030 champions. https://www.usda.gov/foodlossandwaste/champions. Accessed 4 Mar 2023

van den Bos Verma M, de Vreede L, Achterbosch T, Rutten MM (2020) Consumers discard a lot more food than widely believed: estimates of global food waste using an energy gap approach and affluence elasticity of food waste. PLoS One 15(2):e0228369. https://doi.org/10.1371/journal.pone.0228369. Accessed 5 Feb 2024

Vandeplas A, Minten B (2015) Food quality in domestic markets of developing economies: a comparative study of two countries. Agric Econ 46(5):617–628. https://doi.org/10.1111/agec.12158. Accessed 21 Oct 2024

Vos R, Cattaneo A (2021) Poverty reduction through the development of inclusive food value chains. J Integr Agr 20(4):964–978. https://doi.org/10.1016/S2095-3119(20)63398-6. Accessed 24 Oct 2024

Walsh K (2022) Postharvest regulation and quality standards on fresh produce. In: Florkowski WJ, Shewfelt RL, Prussia SE, Banks NH (eds) Postharvest handling: a systems approach. Academic Press, Amsterdam, pp 51–98

Webb P, Shrestha R (2023) Innovations to reduce food loss and waste, promote food safety and access to nutrient-rich foods. Feed the future: food systems for nutrition—innovation (FSN-IL). https://foodsystemsnutrition.org/wp-content/uploads/2023/02/FSN-IL-Bangladesh-Introductory-Webinar-January-29-2023.pdf. Accessed 24 Oct 2024

Wilson NLW, Rickard BJ, Saputo R, Ho S-T (2017) Food waste: the role of date labels, package size, and product category. Food Qual Prefer 55:35–44. https://doi.org/10.1016/j.foodqual.2016.08.004. Accessed 14 Jul 2022

WRI (World Resources Institute) (2020) World's leading food retailers and providers engage nearly 200 suppliers in cutting food loss and waste in half. https://champions123.org/release-worlds-leading-food-retailers-and-providers-engage-nearly-200-suppliers-cutting-food-loss. Accessed 4 Mar 2024

Xue L, Liu G, Parfitt J, Liu X, Van Herpen E, Stenmarck Å, O'Connor C, Östergren K, Cheng S (2017) Missing food, missing data? A critical review of global food losses and food waste data. Environ Sci Technol 51(12):6618–6633. https://doi.org/10.1021/acs.est.7b00401. Accessed 4 Sep 2021

9 Navigating the Complexities: A Deeper Dive into Food Loss and Waste Interventions

9.1 Introduction

Taking action to address food loss and waste (FLW) requires careful selection of specific interventions that are appropriate for the reduction target and food system context. Also important is how intervention success is defined. As discussed throughout this book, FLW reduction is not the ultimate goal. Rather, addressing FLW issues is intended to contribute to sustainable food system transformations that improve food security, reduce environmental impacts, and support livelihoods.

Interventions to prevent FLW can be categorized based on their level of implementation, the involvement of food system actors, and the specific intervention elements (HLPE 2014; Soethoudt et al. 2021). Higher level actions, such as supportive policies and incentives, can create enabling environments to facilitate intervention implementation (HLPE 2014). The involvement and impact of different value chain actors vary depending on the specific intervention and the broader food system context. As food systems transform, the characteristics of value chain actors and their roles in FLW mitigation may change. This chapter summarizes approaches to mitigate FLW, reviews the types of interventions, and examines the evidence on FLW interventions in low- and middle-income countries (LMICs), including their objectives and evaluation methods.

Considering the leverage points and investment priorities, discussed in Chap. 8, we explore the criteria for intervention adoption in which value chain actors are supported and motivated to participate in FLW reduction. Critically, properly matching interventions to food system contexts requires the recognition of the diversity of value chain actor characteristics and their goals (Vernooij et al. 2022). To do so requires, in part, making the "business case" to motivate value chain actors whose livelihoods depend on how the value chain functions. In this chapter, we explore how a value chain approach can be used to design and assess FLW interventions, considering the costs and benefits to different value chain actors and prioritizing inclusive approaches to FLW mitigation.

For FLW interventions to succeed and contribute to healthy, sustainable diets, they must be context-specific, positively impact food security, and last beyond external support. Although FLW is an important indicator for assessing food system sustainability, data limitations hinder its inclusion in current frameworks. This chapter highlights the interconnectedness of FLW with different aspects of sustainable development, extending beyond the traditional focus on economic and environmental impacts.

J. Boiteau, P. Pingali, *Wasted Potential*, Sustainable Development Goals Series,
https://doi.org/10.1007/978-3-031-97411-3_9

9.2 Food Loss and Waste Interventions and Objectives

This book has explored the complex factors contributing to FLW across diverse food value chains and systems. To effectively address FLW, is it critical to design and evaluate interventions that consider the interconnectedness of food systems, including cascading effects and trade-offs. This design requires a two-pronged approach: (1) direct interventions focused on preventing or managing FLW, and (2) broader systemic interventions that address underlying inefficiencies and have FLW reduction as a secondary benefit. Cattaneo et al. (2021) referred to these as first-order and second-order interventions, respectively.

The overarching goal of FLW interventions is to enhance food system sustainability outcomes across multiple dimensions: nutrition and health (including food security), economic and social well-being, and environmental impact. FLW prevention aims to maintain food quality and reduce the quantity of food deemed unacceptable and removed from food value chains. FLW management aims to minimize the negative impacts of physical food loss and waste (pFLW), particularly, environmental consequences. The effectiveness and availability of prevention strategies relative to management strategies can influence both the choice of approach and potential trade-offs between different food system outcomes (Rolker et al. 2022).

Prevention interventions can be categorized into two general approaches: minimizing food surplus and maximizing the recovery of surplus food for human consumption (De Laurentiis et al. 2020). ReFED (2021) defined "surplus food" as food intended for human consumption that is not sold or eaten. To avoid overlap with the definition of pFLW (food intended for human consumption that is never eaten), we refine the definition of surplus food as food that exceeds demand and goes unsold or uneaten within a specific value chain. This distinction allows us to differentiate between strategies to reduce surplus in marketable value chains and those focused on recovering food for consumption through alternative channels.

Whether demand is met or there is a shortfall depends not only on the quantity of food produced and moved through the value chain but also on the quality attributes of the products relative to the quality attributes that are in demand. The reasons for food surplus are often linked to market factors, such as production levels, limiting quality attributes, and processing by-products (ReFED 2021). Strategies to recover surplus food often involve redirection to donation channels, which face their own challenges (Box 9.1). However, the costs associated with food donation may deter value chain actors from redirecting surplus food to these food channels, especially, if the perceived benefits, such as reputational gains or potential tax incentives, are not significant enough to outweigh the costs (Chauhan et al. 2021). Focusing on the balance between supply and demand, as well as the root causes of surplus generation, can lead to more targeted prevention strategies that account for outcome trade-offs and feedback loops.

Box 9.1 Challenges with Food Donation to Support Access to Nutritious Foods

Food redistribution, a strategy to prevent pFLW, involves recovering food from the value chain and redirecting it to people at risk of poverty or social exclusion, often through food banks. Although this approach addresses food security issues, it is important to note that food redistribution channels face management challenges that can introduce potential points of food loss. Food quality may decline, and food may be removed from the value chain as pFLW before reaching consumers or before being eaten. Evaluating the trade-offs between different FLW prevention strategies requires a comprehensive understanding of value chain challenges and the entire FLW pathway.

Akkerman et al. (2023) identified key challenges in food bank supply chain management, including supply-side issues like insufficient and irregular food donations, nutrition quality of donated food, competing uses for surplus food, and donor concerns. Demand-side challenges include reaching beneficiaries, accommodating diverse dietary needs, and addressing the impact of global shocks. Matching supply and demand is further complicated by factors, such as food perishability, food safety, volunteer labor, technology, logistics, and coordination (Akkerman et al. 2023). Although this review focused on three countries with modern food systems, the identified challenges are relevant to FLW pathways we have discussed in earlier chapters.

Additionally, there are concerns about whether food redistribution through donation is an effective strategy to support vulnerable consumers without exacerbating social and economic inequalities. Some have argued that food donation may simply be a mechanism for economic and material efficiency rather than a genuine social benefit (Midgley 2014). Therefore, efforts to develop sustainable and inclusive food systems must also consider broader questions of food access, including affordability and alternative mechanisms like food donation, to address social implications.

A common oversight in the discussion of FLW management interventions is the failure to differentiate between edible and inedible food parts (see Chap. 3). Inedible parts are inherently destined for nonfood uses and, therefore, always require appropriate management to minimize negative environmental and economic impacts. In contrast, diverting edible food parts away from human consumption represents pFLW, which has nutrition and health impacts, in addition to environmental and economic impacts. Therefore, comprehensive assessments of management interventions should consider the distinct trade-offs associated with edible food parts (for example, prevention versus management, different management approaches) and inedible parts (for example, different management approaches). See Box 9.2 for a more detailed discussion on upcycling and valorizing strategies.

Box 9.2 Upcycling and Valorizing Food

Food upcycling and valorizing represent two divergent market-based strategies for addressing FLW. Following the definition of "upcycled food" presented in Chap. 3, upcycling food leverages technology and innovative practices to transform food parts, which would otherwise be wasted, into new food products. These upcycled products are then reintroduced to the market for consumer purchase.

In some cases, inedible plant parts are upcycled into edible food products. However, according to the FLW definitional framework used in this book, this approach does not directly reduce FLW. Inedible plant parts, by definition, are not intended for human consumption, and thus, do not fall within the scope of FLW. Once these parts are transformed into edible food products, they enter the FLW pathway, where they can either be consumed or become pFLW.

Some technologies and practices focus on valorizing food removed from the food value chain (pFLW) for alternative uses, such as animal feed, compost, and energy. While valorization recovers resources, effectively addressing FLW requires weighing the trade-offs between preventing FLW and utilizing pFLW for nonfood uses. This process stands in contrast to the management of inedible food parts that are inherently not used as food. The specific food product, market conditions, and reasons for discarding the food will influence the optimal approach to achieving specific food systems outcomes.

For example, Latka et al. (2022) examined the interdependencies between reducing consumer pFLW and valorizing plant-based food waste as pig feed within the context of the European Union (EU). Their study highlights the trade-offs between these two approaches. While reducing consumer food waste can lead to significant greenhouse gas (GHG) emission savings, compared to valorization, valorizing food waste as animal feed also offers benefits, depending on the market context. Therefore, the choice between these strategies depends on factors, such as the relative cost of animal feed sources and potential impacts on producer incomes. As the EU strives for a more sustainable food system, understanding these trade-offs is crucial for prioritizing effective interventions.

Individual FLW interventions broadly fall into three groupings: technology-based, practice-based, or those that address elements of enabling environments (Table 9.1). Innovators develop technologies, including physical tools, equipment, machinery, or materials, that are distributed by intermediaries and adopted by value chain actors at specific stages. Practices grounded in knowledge sharing include training programs, capacity building, peer-to-peer learning, and participatory approaches to adopt new methods, modify behaviors, and increase awareness of resources and benefits available. Individual decisions to adopt technologies and practices depend on coordination across the value chain and broader food system, which support value chain actors' FLW reduction objectives. Enabling environments, including finance and investment, organization, economics, and policy, facilitate the implementation of technology-based and practice-based interventions at the meso- and macro-levels. Leveraging a socio-technical innovation bundling approach alongside strengthening the enabling environment will be necessary for effectively targeting FLW objectives (Barrett et al. 2020).

9.3 Assessing Intervention Success

As discussed in Chap. 8, determining an optimal level of FLW is complex due to the intricate and sometimes competing relationships between FLW and food system outcomes. Evaluating the success of FLW interventions is similarly challenging, requiring assessment of both FLW reduction (food quality loss [FQL] and/or pFLW) and broader food system impacts. The interpretation of success depends on the specific outcomes selected, the intervention's scale, and the level at which impacts are measured.

Two recent systematic reviews have synthesized the evidence of individual evaluations of FLW interventions (Stathers et al. 2020; Rolker

Table 9.1 Categories of interventions to prevent or manage food loss and waste

Intervention category	Intervention aspects	Intervention types
Technology	• Innovation • Distribution • Adoption and implementation	• Physical tool/equipment • Machine • Material
Practices	• Knowledge sharing • Capacity building • Adoption of best practices/methods • Behavior change	• Training programs • Peer-to-peer learning • Workshops • Education/awareness campaigns
Enabling environment	• Finance and investment • Organization • Economics • Policy	• Infrastructure • Grades and standards • Market demand and access • Contracts and agreements • Finance/credit access

Adapted from De Laurentiis et al. (2020), Soethoudt et al. (2021) and Rolker et al. (2022)
Note: The intervention types are not an exhaustive list, but are those most often reported in the literature

et al. 2022). While Stathers et al. (2020) examined all food groups and Rolker et al. (2022) focused on perishable foods, both reviews found that FLW interventions predominantly rely on technology-based approaches to reduce postharvest loss and prioritize economic and environmental impacts.

Of the 334 studies (comprising 1565 interventions) that Stathers et al. (2020) included in their review, 175 studies (780 interventions) specifically targeted perishable root and tuber crops, fruits, and vegetables. Within this subset, the majority of interventions (696) relied on technological, tool-based, or equipment-based approaches. A smaller proportion focused on modifying handling practices (82 interventions), while a minimal number addressed road infrastructure improvements (2 interventions) (Stathers et al. 2020). When examining FLW measurements within these 780 perishable food interventions, pFLW was assessed in 60% (468 interventions), and FQL was evaluated in 73% (568 interventions) (Stathers et al. 2020). Notably, this emphasis on food quality assessment in intervention studies contrasts with FLW assessment studies, which predominantly focus on pFLW and often overlook food quality indicators (Kitinoja et al. 2018).

Only about 13% of the 334 total studies considered economic, social, or environmental outcomes, with the majority focusing solely on economic impacts, particularly, for durable food commodities (Stathers et al. 2020). Notably, none of the studies addressed gender-specific outcomes, highlighting the critical evidence gap in understanding how gender considerations can be integrated into the design, implementation, and evaluation of FLW reduction strategies and interventions (FAO 2018). While only five studies investigated factors influencing the adoption of FLW reduction interventions for cereals, many others proposed potential barriers and facilitators without providing empirical evidence (Stathers et al. 2020).

Recognizing the historical emphasis on FLW interventions targeting cereals, Rolker et al. (2022) aimed to fill knowledge gaps for perishable foods. Their analysis of 88 studies primarily examined interventions aimed at preventing or managing pFLW in fruits, vegetables (including roots and tubers), animal-source foods (ASF), and mixed food products, considering when food value chain actors would be involved in the intervention. The 59 preventive interventions tended to target earlier value chain stages, from production to processing, while 29 interventions focused on managing existing pFLW were applied at later stages, such as at wholesale, retail, and consumption stages (Fig. 9.1).

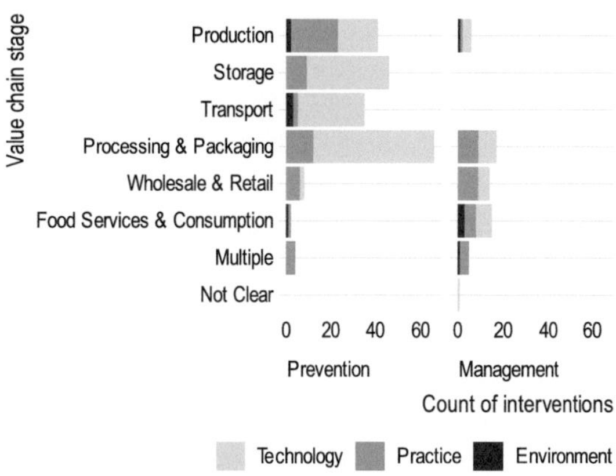

Fig. 9.1 Food loss and waste interventions by value chain stage, type, and aim of intervention

Note: Studies often included multiple interventions. Technology includes machines and materials; practices include methods and people; and environment represents factors for an enabling environment

Data from Rolker et al. (2022)

Similar to Stathers et al. (2020), Rolker et al. (2022) found that preventive interventions focused on technology-based solutions, compared to practice-based approaches (for example, method changes, training), or interventions addressing enabling environmental factors. In contrast, management interventions tended to be more balanced in the application of technology-based and practice-based approaches. No study examined both preventive and management interventions, highlighting a missed opportunity to explore complementary strategies (Rolker et al. 2022).

Preventive interventions primarily focused on economic and social well-being, while management interventions considered economic, social, and environmental impacts (Rolker et al. 2022). Few studies examined nutrition and health impacts. However, lack of sufficient data was also a limitation across studies for comparing FLW outcomes between intervention and control groups. Additionally, stakeholder participation was limited, and many interventions were confined to controlled settings, limiting their real-world applicability. Moreover, only two studies formally assessed barriers and enablers to intervention adoption (Rolker et al. 2022).

The challenges to link FLW outcomes with different food system outcomes depends on where, when, and how FLW targets are defined, implemented, and assessed. A common definition of FLW (Chap. 3) and standardized measurement approaches (Chap. 4) are essential for consistent evaluation of FLW interventions. However, the absence of robust methods to assess the effectiveness of prevention interventions hinders the identification and prioritization of promising strategies (De Laurentiis et al. 2020).

Patinha Caldeira et al. (2019) presented an evaluation framework to consistently assess the performance of FLW prevention interventions. This framework quantifies intervention effectiveness, measuring the reduction in FLW relative to the intended target, and efficiency by evaluating the capacity to achieve the target with minimal input resources (time, cost, effort). To complement this quantitative analysis, additional qualitative criteria are incorporated to assess the quality of intervention design, long-term sustainability, transferability, scalability, and intersectoral collaboration. By examining both quantitative and qualitative aspects, this framework aims to provide a comprehensive evaluation to identify facilitators and barriers to intervention success.

To assess successful FLW reduction interventions through a food systems lens, government and civil society evaluators must consider not only their impact on loss reduction but also potential trade-offs with other food system outcomes. These outcomes often focus on net economic or environmental benefits, which can be calculated based on avoided costs or impacts associated with producing and distributing food up to the point of where pFLW is avoided, diverting food to nonfood destinations, and implementing the intervention itself (Patinha Caldeira et al. 2019; De Laurentiis et al. 2020). For example, expanding cold chain networks to prevent FLW incurs economic and environmental costs (for example, equipment, energy consumption, and emissions). Technological advancements can shift cost–benefit calculations over time, making the trade-offs between FLW reduction and other factors more dynamic. This is exemplified by ongoing efforts to scale "clean cold chains," powered by renewable or alternative energy sources (Kitinoja et al. 2019). Similar debates of the trade-offs with FLW and different packaging are still unresolved (Box 9.3).

Although often underemphasized in favor of economic and environmental concerns, diet and nutrition outcomes are important, as FLW pathways are linked to various aspects of food security (Chaps. 5–8). Recognizing that various indicators within and across nutrition, environment, and economic categories have an impact, entities monitoring intervention implementation can identify potential synergies and trade-offs in FLW reduction strategies.

Box 9.3 A Closer Look at the Environmental Impact of Food Packaging

As food production and distribution systems become more complex, the importance of effective packaging increases. However, the life cycle of packaging materials, from production to disposal, has important environmental implications (Grönman et al. 2013). While various materials, including glass, metal, paper, and cardboard, are used for food packaging, plastics have become increasingly prevalent due to their versatility and cost-effectiveness (Otto et al. 2021).

The share of the environmental impact of food packaging relative to the whole food product–package combination is generally small, but does depend on the product and type of packaging (Grönman et al. 2013; Otto et al. 2021). Still, food and beverage packaging accounts for between 10% and 20% of plastics products, often with very short lifespans and containing chemicals associated with health and environmental hazards (Yates and Deeney 2024). Identifying sustainable packaging solutions requires a careful assessment of the trade-offs between reducing FLW and minimizing the environmental impact of packaging materials, from production to disposal, and prioritizing packaging for products with high environmental footprints (Otto et al. 2021; Yates and Deeney 2024). However, to make informed decisions, more data on FLW and the environmental impact of different packaging materials is needed for civil society organizations and governments to develop evidence-based recommendations and policies that promote safe, sustainable, and effective packaging solutions.

While some packaging is recycled within countries, a significant portion is exported or disposed of through incineration or landfilling. For example, in Europe, 26.3% of post-consumer plastic was recycled inside Europe, and 6.2% was exported in 2018, representing 7.7 and 1.8 MT of plastic, respectively (PlasticsEurope 2019). These disposal methods can have significant environmental and social implications, especially in low-income regions where open burning is common (Periathamby and Law 2020; Velis and Cook 2021).

9.4 Connecting Food Loss and Waste Intervention Benefits Across the Value Chain

To effectively scale FLW interventions and improve food systems, enabling environments are necessary for value chain actors to be sufficiently motivated to adopt and sustain technology and practice-based changes. Desirable impacts of addressing FLW include ensuring adequate supply and reducing the cost of nutritious food and supporting the livelihoods of value chain actors. These impacts, however, manifest at various stages and scales. To realize the broader food systems benefits of FLW reduction, evaluations should consider the specific impacts on value chain actors directly involved in prevention and management strategies and to ensure an enabling environment.

Individual-level interventions to address FLW often require broader value chain changes to achieve sustainable benefits. As discussed throughout this book, many drivers of FLW stem from a lack of coordination within the food value chain. Moreover, FLW at one stage often results from the actions and decisions of actors at other stages (HLPE 2014). The absence of concerted action can itself contribute to FLW, as discussed in Chap. 8. For example, when producers lack financial incentives for producing higher quality produce, they may neglect grading and sorting processes. Consequently, a value chain approach must consider the motivations and responses of each actor in FLW interventions, as well as their potential impact on the overall system.

The success and sustainability of FLW interventions depends on how the costs and benefits are distributed. While the costs and benefits of FLW interventions can occur at individual, value chain, and societal levels, they may not be evenly distributed. The optimal level of FLW for sustainable food systems may diverge from individual perceptions of acceptable levels of loss, particularly, among those who bear the highest costs. Further, value chain actors respond to changes in other parts of the system, necessitating a value chain approach to FLW intervention design. This approach aims to balance societal benefits with individual incentives, considering who gains and who loses, with particular attention to whether or not benefits accrue to marginalized value chain actors and consumers. Box 9.4 provides insights into the complex interplay between FLW and value chain actors, highlighting the importance of considering feedback loops and cascading effects.

Box 9.4 Modeling Approaches to Assessing Food Loss and Waste Reduction Impacts Across the Value Chain

To effectively reduce food loss and waste, a multifaceted approach is necessary, involving individual actions, value chain coordination, and supportive policies. Value chain actors must be motivated to implement and sustain actions that reduce food loss. For producers to retailers, this often involves balancing costs and benefits, and ensuring that the sale price exceeds the purchase price. For consumers, lower food prices can enhance affordability and access.

To model the economic impacts of FLW reduction, de Gorter et al. (2021) evaluated three economic models to examine how public and private interventions influence food markets and trigger cascading effects across the value chain (production; transportation, handling and storage; processing; food service; retail; and consumption). The study highlights that the potential synergies and conflicts between policy goals, such as food security, environmental impact, economic impact, and social impact, are influenced by market structure. These findings underscore the importance of adopting a value chain perspective to fully comprehend the interactions between stages and the potential trade-offs between policy objectives.

Table 9.2 Categories of adoption criteria for food loss and waste interventions

Logistical	• Availability and (physical) accessibility of the intervention technology or services • Availability of extension services to facilitate knowledge sharing
Sociocultural	• Acceptable to value chain actor and wider society • Awareness of the intervention and its benefits • Participatory approach to involve local stakeholders in the decision-making process
Economic	• Affordable to all people targeted for the intervention • Resources available at the local level
Technical	• Technically feasible to implement in the context • Adaptable to current and changing environment/infrastructure • Scalable for widespread adoption • Time to impact (initial effect and duration)

Source: Adapted from Soethoudt et al. (2021)

The success of FLW interventions is determined by achieving performance targets and ensuring sustained adoption. However, the implementation of FLW interventions, how strategies are introduced, adopted, and sustained, is rarely considered in overall intervention design (Rolker et al. 2022). While a priori evaluations often narrowly focus on economic impact and affordability, adoption failures can stem from a broader range of factors, including stakeholder engagement, technical feasibility, market dynamics, and environmental and infrastructure constraints (Soethoudt et al. 2021). Soethoudt et al. (2021) reviewed the literature to identify key factors and propose 11 criteria likely to influence the adoption of FLW interventions in LMIC contexts (Table 9.2). To learn more about the criteria and explore case studies on intervention evaluation, we encourage readers to consult Soethoudt et al. (2021).

While many FLW interventions prioritize technology-based solutions, experts argue that addressing information and knowledge gaps is an important starting point to prioritize (HLPE 2014). By providing information about FLW, potential solutions, and their associated benefits, value chain actors can make informed decisions and adopt effective strategies. This communication aligns with the criteria for successful intervention adoption, which emphasize the importance of information and knowledge dissemination (Soethoudt et al. 2021). Achieving a common conceptualization of how FLW is defined and measured is critical to building evidence of the benefits of specific interventions and consistently communicating the FLW issue and intervention benefits to value chain actors.

As food systems transform, shifts from informal to formal value chains, increasing value chain length, and the consolidation of actors influence market dynamics, motivations, and power relationships. To ensure successful FLW interventions that support inclusive food systems, the design and evaluation of FLW interventions, including those focused on enabling environments, must consider how value chain actors are intentionally included or excluded.

The informal sector, operating largely outside formal systems, plays a significant role in food systems, particularly, in LMICs. Midstream and downstream food value chain stages are dominated by small, informal businesses that play a critical role in distributing perishable and nutritious foods from farms to retail outlets. Research by de Steenhuijsen Piters et al. (2021) indicated that over 90% of domestically produced fruits and vegetables in LMICs are traded through the informal midstream sector. Women, in particular, are key contributors to food security, both within households and through their involvement in agricultural production and marketing (Doss et al. 2018). To effectively address FLW in a way that leads to positive impacts supporting sustainable food systems that deliver healthy diets to all, FLW interventions must explicitly incorporate strategies to include informal value chain actors, especially women, in the design, monitoring, and evaluation.

Small- and medium-sized value chain actors, particularly, those involved in perishable foods, face challenges due to economies of scale and limited bargaining power. Although value chain actors may express interest in specific technologies, scaling these interventions requires careful consideration of factors, such as production levels, market demand, and the enabling environment. For instance, Debnath et al. (2021) found that Indian value chain actors were not interested in investing in local milk chillers and processing equipment, as the limited scalability of milk production hindered the potential benefits of reduced spoilage. Similarly, Obanubi et al. (2020) highlighted the impact of business models, scale, and the enabling environment on the financial viability of technology-based FLW interventions in Nigeria's tomato value chains. While much of the focus on FLW prevention is to support value chains for growing urban populations, it is also important to address rural food security and market inclusion (Box 9.5).

As discussed throughout this chapter, most FLW interventions rely heavily on technology-based solutions. This narrow focus limits the integration of practice-based interventions and socioeconomic considerations necessary for creating inclusive enabling environments. For example, Benyam et al. (2021) found that the adoption of digital agricultural technologies to reduce FLW is limited by factors, such as investment costs and the digital divide.

More data across value chain stages is needed to design more inclusive interventions. For example, gender-sensitive data collection, such as gender-sensitive mapping, can help identify and address gender-specific barriers to FLW reduction (Doss et al. 2018; FAO 2018). By understanding social norms and traditions, FLW interventions can be tailored to the specific needs and contexts of value chain actors. Further, understanding how the benefits and costs of FLW interventions change with scale is essential for anticipating the long-term sustainability and impact of these interventions.

Box 9.5 Inclusion of Rural Markets to Enhance Stable Access to Nutritious Foods

As food systems transform, the transitional stages are marked by rapidly expanding and changing food value chains that aim to reach growing urban markets and international markets (Chap. 2). At the same time, goals for developing sustainable and equitable food systems must consider the ongoing trade-offs that emerge from transitioning food systems, particularly, with regard to supporting rural value chains through the transformation.

Cooper et al. (2021) used a systems dynamics modeling approach to examine the potential of producer aggregation schemes to increase the availability of nutritious fruits and vegetables in smaller Indian markets while improving producer profitability. Their findings suggest that a combination of strategies, including aggregation, cold storage, and demand-side interventions, is necessary to achieve this goal. Relying solely on aggregation proved insufficient. This study emphasizes the importance of bundling interventions and considering the role of technology, practices, and enabling environments, particularly, in smaller markets that may be overlooked compared to more profitable markets in more distant urban areas.

9.5 Connecting Sustainable Development to Food Loss and Waste

This chapter distinguishes between preventing and managing FLW, as well as managing inedible food parts. Although Sustainable Development Goal (SDG) Target 12.3 focuses on FLW reduction, these three interconnected pathways will collectively contribute to achieving sustainable consumption and production patterns (SDG 12). Additionally, the objectives for reducing FLW and managing pFLW are linked to broader food system outcomes. Progress in addressing FLW can contribute to multiple SDGs, including hunger, food security, and nutrition (SDG 2), waste management, water and land resources, climate change (SDGs 6, 11, 13, 14, and 15), and poverty, livelihoods, and equality (SDGs 1, 8, and 10) (FAO 2019).

At the same time, policies targeting other SDGs have the potential to reduce FLW as a secondary benefit when the goals address underlying inefficiencies. For example, progress on SDGs, including gender equality (SDG 5); affordable and clean energy (SDG 7); infrastructure, industry, and innovation (SDG 9); and partnerships (SDG 17) may support FLW interventions through building enabling environments, depending on the context and linkage (FAO 2019). Therefore, addressing FLW is an iterative process by which changes in other food system aspects can enable the feasibility, effectiveness, and sustainability of FLW-specific technologies and practices.

The transformation to sustainable food systems is required for achieving the SDGs. The recent Food Systems Countdown Initiative (see Chap. 2) uses an indicator framework to track progress toward global development, health, and sustainability goals, encompassing diets, nutrition, and health; environment, natural resources, and production; livelihoods, poverty, and equity; and governance and resilience (Schneider et al. 2023). Although the initiative aims to align indicators with the SDG indicator framework, the SDG 12.3 FLW indices (Food Loss Index and Food Waste Index) are currently not included in the Food Systems Countdown due to data limitations. Addressing these data gaps is essential to fully understand the linkages between FLW and other sustainable development indicators and identify synergistic opportunities for progress.

9.6 Conclusions

The chapter examined the complexities of FLW interventions, emphasizing the need for a comprehensive approach that considers the interconnectedness of food systems. Effectively addressing FLW requires both direct interventions targeting prevention and management as well as broader systemic changes addressing underlying inefficiencies. Prevention aims to maintain food quality and reduce waste of edible food parts, while management focuses on minimizing the negative impacts of edible and inedible waste. The balance between supply and demand, as well as the root causes of surplus generation, are critical factors to consider in designing effective prevention strategies. Further, the differentiation between edible and inedible food parts is necessary to more consistently articulate the types of interventions and their associated trade-offs.

To evaluate the effectiveness of interventions, robust methods are needed to assess both FLW reduction and broader food system impacts. While technology-based interventions have dominated the field, a more balanced approach is needed, incorporating practice-based interventions and addressing enabling environmental factors. To ensure the success and sustainability of FLW interventions, entities designing evaluations must consider factors including stakeholder engagement, social and cultural, and the specific context of implementation. Therefore, a comprehensive evaluation approach should consider economic, environmental, and social outcomes, as well as potential trade-offs between different intervention strategies.

In both the design and assessment of FLW interventions, it is crucial for intervention evaluations to consider the entire food value chain. Although individual-level interventions can address FLW, broader value chain changes are necessary for sustainable impact. However, the distribution of costs and benefits from FLW interventions can be uneven, with potential conflicts between societal goals and individual incentives. Therefore, a value chain approach is essential to balance these interests and ensure that FLW interventions are inclusive and sustainable. This approach involves considering the motivations and responses of each actor, as well as their potential impact on the overall system. The complex interplay between FLW and value chain actors informs the design of interventions that address information and knowledge gaps, promote the adoption of effective strategies, and support the inclusion of informal value chain actors, particularly women.

The ultimate goal of FLW interventions is to enhance food system sustainability across multiple dimensions, including nutrition, health, economic well-being, and environmental impact. Understanding of the interconnectedness of actions on FLW with other aspects of sustainable development is important to identify synergies in supporting sustainable food systems transformations. However, data gaps currently prevent FLW from being included as an indicator in comprehensive food system indicator frameworks. Filling FLW data gaps requires harmonizing the FLW definition, developing a data value chain that is inclusive across value chains and actors, and comprehensively evaluating progress on FLW reduction and management.

References

Akkerman R, Buisman M, Cruijssen F, de Leeuw S, Haijema R (2023) Dealing with donations: supply chain management challenges for food banks. Int J Prod Econ 262:108926. https://doi.org/10.1016/j.ijpe.2023.108926. Accessed 25 Nov 2024

Barrett CB, Benton T, Fanzo J, Herrero M, Nelson RJ, Bageant E, Buckler E, Cooper K, Culotta I, Fan S, Gandhi R, James S, Kahn M, Lawson-Lartego L, Liu J, Marshall Q, Mason-D'Croz D, Mathys A, Mathys C, Mazariegos-Anastassiou V, Miller A, Misra K, Mude AG, Shen J, Majele Sibanda L, Song C, Steiner R, Thornton P, Wood S (2020) Socio-technical innovation bundles for agri-food systems transformation. Report of the international expert panel on innovations to build sustainable, equitable, inclusive food value chains. Cornell Atkinson Center for Sustainability and Springer Nature, Ithaca/London. https://www.nature.com/documents/Bundles_agrifood_transformation.pdf. Accessed 24 Aug 2020

Benyam A(A), Soma T, Fraser E (2021) Digital agricultural technologies for food loss and waste prevention and reduction: global trends, adoption opportunities and barriers. J Clean Prod 323(129099). https://doi.org/10.1016/j.jclepro.2021.129099. Accessed 4 Oct 2021

Cattaneo A, Sánchez MV, Torero M, Vos R (2021) Reducing food loss and waste: five challenges for policy and research. Food Policy 98:101974. https://doi.org/10.1016/j.foodpol.2020.101974. Accessed 3 Sep 2021

Chauhan C, Dhir A, Akram MU, Salo J (2021) Food loss and waste in food supply chains. A systematic literature review and framework development approach. J Clean Prod 295:126438. https://doi.org/10.1016/j.jclepro.2021.126438. Accessed 9 Nov 2021

Cooper GS, Shankar B, Rich KM, Ratna NN, Alam MJ, Singh N, Kadiyala S (2021) Can fruit and vegetable aggregation systems better balance improved producer livelihoods with more equitable distribution? World Dev 148:105678. https://doi.org/10.1016/j.worlddev.2021.105678. Accessed 12 Dec 2021

de Gorter H, Drabik D, Just DR, Reynolds C, Sethi G (2021) Analyzing the economics of food loss and waste reductions in a food supply chain. Food Policy 98:101953. https://doi.org/10.1016/j.foodpol.2020.101953. Accessed 31 Aug 2021

De Laurentiis V, Caldeira C, Sala S (2020) No time to waste: assessing the performance of food waste prevention actions. Resour Conserv Recycl 161:104946. https://doi.org/10.1016/j.resconrec.2020.104946. Accessed 4 Nov 2024

de Steenhuijsen Piters B, Dijkxhoorn Y, Hengsdijk H, Brouwer I, Tichař T, Carrico C (2021) Synthesis report of the global fruits and vegetables scoping study: assessing opportunities for philanthropic investment. Wageningen economic research, report 2021-113, Wageningen. https://research.wur.nl/en/publications/c425c239-a667-4539-8dfa-5cd4ddc38311. Accessed 10 Jan 2024

Debnath KB, Wang X, Peters T, Menon S, Awate S, Patwardhan G, Wadkar N, Patankar M, Shendage P (2021) Rural cooling needs assessment towards designing community cooling hubs: case studies from Maharashtra, India. Sustainability 13(10):5595. https://doi.org/10.3390/su13105595. Accessed 20 Sep 2021

Doss C, Meinzen-Dick R, Quisumbing A, Theis S (2018) Women in agriculture: four myths. Glob Food Secur 16:69–74. https://doi.org/10.1016/j.gfs.2017.10.001. Accessed 4 Sep 2021

FAO (Food and Agriculture Organization of the United Nations) (2018) Gender and food loss in sustainable food value chains: a guiding note, Rome. https://openknowledge.fao.org/server/api/core/bitstreams/0e730b95-9ead-4ced-9fd9-23cdcf01a189/content. Accessed 3 Sep 2021

FAO (Food and Agriculture Organization of the United Nations) (2019) The state of food and agriculture 2019. Moving forward on food loss and waste reduction. FAO, Rome. https://openknowledge.fao.org/server/api/core/bitstreams/11f9288f-dc78-4171-8d02-92235b8d7dc7/content. Accessed 30 Aug 2021

Grönman K, Soukka R, Järvi-Kääriäinen T, Katajajuuri J-M, Kuisma M, Koivupuro H-K, Ollila M, Pitkänen M, Miettinen O, Silvenius F, Thun R, Wessman H, Linnanen L (2013) Framework for sustainable food packaging design. Packag Technol Sci 26(4):187–200. https://doi.org/10.1002/pts.1971. Accessed 29 May 2024

HLPE (2014) Food losses and waste in the context of sustainable food systems. A report by the high level panel of experts on food security and nutrition of the Committee on World Food Security, Rome. https://openknowledge.fao.org/server/api/core/bitstreams/b1949fae-23d4-473c-8b87-8c4359b74d6c/content. Accessed 25 Aug 2021

Kitinoja L, Tokala VY, Brondy A (2018) A review of global postharvest loss assessments in plant-based food crops: recent findings and measurement gaps. J Postharvest Technol 6(4):1–15. http://jpht.in/MenuscriptFile/ed17639f-edd5-4802-b4ad-d5a1bdbadb28.pdf. Accessed 14 Dec 2024

Kitinoja L, Tokala VY, Mohammed M (2019) Clean coldchain development and the critical role of extension education. Agric Dev 36:3. https://www.cabidigitallibrary.org/doi/pdf/10.5555/20193201962. Accessed 14 Oct 2021

Latka C, Parodi A, van Hal O, Heckelei T, Leip A, Witzke H-P, van Zanten HHE (2022) Competing for food waste—policies' market feedbacks imply sustainability tradeoffs. Resour Conserv Recycl 186:106545. https://doi.org/10.1016/j.resconrec.2022.106545. Accessed 8 May 2023

Midgley JL (2014) The logics of surplus food redistribution. J Environ Plan Manag 57(12):1872–1892. https://doi.org/10.1080/09640568.2013.848192

Obanubi M, Okoruwa A, Morris S, Nordhagen S, Djimeu Wouabe E, Yalch T (2020) Business opportunities to reduce post-harvest loss of nutritious food: modelling the return on investment for field-ready technologies in Nigeria. GAIN working paper no. 16, Global Alliance for Improved Nutrition, Geneva. https://doi.org/10.36072/wp.16. Accessed 20 May 2024

Otto S, Strenger M, Maier-Nöth A, Schmid M (2021) Food packaging and sustainability—consumer perception vs. correlated scientific facts: a review. J Clean Prod 298:126733. https://doi.org/10.1016/j.jclepro.2021.126733. Accessed 29 May 2024

Patinha Caldeira C, De Laurentiis V, Sala S (2019) Assessment of food waste prevention actions: development of an evaluation framework to assess performance of food waste prevention actions. EUR 29901 EN. Publications Office of the European Union, Luxembourg. https://publications.jrc.ec.europa.eu/repository/handle/JRC118276. Accessed 18 Nov 2024

Periathamby A, Law HJ (2020) Do we need landfills? Waste Manag Res 38:1075–1077. https://doi.org/10.1177/0734242x20943036. Accessed 29 May 2024

PlasticsEurope (2019) Plastic—the facts 2019. An analysis of European plastic production, demand and waste data. https://plasticseurope.org/wp-content/uploads/2021/10/2019-Plastics-the-facts.pdf. Accessed 29 May 2024

ReFED (2021) Roadmap to 2030: reducing U.S. Food Waste by 50% and the ReFED Insights Engine. At-A-Glace. https://refed.org/uploads/refed_roadmap2030-FINAL.pdf. Accessed 11 Nov 2021

Rolker H, Eisler M, Cardenas L, Deeney M, Takahashi T (2022) Food waste interventions in low- and-middle-income countries: a systematic literature review. Resour Conserv Recycl 186:106534. https://doi.org/10.1016/j.resconrec.2022.106534. Accessed 17 Apr 2023

Schneider KR, Fanzo J, Haddad L, Herrero M, Moncayo JR, Herforth A, Remans R, Guarin A, Resnick D, Covic N, Béné C, Cattaneo A, Aburto N, Ambikapathi R, Aytekin D, Barquera S, Battersby J, Beal T, Molina PB, Cafiero C, Campeau C, Caron P, Conforti P, Damerau K, Di Girolamo M, DeClerck F, Dewi D, Elouafi I, Fabi C, Foley P, Frazier TJ, Gephart J, Golden C, Fischer CG, Hendriks S, Honorati M, Huang J, Kennedy G, Laar A, Lal R, Lidder P, Loken B, Marshall Q, Masuda YJ, McLaren R, Miachon L, Muñoz H, Nordhagen S, Qayyum N, Saisana M, Suhardiman D, Sumaila UR, Cullen MT, Tubiello FN, Vivero-Pol J-L, Webb P, Wiebe K (2023) The state of food systems worldwide in the countdown to 2030. Nat Food 4:1090–1110. https://doi.org/10.1038/s43016-023-00885-9. Accessed 16 Jan 2024

Soethoudt JM, Pedrotti M, Bos-Brouwer HEJ, Castelein RB (2021) Adoption of food loss and waste-reducing interventions in low- and middle-income countries. Wageningen Food & Biobased Research, Wageningen. https://research.wur.nl/en/publications/2856b65d-71bb-424d-b1d2-606cadb86860. Accessed 27 Sep 2021

Stathers T, Holcroft D, Kitinoja L, Mvumi BM, English A, Omotilewa O, Kocher M, Ault J, Torero M (2020) A scoping review of interventions for crop postharvest loss reduction in sub-Saharan Africa and South

Asia. Nat Sustain 3:821–835. https://doi.org/10.1038/s41893-020-00622-1. Accessed 13 Nov 2024
Velis CA, Cook E (2021) Mismanagement of plastic waste through open burning with emphasis on the Global South: a systematic review of risks to occupational and public health. Environ Sci Technol 55(11):7186–7207. https://doi.org/10.1021/acs.est.0c08536. Accessed 31 May 2024
Vernooij DM, Oostewechel RJA, Verschoor JA, Axmann HB, Pereira da Silva FIDG, Castelein RB (2022) Understanding the mechanisms to strengthen food systems through value chain-driven action. Wageningen Food & Biobased Research, Wageningen. https://research.wur.nl/en/publications/e07b5475-c11d-42cf-b0f5-c4b0ab73fcbc. Accessed 9 Jan 2024
Yates J, Deeney M (2024) Policy brief: impacts of plastics across the food system. https://ikhapp.org/material/policy-brief-impacts-of-plastics-across-the-food-system/. Accessed 31 May 2024

10 The Way Forward: Setting a Food Loss and Waste Policy Agenda

10.1 Introduction

Effectively addressing FLW requires defining success metrics and carefully selecting appropriate interventions tailored to reduction targets and food system contexts. As explored throughout this book, reducing FLW is not an end goal but a means to achieve broader sustainable food systems transformations. These transformations aim to enhance food security and nutrition, mitigate environmental impacts, and support livelihoods. Designing and implementing relevant FLW interventions requires a collaborative effort between the public sector and private value chain actors to align their respective strategies and decision-making processes. This chapter presents a FLW policy agenda, which addresses the demand for nutritious foods, supports innovations in value addition, enhances market coordination, and considers environmental trade-offs, while prioritizing food security outcomes (Fig. 10.1).

Considering the leverage points and investment priorities discussed in Chap. 8, we explore the mechanisms and conditions in which value chain actors are supported and motivated to participate in FLW reduction. Critically, properly matching interventions to food system contexts requires the recognition of the diversity of value chain actor characteristics and goals (Vernooij et al. 2022). This requires, in part, making the business case to motivate value chain actors whose livelihoods depend on how the value chain functions. This chapter explores the strategies and policies needed to create enabling environments that foster the integration of FLW reduction into core business practices.

For FLW interventions to effectively contribute to healthy, sustainable diets, they must be contextually relevant, positively impact food security, and be sustainable beyond external support. This contrasts with costly strategies that may not be practical or sustainable in specific contexts (Soethoudt et al. 2021). This chapter identifies policy approaches that can address barriers to implementing various FLW reduction strategies across different food systems, aligning with investment priorities to prevent and manage FLW.

J. Boiteau, P. Pingali, *Wasted Potential*, Sustainable Development Goals Series,
https://doi.org/10.1007/978-3-031-97411-3_10

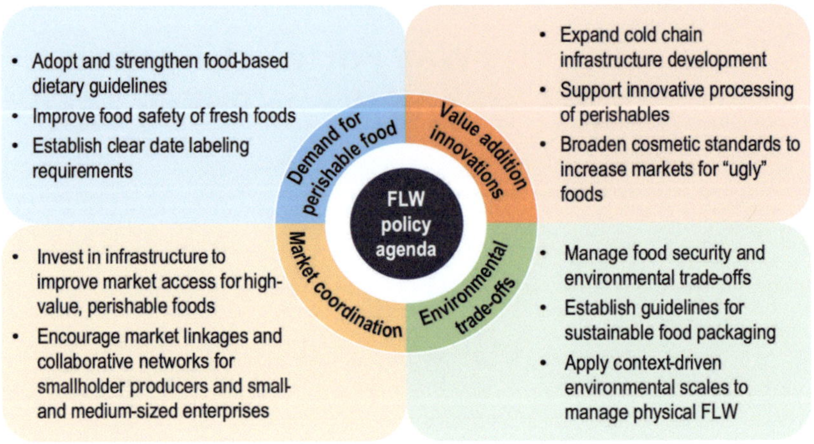

Fig. 10.1 Policy agenda to address food loss and waste for sustainable and healthy food systems

10.2 Policies for Incentivizing Supply: Increasing Demand for Safe, Nutritious Foods

Reducing FLW is a critical step toward building sustainable food systems that deliver healthy diets. Following the FLW pathways, discussed throughout this book, minimizing FLW of perishable foods aims to increase the availability, physical accessibility, and affordability of nutritious food. Ultimately, the goal is to ensure that adequate quantities of quality food reach all consumers and are consumed, thereby contributing to healthier diets in a sustainable way.

The potential price and income effects of increasing food supplies are important considerations for the success of sustained FLW interventions. To effectively increase food availability and affordability, FLW reduction strategies must target losses that occur before the consumption stage (FAO 2019). However, there is a concern of negative impacts from feedback loops, such as decreased demand and lower incomes for value chain actors (FAO 2019). Producers, in particular, often face risks, including information asymmetry, product specialization, and power imbalances, which can lead to overproduction and loss of perishable foods that cannot be stored for long periods (Messner et al. 2021). To mitigate price decreases below profitable levels, producers may choose to leave harvest-ready produce unharvested, further exacerbating FLW of nutritious foods (Gillman et al. 2019; Soma et al. 2021). Therefore, policies are necessary to address these risks by encouraging demand for healthy diets and creating an enabling environment for scaling up FLW reduction interventions. To encourage healthier diets, the private sector needs to market and sell nutritious foods, while the public sector must use procurement, incentives, and regulations to stimulate demand at the individual and household levels (Carducci and Oh 2024).

10.2.1 Adopt and Strengthen Food-Based Dietary Guidelines

Focusing on nutritious, perishable food value chains offers the most promising approach to increase the supply and affordability of these foods, thereby supporting healthy diets. Globally, and in many countries, there is not enough nutritious food available to meet current population nutrient requirements (Siegel et al. 2014; Krishna Bahadur et al. 2018; Wang et al. 2023). The distribution of micronutrient-rich food availability often correlates with the level of food system modernization, with more modern food systems generally offering greater availability to these foods (Marshall et al. 2021; Wang et al. 2023). In contrast, calorie and protein availability is generally adequate or even in surplus in many regions, including the United States, Canada, and Europe (Wang et al. 2023). Although progress has been made in increasing the production of micronutrient-rich foods (Beal et al. 2017; Schmidhuber et al. 2018; Bell et al. 2021), future projections suggest that micronutrient availability gaps will persist in many regions through 2050, particularly, as income growth and climate change impact food systems (Nelson et al. 2018).

With insufficient supply, a healthy diet—one that provides adequate nutrients and long-term health—is unaffordable for three billion people, primarily in Southern Asia and sub-Saharan Africa, regions with predominantly traditional and early transitional food systems (Herforth et al. 2020; Marshall et al. 2021). The low affordability of perishable produce is linked to lower consumption (Miller et al. 2016). Over 80% of the cost of a healthy diet comes from perishable, nutrient-rich foods, like animal-source foods (ASF), fruits, and vegetables (Herforth et al. 2020). In contrast, a calorie-sufficient, but micronutrient-deficient, diet based on starchy staples is three to eight times less expensive (Herforth et al. 2020).

Policies that stimulate demand for healthy diets require foundational guidelines that specify the particular food groups, nutrients, and dietary patterns needed at a population-level. These guidelines inform other critical policy areas, including food production priorities, marketing and labeling regulations, public food procurement practices (Box 10.1), and public nutrition education initiatives within a national context. Although food-based dietary guidelines exist for 90 countries, which encourage the consumption of fruits, vegetables, and ASF, there is a need to refine current and future guidelines to better align with WHO global guidance, incorporate environmental sustainability, and address sociocultural factors influencing dietary transitions (Herforth et al. 2019). Therefore, adopting and strengthening food-based dietary guidelines is a necessary component of comprehensive interventions to increase demand for nutritious, often perishable foods, which can incentivize FLW reductions along the value chain.

Box 10.1 Food Loss and Waste in Public Food Procurement Systems

As part of Sustainable Development Goal (SDG) 12 to ensure sustainable consumption and production patters, Target 12.7 calls for the promotion of public procurement practices that are sustainable, in accordance with national policies and priorities (UN 2015). Public food procurement, also referred to as institutional food procurement, includes school meals programs and purchase of food for public hospitals, prisons, universities, public building cafeterias, and other social programs (Swensson and Tartanac 2020). Similar to private procurement, public food procurement can suffer from organizational drivers of FLW, including inflexible procurement requirements and contracts, as well as issues of demand fluctuation and inventory management (Flanagan et al. 2019; Swensson and Tartanac 2020). Unlike private procurement, public food procurement is subject to specific public procurement regulatory frameworks that govern the entire process, including what, how, and from whom to purchase (Swensson and Tartanac 2020).

The lack of a supportive regulatory framework is a major barrier to leveraging public food procurement for driving more sustainable diets (Swensson and Tartanac 2020). FLW is a recognized issue in public food procurement, often framed primarily in terms of environmental sustainability within recommended tendering and contracting processes (European Commission et al. 2019; Day Farnsworth et al. 2019). However, a food systems approach requires considering the nutritional, social, and environmental impacts of food procurement, including the potential trade-offs between the outcomes associated with specific procurement criteria (Casonato et al. 2024). Such considerations are important, particularly, for programs serving diverse populations with varying nutritional needs, as they can lead to different trade-offs between environmental, nutritional, and socioeconomic impacts (Casonato et al. 2024). Therefore, evidence-based public food procurement policies that define what food is to be purchased will benefit from improved understanding of FLW and robust data generation to achieve sustainable food procurement programs that deliver diverse, nutritious, and culturally appropriate food.

10.2.2 Improve Food Safety of Fresh Foods

In alignment with public health goals, public policies, which prioritize inclusive approaches to ensure that safe food reaches all consumers, must address two key FLW pathways. First, preventing FQL ensures that safe food remains safe. Second, identifying and removing unsafe food from the food value chain, even if it generates pFLW, is necessary for protecting public health. Consumer demand for improved food safety can influence decision-making, such as opting for processed foods over fresh, nutritious foods, due to perceived food safety concerns associated with fresh foods (Carducci and Oh 2024). Governments and the private sector share an interest in building consumer trust and stimulating demand for food safety as a required quality attribute by improving food safety through FQL prevention and the identification of unsafe food.

Institutional policies that establish and enforce evidence-based food safety standards are necessary for setting clear guidelines on food safety. As food systems transform, food safety standards and practices may change and require additional training and certification, such as updated safety standards or expanding to export markets (Pingali 2015). Public investments may be required to support specific value chain actors in complying with these standards, as noncompliance can hinder market access, particularly for high-value, nutritious foods. To promote participation, policy interventions that facilitate knowledge sharing and training on food safety standards and practices can aim to create equitable enabling environments.

Further, investment in public infrastructure, such as access to clean water and hygienic market facilities, plays a role in reducing FQL and maintaining food safety. Clean water is essential for sanitizing food, equipment, and surfaces throughout the value chain, minimizing the risk of foodborne hazards. Additionally, policies that enforce hygienic market facilities are important for maintaining food safety standards where consumers interface with food value chain actors. Policies include promoting best practices for handling animals, livestock, and fresh produce, as well as maintaining sanitary conditions in market areas.

10.2.3 Establish Clear Date Labeling Requirements

Food labeling on retail and consumer packaging provides information to consumers. Certain label information may be mandatory or voluntary, and specific nutrition and health claims may be regulated (Cheftel 2005). Packaging design can also influence consumer perceptions and purchasing decisions, potentially leading to increased FLW, if it obscures important product information or observable attributes. Therefore, innovative food packaging and labeling can work together to provide consumers with clear and accurate food safety information.

Issues with date labels and damaged food packaging can contribute to FLW, based on unclear or incorrect information that consumers associate with FQL, particularly food safety risks. Confusion of date labels is often a cause of pFLW, when food that is safe and edible is discarded, particularly, for perishable foods (Leib et al. 2016). Consumers often struggle to understand vague and unclear date label terminology, making it difficult to determine whether the label is related to food safety, quality, or neither (Wilson et al. 2017; Neff et al. 2019). Consumers may also interpret packaging errors or damaged packaging (for example, dented cans) as contamination cues, even when there is no FQL (White et al. 2016; Aschemann-Witzel et al. 2018). Therefore, public institutions should collaborate to develop consistent policies for clear date labeling requirements that differentiate between safety-based and quality-based labels to contribute to building consumer demand and confidence in foods that are safe to consume, to improving decision-making around food purchases, and to avoiding unnecessary household-level FLW.

10.3 Policies for Innovations: Supporting Value Addition to Keep Nutritious, Perishable Foods in the Food Value Chain

Time and temperature are critical factors in preventing FLW, particularly, for perishable foods. While production diversification into high-value, nutritious foods is already happening in many food systems, the availability and adoption of technologies and best practices to maintain food quality along the value chain remain significant challenges. This section explores three key areas of FLW reduction technologies: cold chain, processing, and cosmetic standards. The viability of investing in these technologies depends on factors, such as business models (for example, ownership, leasing, service-based) and market context (Obanubi et al. 2021). To encourage innovation, policies should create an enabling environment that supports the development and adoption of technologies that address time and temperature issues to preserve food quality.

10.3.1 Expand Cold Chain Infrastructure Development

Cold chain technology has been instrumental in reducing FLW of perishable fresh foods, particularly, as value chains extend and consumer demand for out-of-season produce increases. By reducing FQL and enabling the transportation and storage of produce over longer distances, cold chains add value and expand market opportunities. However, policies should also encourage demand for seasonal produce, when feasible, to minimize the need for extended cold storage.

Cold chain interventions, both mechanical and nonmechanical, aim to slow the degradation of perishable foods along the value chain. While mechanical technologies, like refrigeration and freezing, are effective, they require significant investments in infrastructure and energy, which is expensive to establish, operate, and maintain. Nonmechanical technologies, such as shaded coverings and evaporative cooling, offer more affordable options, especially in rural areas (Brondy 2019; Soethoudt et al. 2021). However, the cost of implementing and maintaining cold chain solutions, whether mechanical or nonmechanical, is often passed on to consumers, particularly, for high-value products that are prioritized for cold storage (Brondy 2019). Therefore, inclusive policies should prioritize ensuring equitable access to cold storage benefits, especially, in relation to food security and livelihoods of marginalized groups.

Policies that promote knowledge sharing and education on the benefits and best practices of cold chain technologies can empower value chain actors to make informed decisions and maximize their investments. However, cold chain technologies cannot completely eliminate FQL. Ideal cooling conditions vary by product and require specific temperature and humidity settings (Brondy 2019). Therefore, the choice of cold chain technology will depend on factors, such as availability, accessibility, and the specific foods.

The effectiveness of cold chains in slowing food quality loss depends on their integration throughout the entire value chain, from harvest to consumption. Although cold storage is often the primary focus of cold chains, cooling at early stages, especially for ASF, is important for food safety and quality preservation. For example, cold chain infrastructure is needed on fishing vessels and at landing sites to prevent deterioration and spoilage (Odoli et al. 2019). Additionally, maintaining consistent temperatures during transportation is vital to prevent fluctuations that can accelerate food degradation. Cold chain infrastructure at later stages, including retail, food service, and consumer storage, employs different refrigeration and freezing equipment to maintain food quality.

Public infrastructure, particularly electrification, is essential for the integration of cold chain technologies. Value chain actors, who act as buyers at different scales, may face limitations in their ability to procure and store perishable goods due to constraints in cold chain infrastructure and timing. Therefore, public policies should consider the flow of food and market connections when designing electricity distribution utilities. Additionally, policies that facilitate the adoption of off-grid solar solutions (Box 10.2), such as energy buy-back or credit programs in which excess electricity produced is sold to the energy grid for profit or credit, can incentivize private investment and increase the availability of clean, affordable energy for cold chain operations.

Box 10.2 Spotlight on the Availability of Off-Grid Solar Innovations for Food Value Chains

In 2020, globally, 733 million people were living without access to electricity (Lighting Global/ESMAP et al. 2022). Off-grid solar technologies and markets are rapidly expanding, offering promising solutions for achieving universal access to electricity in many regions. Among the several technologies, off-grid solar appliances are energy-efficient, solar-powered appliances that include household/small business appliances and larger commercial appliances (Lighting Global/ESMAP et al. 2022).

Uninterrupted energy supplies are essential for operating food appliances, including refrigerators and cooking equipment like pressure cookers, as well as commercial food processing. Despite advancements in solar technology, the availability of off-grid solar appliances varies widely, ranging from conceptual designs to commercially available products (Lighting Global/ESMAP et al. 2022). Solar inverters, which can serve as backup power sources, are commercially available and can be useful for powering appliances during power outages. Off-grid solar technologies for small-scale commercial food applications are still under development. While early versions of walk-in cold storage solutions exist, other products like milk chillers, ice makers, electric chiller vehicles, and agro-processing equipment are still in the pilot phase (Lighting Global/ESMAP et al. 2022).

Despite recent advancements, challenges hinder widespread and sustainable adoption of off-grid solar technologies for food applications. Although recent innovations have made solar-powered refrigerators and freezers more efficient and affordable, they still remain relatively expensive for household and small business use (Lighting Global/ESMAP et al. 2022). Moreover, concerns have been raised about the environmental impact of off-grid solar products, particularly, with regard to their short lifespan, end-of-life disposal, and the potential for lead exposure as these technologies become more widespread (Kinally et al. 2022).

10.3.2 Encourage Processing of Perishable Foods

The seasonal production of nutritious, perishable foods and their short shelf life pose significant challenges in ensuring year-round availability and access. Food processing is a valuable strategy for extending the shelf life of these products. Although processing can impact certain quality attributes, it can also prevent pFLW by preserving food beyond its natural shelf life. Policies that stimulate demand for nutritious foods can also encourage the demand for safe and nutritious processed food products that align with food-based dietary guidelines.

At the earlier stages of food system transformation, shelf-stable products, including dried, canned, or jarred foods, provide consumers access to perishable foods, especially in regions with limited cold chain infrastructure. Innovations in processing facilities, such as multi-seasonal use facilities (Machala et al. 2022), offer strategies to overcome technical, financial, and operational barriers often faced by small- and medium-sized enterprises (SMEs) involved in seasonal produce value chains. Public policies that invest in infrastructure to support processing facilities, including reliable energy and transportation infrastructure, can help reduce barriers to entry and enable the scalability of processing approaches.

As cold chain infrastructure expands, particularly, in high-value markets, refrigeration and freezing technologies can be used to extend the shelf life of perishable products even further. Increasing the access and affordability of cold chain infrastructure can support minimal processing methods for fresh foods, such as trimming and cutting fruits and vegetables—and thus, adding value by making these products more convenient to use and consume, while also maximizing nutrient content relative to other processing methods. Attention should be paid to what other ingredients are added to products, as these can alter the nutrient profile of the final product, particularly, in terms of sugar, salt, and fat content.

Although certain food qualities may be desirable for fresh consumption, they might not be ideal for processing. For example, the increasing demand for tomato-based products in India has led to a need for larger volumes of consistently high quality tomatoes (Subramanian 2016). To address this demand, there has been growing interest in developing dual-purpose cultivars that are suitable for both fresh consumption and processing (Subramanian 2016). Therefore, policies that support research and innovation in seed development and processing technologies can create new opportunities for both fresh and processed food markets.

10.3.3 Broaden Cosmetic Standards to Increase Markets for "Ugly" Foods

Typically, in modern food systems, institutional policies set cosmetic food quality standards to add value. Cosmetic attributes generally refer to attributes that are based on the natural variations in the size, shape, and color of foods that do not affect the taste, nutritive value, or food safety (de Hooge et al. 2018). Suboptimal foods, sometimes referred to as "ugly" foods, are those with cosmetic attributes deviating from average or common cosmetic attributes (Xu et al. 2021). Export markets in transitioning food systems must then adapt to these standards, which can present challenges related to logistics, pricing, and competition (de Hooge et al. 2018). These challenges arise as businesses seek to integrate new quality standards into existing models and market conditions.

However, simply removing cosmetic quality standards can have unintended consequences for value chain actors, as some cosmetic attributes also serve functional purposes in packaging, transport, and processing (Raak et al. 2017; de Hooge et al. 2018). When cosmetic attributes overlap with functional quality attributes, FLW reduction strategies must consider the potential trade-offs in value chain efficiency. To address these trade-offs, interventions may focus on value chain improvements to separate cosmetic and functional attributes. Technological innovations or shifts in consumer demand can expand the range of acceptable quality standards while maintaining efficiency and minimizing food quality loss along the value chain.

One strategy to mitigate FLW stemming from implementation of cosmetic standards involves expanding market opportunities for "ugly" produce. This expansion can be achieved through price differentiation within existing retail outlets or by creating alternative channels, such as dedicated retailers who specialize in cosmetically imperfect produce. However, using price differentiation has several challenges (Box 10.3). These suboptimal products may have minor blemishes, be close to their expiration date, or have minor packaging damage that does not compromise food safety (Aschemann-Witzel et al. 2015; Hartmann et al. 2021). Implementing price-differentiation strategies requires clear and effective communication with consumers. This communication should positively frame "ugly" produce, emphasizing its safety and acceptability for consumption (Hartmann et al. 2021).

Changes in market conditions can influence perceptions of what constitutes suboptimal food (see Chap. 8). Additionally, the demand for extrinsic attributes like organic, fair trade, and animal welfare certification can further complicate consumer decision-making and influence their perception of acceptable food quality attributes. One strategy to address this is to position suboptimal foods as environmentally friendly options. Policies that support market diversification can create channels for both optimal and suboptimal foods, serving consumers at different income levels while promoting environmentally friendly FLW reduction strategies. An alternative to creating separate market segments for suboptimal foods involves using evidence-based approaches to broaden the acceptable range of cosmetic qualities, such as identifying consumer tolerance for mixing "ugly" produce with standard produce (Qi et al. 2022). Successfully increasing the acceptability of suboptimal foods will require public policies that support ensuring food safety can give consumers confidence that foods in the marketplace, regardless of shape, size, or color, are safe to consume. This approach also called for clear product labeling and messaging to consumers.

Box 10.3 Pricing Challenges for "Ugly" Foods

Determining the appropriate pricing for suboptimal foods presents a challenge for retailers. They must satisfy consumer demands for value and quality food while also striving to minimize FLW. To mitigate FLW, retailers may use price differentiation strategies, offering varying prices based on specific quality attributes. However, the chosen attribute(s) for price reduction can impact consumer perception and potential for success. For example, price reductions based solely on cosmetic imperfections may be more readily accepted than those based on attributes that directly affect food safety or sensory attributes, such as food with reduced storage life.

Retailers encounter several challenges and opportunities when offering price-reduced "ugly" foods (Aschemann-Witzel et al. 2017). While offering price reductions can generate revenue and avoid the costs associated with the disposal of food, they also present inherent costs for retailers, including inventory management, dedicated storage space, and marketing efforts to promote these products. Moreover, the introduction of "ugly" foods may inadvertently decrease sales of higher-priced, "normal" items that meet more stringent cosmetic standards. While reducing FLW may be a sustainable practice, the impact on retailer brand image and consumer perception depends on the target market and the extent to which environmental and social considerations contribute to brand value. Price reductions for "ugly" foods may inadvertently diminish overall consumer perception of product quality offered by the retailer (Aschemann-Witzel et al. 2017).

Consumer concerns regarding the quality of "ugly" foods remain a significant obstacle to the successful development of market channels for these products (Hartmann et al. 2021). When consumers lack familiarity with "ugly" foods, or have limited knowledge about food production and quality, they may be unclear and concerned about which specific quality attributes are suboptimal. Addressing these concerns requires increasing the availability of "ugly" foods to enhance consumer familiarity, providing education on food quality evaluation, and effectively communicating quality information through clear labeling and informative messaging. Offering taste samples and presenting "ugly" foods in an appealing and visually attractive way may also improve consumer acceptance. However, an outstanding challenge is the lack of field experiments that translate consumer preferences for "ugly" foods into real-world market settings, limiting the development of effective marketing strategies and evidence-based policies (Hartmann et al. 2021).

10.4 Policies for Market Coordination: Facilitating Efficient Transactions and Improving Market Linkages

Addressing the complex interconnections within FLW pathways requires collaboration across the entire food value chain. Implementing policies that support the inclusion of SMEs in high-value, perishable food markets can create economic opportunities and incentivize investments in food quality and standards (Vorley et al. 2009). Improvements in public infrastructure, such as transportation and storage facilities, are essential for expanding market access and enabling FLW reduction initiatives. Transparent governance and equitable partnerships can further strengthen market linkages and mitigate risks for all stakeholders.

10.4.1 Invest in Infrastructure to Improve Market Access for High-Value, Perishable Foods

Improving market access for high-value, perishable foods can indirectly contribute to FLW reduction by facilitating the timely movement of products from farm to consumer. As discussed throughout this book, factors such as timing, distance, and transportation conditions significantly impact FLW. To optimize these conditions, interventions are needed at multiple food value chain stages, from production to consumption.

Public investments in durable quality roads are important for reducing transportation time and minimizing physical damage to food products (Akande and Diei-Ouadi 2010; Odoli et al. 2019). Road infrastructure networks, including motorways, rural roads, and urban roads, significantly impact the efficiency and environmental emissions of transportation (Soysal et al. 2014). Policies that consider the trade-offs between optimizing transportation networks and minimizing environmental impacts, while accounting for the relationship between road conditions, vehicle types, and load size, can help reduce transportation-related FLW.

Improvements in communication infrastructure, such as cellular networks, will be important for facilitating the dissemination of information on FLW reduction approaches and technologies, as well as market prices (Affognon et al. 2015; Pingali 2015). By improving information flow, value chain actors may be empowered to make informed decisions, diversify targeted markets, and optimize storage and transport strategies (Akande and Diei-Ouadi 2010). Additionally, increased access to reliable communication can foster collaboration and networking among value chain actors to improve the organization and efficiency between value chain stages.

Expanding access to financial services is essential for creating enabling environments that promote equitable access to FLW reduction technologies and knowledge. By supporting a diverse range of financial providers, policies can facilitate investment in FLW reduction initiatives and connect value chain actors to markets (Soethoudt et al. 2021). Given the limited data available on FLW, it is difficult to define the equity challenges and identify solutions related to financial barriers of specific groups. Therefore, public investment in financial institutions should prioritize equity indicators to ensure that marginalized groups have opportunities to benefit from FLW reduction strategies.

10.4.2 Encourage Market Linkages and Collaborative Networks for Smallholder Producers and Small- and Medium-Sized Enterprises

Policies to create incentive structures that support collaborative partner networks are important, particularly, for perishable foods that require traceability and have higher food safety risks (Vorley et al. 2009). By supporting market linkages and equitable access to FLW reduction strategies for smallholder producers and SMEs, policies can foster competition, leading to higher producer incomes and lower consumer prices (Pingali 2015).

Institutional interventions that enable smallholder producers to adopt and scale FLW reduction technologies and best practices are essential for their participation in high-value, perishable food markets, which demand good quality produce. Aggregation models, such as producer organizations and cooperatives, can facilitate access to finance, reduce risks associated with investing in FLW reduction strategies, and enhance bargaining power. Additionally, lead farmer models can provide valuable knowledge, technology, and market access to networks of smallholder producers. By empowering smallholders, these strategies can help them participate in high-value food markets and reduce FLW.

Policies that address power imbalances related to pricing, quality standards, and contractual agreements will require transparent governance and clear communication to help mitigate information asymmetry and align incentives among value chain actors. Consistent pricing systems that reward quality can incentivize investments in FQL reduction (FAO 2014). Further, public quality standards can establish minimum benchmarks for food safety, nutrition, and other attributes used for marketing. Formalizing food markets for perishable foods, through registration and licensing of SMEs, can support improvements in food quality and FLW reduction, when combined with policies that ensure equitable access to resources for meeting regulatory requirements.

Private standards, primarily set by food processing and retail sectors, aim to differentiate products and meet specific consumer preferences. However, these standards can lead to the rejection of food that does not meet strict criteria for limiting attributes, even when the food is still safe to consume and may be acceptable to a consumer (De Hooge et al. 2018). To minimize waste, policies can encourage the adoption of innovative contract arrangements, such as "whole-crop contracts," in which retailers commit to purchasing entire yields, including suboptimal produce (Hezarkhani et al. 2024). This approach can help reduce food loss and provide more stable markets for producers.

10.5 Policies for the Environment: Examining Environmental Trade-Offs in Food Loss and Waste Reduction and Management

The environmental impacts of FLW are multifaceted, arising from both prevention and management strategies. When assessing the environmental impact of FLW prevention interventions, it is essential for evaluations to consider the trade-offs related to increased production to meet growing demand for nutritious food, methods for preventing FLW, and diversion of food intended for human consumption to nonfood uses. Beyond these considerations, evaluations of the environmental impact of pFLW management must examine the trade-offs between different nonfood destinations.

10.5.1 Manage Food Security and Environmental Trade-Offs

Food system transformations have not consistently led to the availability, accessibility, and consumption of healthy diets for all (Ambikapathi et al. 2022). At the global level, the quantity and quality of food made available and accessible for human consumption are based on food production and FLW. At the regional level, food production, trade, FLW, and other stock changes determine the quantity and quality of food made available and accessible for human consumption. While aiming for zero FLW is neither realistic nor optimal (Kader 2005; Sheahan and Barrett 2017), efficient food systems require a balance between producing sufficient food to meet global demand and effectively managing the postproduction movement of food to minimize waste (Ellison et al. 2019).

Nutritious, perishable foods, especially, fruits, vegetables, and ASF, are often high-value commodities that have the most potential for FLW. Across all food systems, increasing demand for fruits and vegetables can support healthier diets. However, policies promoting ASF should be carefully considered to avoid promoting overconsumption, which may have negative health and environmental consequences. Therefore, determining optimal production levels and FLW reduction strategies require comprehensive understanding of the specific food value chain and dietary context.

The consumer stage presents a unique opportunity for FLW prevention and management, as it is the final point of control within the food value chain. Institutional policies should focus on collecting data on FLW at consumer stages to better understand the complex drivers of loss at this stage. To identify optimal FLW reduction strategies, it is important to consider consumer-level food waste to avoid shifting the burden of waste from earlier stages of the food chain onto consumers. Therefore, a value chain approach will be useful for examining FLW pathways and the impacts of different interventions and accounting for interactions between each stage. This includes evaluating redistribution programs aimed at reducing FLW, with consideration of the unique characteristics and needs of target consumers.

10.5.2 Establish Guidelines for Sustainable Food Packaging

Food packaging protects food from quality loss to extend the product's shelf life. The widespread and increasing use of plastics raises concerns about food safety (Chap. 7) and environmental sustainability (Chap. 9). Maximizing the benefits of packaging while minimizing negative impacts calls for innovative packaging materials, combined with appropriate handling and packing techniques and promotion of sustainable end-of-life package management. Therefore, evidence-based policies on sustainable food packaging must address potential impacts of human health, including worker safety, as well as the entire lifecycle of packaging, including disposal and export, to ensure equitable distribution of the environmental and social costs. Such guidelines can help drive private sector investment to develop packaging materials, designs, and practices, which are sustainable throughout their lifecycle, from formation to disposal.

Retail and consumer packaging is more visible to the public than bulk packaging used throughout the value chain. Although packaging is often associated with processed foods, bulk packaging is essential for protecting both fresh and processed products during transport and storage. Smaller retail and consumer packaging can help preserve the quality of perishable foods, which continue to undergo biochemical changes after harvest.

Bulk packaging protects food during handling and transportation, minimizing damage from factors like light and moisture. However, damaged packaging can compromise food safety and quality. Innovations in packaging materials, such as transitioning from raffia baskets to stackable plastic crates, can enhance efficiency and reduce damage. Additionally, advanced technologies like modified atmosphere packaging and coatings can further protect food. To ensure safe and efficient handling, institutional policies should consider the impact of packaging innovations on labor conditions, particularly regarding manual labor and exposure to hazards, such as manually lifting and handling heavy loads. Policies that prioritize worker safety and well-being support transformations toward more sustainable and equitable food systems.

Food packaging also serves as a platform for food labeling. Improvements in bulk packaging can make it suitable for labeling, providing valuable information for value chain actors and facilitating traceability, a tool that can be used for monitoring food safety. Public investments in infrastructure, such as improved Internet connectivity, can support the implementation of labeling systems for bulk packaging traceability, streamlining processes, and reducing costs.

The extent to which retail and consumer food packaging is used for fresh foods varies significantly across different markets (Chap. 6). Although food packaging has become a significant sustainability concern, the role of consumer packaging in reducing FLW is relatively less visible (Croad et al. 2024). For example, public perception of sustainable packaging tends to focus on factors like recyclability and natural materials (Otto et al. 2021), rather than its impact on food waste reduction.

The increasing use of single-use and flexible plastics in retail and consumer food packaging is a major environmental concern. Mitigating the negative environmental impacts of plastic waste requires reductions in plastic waste disposal. This can be achieved by minimizing plastic packaging usage, maximizing reuse, and enhancing recyclability to establish a circular economy for packaging materials. Although much attention has focused on plastic bags, with policies, such as bans and fees/taxes, encouraging consumers to adopt reusable alternatives (Dey et al. 2021; Borg et al. 2022), a more comprehensive approach is needed. Such an approach requires shifting the focus beyond point-of-sale bags to addressing the use of plastic packaging for individual food items. Prioritizing which plastic packaging streams to address requires an understanding of FLW pathways and the associated trade-offs related to packaging. For example, packaging for fresh meats is important in preventing cross-contamination and minimizing foodborne hazards, whereas single-use plastic bags for unpackaged produce may be less essential. Further research is needed to identify the most effective strategies for reducing different types of plastic food packaging usage while simultaneously increasing reuse and recyclability.

Looking toward the future, efforts are ongoing to develop sustainable alternatives to single-use plastic food packaging. Effectively transitioning away from single-use plastic will require a comprehensive understanding of packaging functions, including how packaging minimizes FLW, while also considering the value chain actors' and consumers' preferences for factors like information display, brand recognition, aesthetic appeal, convenience, and traceability (Diprose et al. 2023). Ongoing research in green chemistry aims to develop bio-based alternatives to conventional plastics, derived from renewable resources, and designed for easier recycling and enhanced biodegradability (Sheldon and Norton 2020). Therefore, evaluating the trade-offs between the health and environmental impacts of packaging and FLW will be an ongoing iterative process.

10.5.3 Apply Context-Driven Environmental Scales to Manage Physical FLW

The environmental and climate change impacts of FLW come from the resources used in both the production and postproduction stages of food that, ultimately, is not eaten, as well as the management of pFLW (Muth et al. 2019). These impacts can affect climate, water, air, and land, and often are interconnected (Muth et al. 2019). Addressing the environmental impacts associated with food production and postproduction can be second-order policies, which not only improve the sustainability of food that is eaten but also reduce the potential environmental burden related to pFLW. Policies directly targeting the environmental impact of pFLW must focus on effective management. Additionally, the management of inedible food parts can have environmental implications. These parts may be managed separately or alongside pFLW, depending on the specific context.

When pFLW and inedible food parts are disposed of (incinerated without energy capture or sent to landfill), this creates a linear path for food nutrients. Environmental policies that promote a circular economy in food systems prioritize efficient strategies to recycle pFLW into circular pathways, in which food nutrients are used for nonfood purposes (for example, anaerobic digestion, composting, and animal feed) (Wang et al. 2021). For example, using pFLW as animal feed recycles nutrients back into feed production, and composting and anaerobic digestion recycles nutrients back to the soil.

Evidence-based guidance prioritizes pFLW management strategies with the lowest environmental impact. The US Environmental Protection Agency's Wasted Food Scale, an update to the Food Recovery Hierarchy, ranks common pFLW management pathways in the United States, prioritizing strategies that avoid landfills and incineration (US EPA 2023). Assessing the generalizability of such environmental scales can inform the development of tailored strategies for different contexts, as the choice of pFLW management approach can have different climate change impacts (Box 10.4).

Box 10.4 Environmental Benefits of Preventing Physical Food Loss and Waste from Going to the Landfill

In their systematic review, Wang et al. (2021) found that managing pFLW through animal feed and composting can effectively minimize GHG emissions, compared to anaerobic digestion, incineration, and landfilling. Although anaerobic digestion, incineration, and landfilling can reduce GHG emissions through energy recovery, globally, landfills are common destinations with low gas capture rates, making their use a less environmentally friendly option. Additionally, several challenges exist for prioritizing pFLW for nonfood uses, including food safety, nutrient content concerns for animal feed, and technical limitations for composting and anaerobic digestion.

Livestock production and feed requirements vary significantly, based on species, feed sources, land use, and the extent to which livestock consume resources that could be used for human food (Mottet et al. 2017; Parlasca and Qaim 2022). Assessing the sustainability of using recycled pFLW as animal feed requires evaluating its suitability for animal consumption and determining the environmental benefits of diverting it from other nonfood uses or disposal (Pinotti et al. 2021). For example, leftovers from cereal-based food processing (such as bread, pasta, confectionery products, and salty snacks) can offer a sustainable alternative to traditional feed ingredients, while posing a low risk to animal health (Pinotti et al. 2021). Factors such as the distance, quantity, and intensity involved in processing and transporting discarded food for nonfood purposes are also important considerations (Parlasca and Qaim 2022).

As the demand for sustainable fuels grows amid depleting fossil fuel resources, Sridhar et al. (2021) reviewed technologies converting pFLW into fuels, identifying environmental and logistical hurdles. Various approaches, such as composting, anaerobic digestion, pyrolysis, biochemical conversion, incineration, and landfilling, can generate energy from pFLW. However, technological feasibility, economic viability, and societal behaviors and attitudes pose significant challenges. Ensuring safety is crucial, considering the nutrient requirements for conversion processes and the potential presence of hazardous, bacterial, or chemical contaminants. To make informed decisions within a circular economy framework, further research is needed to assess the environmental trade-offs and feasibility of different pFLW-to-fuel strategies.

A key concern with current FLW management hierarchies is their failure to account for rebound effects, which occur when benefits from reduced FLW are offset by additional consumption, either directly on the same products or indirectly on other products or services (Albizzati et al. 2022). For example, Hegwood et al. (2023) examined the direct rebound effects of avoided FLW, where reduced waste lowers prices, leading to increased food consumption and offsetting the environmental benefits of avoided waste, creating a trade-off between improved food security (higher calorie intake at lower prices) and potential negative environmental impacts. Therefore, to effectively address FLW, comprehensive policies must incorporate strategies to mitigate potential rebound effects. To do so necessitates a nuanced understanding of the trade-offs between food security and environmental impacts, and consideration of which environmental impacts are assessed, which foods become more accessible to promote dietary diversity and consumption of healthy diets, and the specific FLW pathways and prevention approaches discussed throughout this book.

10.6 Conclusion

Throughout this book, we have examined the complexities of FLW pathways, considering both quality loss and physical removal of food from value chains. We have addressed challenges in defining FLW and highlighted data gaps. To inform evidence-based guidelines for policies and initiatives, we propose a comprehensive framework and a data value chain approach and call on the UN to take a more coordinated lead. We then explored the connections between food quality, pFLW, and the dimensions of food security, while considering environmental and socioeconomic trade-offs. By identifying the drivers of FLW at various levels, stronger collaboration between UN agencies and national governments can align policies and regulations for FLW reduction and management goals with broader sustainable food system objectives.

Looking ahead, comprehensive FLW reduction strategies must use several types of interventions applied at different levels to leverage both technical and behavioral changes and incentivize actors across the value chain. Food systems stakeholders have many objectives that, at times, overlap, and at times, conflict. Multiple FLW interventions can work together within a FLW strategy, but also require assessment of potential trade-offs and issues of inequity. Therefore, a comprehensive policy agenda is required to create an inclusive and enabling environment in which FLW reduction strategies and FLW management approaches can be optimized to support food security, environment, and socioeconomic outcomes.

Policies that encourage consumer demand for healthy diets can incentivize value chain actors to increase the supply of nutritious, perishable foods by implementing FLW reduction strategies. Strong food safety standards and traceability systems are important for building consumer trust and minimizing FQL, and later, pFLW. Moreover, creating market opportunities for diverse cosmetic standards can prevent the unnecessary disposal of edible food and improve accessibility to various consumer needs and income levels.

Reducing FLW requires collaboration across the entire food value chain. Expanding access to financial services can empower smaller producers and SMEs to invest in FLW reduction technologies and practices. Investing in infrastructure, like transportation and storage facilities, is essential for expanding market access and reducing FLW. This will require transparent governance and equitable partnerships to strengthen market connections and mitigate risks across value chain actors.

Finally, balancing food security and environmental impacts of FLW is crucial as urban populations grow and diet preferences shift. Sustainable food systems transformations must address these challenges by innovating to sustainably increase production of nutritious, perishable foods and reduce FLW to make food value chains more efficient. This will involve identifying optimal FLW levels and implementing circular economy approaches to minimize the environmental impact of food waste. A comprehensive approach to FLW is necessary for FLW reduction and management initiatives to effectively support sustainable food systems and ensure healthy diets for all.

References

Affognon H, Mutungi C, Sanginga P, Borgemeister C (2015) Unpacking postharvest losses in sub-Saharan Africa: a meta-analysis. World Dev 66:49–68. https://doi.org/10.1016/j.worlddev.2014.08.002. Accessed 20 Oct 2021

Akande G, Diei-Ouadi Y (2010) Post-harvest losses in small-scale fisheries: case studies in five sub-Saharan African countries. FAO fisheries and aquaculture technical paper no. 550. FAO, Rome. https://openknowledge.fao.org/items/de4209a3-e521-4587-8252-4f27429702ff. Accessed 18 Aug 2023

Albizzati PF, Rocchi P, Cai M, Tonini D, Astrup TF (2022) Rebound effects of food waste prevention: environmental impacts. Waste Manag 153:138–146. https://doi.org/10.1016/j.wasman.2022.08.020. Accessed 30 Oct 2023

Ambikapathi R, Schneider KR, Davis B, Herrero M, Winters P, Fanzo JC (2022) Global food systems transitions have enabled affordable diets but had less favourable outcomes for nutrition, environmental health, inclusion and equity. Nat Food 3:764–779. https://doi.org/10.1038/s43016-022-00588-7. Accessed 6 Nov 2023

Aschemann-Witzel J, De Hooge I, Amani P, Bech-Larsen T, Oostindjer M (2015) Consumer-related food waste: causes and potential for action. Sustainability 7(6):6457–6477. https://doi.org/10.3390/su7066457. Accessed 18 Jan 2024

Aschemann-Witzel J, Jensen JH, Jensen MH, Kulikovskaja V (2017) Consumer behaviour towards price-reduced suboptimal foods in the supermarket and the relation to food waste in households. Appetite 116:246–258. https://doi.org/10.1016/j.appet.2017.05.013. Accessed 26 Feb 2024

Aschemann-Witzel J, Giménez A, Ares G (2018) Consumer in-store choice of suboptimal food to avoid food waste: the role of food category, communication and perception of quality dimensions. Food Qual Prefer 68:29–39. https://doi.org/10.1016/j.foodqual.2018.01.020. Accessed 18 Jan 2024

Beal T, Massiot E, Arsenault JE, Smith MR, Hijmans RJ (2017) Global trends in dietary micronutrient supplies and estimated prevalence of inadequate intakes. PLoS One 12(4):e0175554. https://doi.org/10.1371/journal.pone.0175554. Accessed 6 Nov 2023

Bell W, Lividini K, Masters WA (2021) Global dietary convergence from 1970 to 2010 altered inequality in agriculture, nutrition and health. Nat Food 2:156–165. https://doi.org/10.1038/s43016-021-00241-9. Accessed 6 Nov 2023

Borg K, Lennox A, Kaufman S, Tull F, Prime R, Rogers L, Dunstan E (2022) Curbing plastic consumption: a review of single-use plastic behaviour change interventions. J Clean Prod 344:131077. https://doi.org/10.1016/j.jclepro.2022.131077. Accessed 19 Dec 2024

Brondy A (2019) Growing the cold chain in developing and emerging economies. Agric Dev 36:4–9. https://www.cabidigitallibrary.org/doi/pdf/10.5555/20193201960. Accessed 14 Oct 2021

Carducci B, Oh C (2024) Increasing the demand for healthy diets evidence on approaches across the food system in Feed the Future contexts. Prepared for the United States Agency for International Development by the Board for International Food and Agriculture Development (BIFAD) Support Contract, contract number GS00Q14OADU138, Task Order 7200AA21M00003. https://www.usaid.gov/bifad/document/increasing-demand-health-diets-evidence-approaches-across-food-system-feed-future-contexts. Accessed 13 Dec 2024

Casonato C, Sanyé-Mengual E, Vittuari M, Sala S (2024) Life cycle assessment to support public procurement of food: a review. Clean Environ Syst 15:100239. https://doi.org/10.1016/j.cesys.2024.100239. Accessed 29 Nov 2024

Cheftel JC (2005) Food and nutrition labelling in the European Union. Food Chem 93(3):531–550. https://doi.org/10.1016/j.foodchem.2004.11.041. Accessed 30 May 2024

Croad T, Campbell H, Mirosa M (2024) Investigating systemic and social dynamics of food loss and waste: an application of waste regime theory to food production in Aotearoa New Zealand. Clean Waste Syst 7:100125. https://doi.org/10.1016/j.clwas.2023.100125. Accessed 19 Dec 2023

Day Farnsworth L, Delwiche A, McKinney C (2019) The good food purchasing program: a policy tool for promoting supply chain transparency and food system change. In: Thottahil SE, Goger AA (eds) Institutions as conscious food consumers. Academic Press, London, pp 103–126. https://linkinghub.elsevier.com/retrieve/pii/B9780128136171000058. Accessed 2 Dec 2024

de Hooge IE, van Dulm E, van Trijp HCM (2018) Cosmetic specifications in the food waste issue: supply chain considerations and practices concerning suboptimal food products. J Clean Prod 183:698–709. https://doi.org/10.1016/j.jclepro.2018.02.132. Accessed 8 Mar 2024

Dey A, Dhumal CV, Sengupta P, Kumar A, Pramanik NK, Alam T (2021) Challenges and possible solutions to mitigate the problems of single-use plastics used for packaging food items: a review. J Food Sci Technol 58:3251–3269. https://doi.org/10.1007/s13197-020-04885-6. Accessed 19 Dec 2021

Diprose G, Lee L, Blumhardt H, Walton S, Greenaway A (2023) Reducing single use packaging and moving up the waste hierarchy. Kōtuitui N Zeal J Soc Sci Online 18(3):268–289. https://doi.org/10.1080/1177083X.2022.2154230. Accessed 19 Dec 2024

Ellison B, Muth MK, Golan E (2019) Opportunities and challenges in conducting economic research on food loss and waste. Appl Econ Perspect Policy 41(1):1–19. https://doi.org/10.1093/aepp/ppy035. Accessed 13 Nov 2023

European Commission: Joint Research Centre, Rodriguez Quintero R, Gama Caldas M, Boyano A, Wolf O, Espinosa N, Neto B (2019) EU GPP criteria for food procurement, catering services and vending machines. Publications Office. https://data.europa.eu/doi/10.2760/748165. Accessed 2 Dec 2024

FAO (Food and Agriculture Organization of the United Nations) (2014) Food loss assessments: causes and solutions. Case studies in small-scale agriculture and fisheries subsectors. Kenya: banana, maize, milk, fish. FAO, Rome. https://www.fao.org/fileadmin/user_upload/save-food/PDF/Kenya_Food_Loss_Studies.pdf. Accessed 19 Jan 2022

FAO (Food and Agriculture Organization of the United Nations) (2019) The state of food and agriculture 2019. Moving forward on food loss and waste reduction. FAO, Rome. https://openknowledge.fao.org/server/api/core/bitstreams/11f9288f-dc78-4171-8d02-92235b8d7dc7/content. Accessed 30 Aug 2021

Flanagan K, Robertson K, Hanson C (2019) Reducing food loss and waste: setting a global action agenda. Report. World Resources Institute, Washington, DC. https://doi.org/10.46830/wrirpt.18.00130. Accessed 1 Jun 2022

Gillman A, Campbell DC, Spang ES (2019) Does on-farm food loss prevent waste? Insights from California produce growers. Resour Conserv Recycl 150:104408. https://doi.org/10.1016/j.resconrec.2019.104408. Accessed 8 Sep 2021

Hartmann T, Jahnke B, Hamm U (2021) Making ugly food beautiful: consumer barriers to purchase and marketing options for suboptimal food at retail level—a systematic review. Food Qual Prefer 90:104179. https://doi.org/10.1016/j.foodqual.2021.104179. Accessed 8 Mar 2024

Hegwood M, Burgess MG, Costigliolo EM, Smith P, Bajželj B, Saunders H, Davis SJ (2023) Rebound effects could offset more than half of avoided food loss and waste. Nat Food 4:585–595. https://doi.org/10.1038/s43016-023-00792-z. Accessed 24 Oct 2023

Herforth A, Arimond M, Álvarez-Sánchez C, Coates J, Christianson K, Muehlhoff E (2019) A global review of food-based dietary guidelines. Adv Nutr 10(4):590–605. https://doi.org/10.1093/advances/nmy130. Accessed 5 Oct 2022

Herforth A, Bai Y, Venkat A, Mahrt K, Ebel A, Masters WA (2020) Cost and affordability of healthy diets across and within countries. Background paper for the state of food security and nutrition in the world 2020. FAO, Rome. https://doi.org/10.4060/cb2431en. Accessed 16 Aug 2022

Hezarkhani B, Bouchery Y, Demirel G (2024) Battling food loss and empowering growers with ugly veg: the role of whole-crop contracts. SSRN J. https://doi.org/10.2139/ssrn.4845048. Accessed 31 May 2024

Kader AA (2005) Increasing food availability by reducing postharvest losses of fresh produce. Acta Hortic 682:2169–2176. https://doi.org/10.17660/ActaHortic.2005.682.296. Accessed 10 Nov 2021

Kinally C, Antonanzas-Torres F, Podd F, Gallego-Schmid A (2022) Off-grid solar waste in sub-Saharan Africa: market dynamics, barriers to sustainability, and circular economy solutions. Energy Sustain Dev 70:415–429. https://doi.org/10.1016/j.esd.2022.08.014. Accessed 2 Dec 2024

Krishna Bahadur KC, Dias GM, Veeramani A, Swanton CJ, Fraser D, Steinke D, Lee E, Wittman H, Farber JM, Dunfield K, McCann K, Anand M, Campbell M, Rooney N, Raine NE, Acker RV, Hanner R, Pascoal S, Sharif S, Benton TG, Fraser EDG (2018) When too much isn't enough: does current food production meet global nutritional needs? PLoS One 13(10):e0205683. https://doi.org/10.1371/journal.pone.0205683. Accessed 2 Sep 2021

Leib EB, Rice C, Neff R, Spiker M, Schklair A, Greenberg S (2016) Consumer perceptions of date labels: National Survey. Center for Health Law and Policy Innovation, Harvard Law School. https://chlpi.org/wp-content/uploads/2013/12/Consumer-Perceptions-on-Date-Labels_May-2016.pdf. Accessed 8 Jul 2022

Lighting Global/ESMAP, GOGLA, Efficiency for Access, Open Capital Advisors (2022) Off-grid solar market trends report 2022: state of the sector. World Bank, Washington, DC. http://www.osti.gov/servlets/purl/1220825/. Accessed 20 May 2024

Machala ML, Tan FL, Poletayev A, Khan MI, Benson SM (2022) Overcoming barriers to solar dryer adoption and the promise of multi-seasonal use in India. Energy Sustain Dev 68:18–28. https://doi.org/10.1016/j.esd.2022.02.001. Accessed 3 Apr 2023

Marshall Q, Fanzo J, Barrett CB, Jones AD, Herforth A, McLaren R (2021) Building a global food systems typology: a new tool for reducing complexity in food systems analysis. Front Sustain Food Syst 5:746512. https://doi.org/10.3389/fsufs.2021.746512. Accessed 2 Mar 2022

Messner R, Johnson H, Richards C (2021) From surplus-to-waste: a study of systemic overproduction, surplus and food waste in horticultural supply chains. J Clean Prod 278:123952. https://doi.org/10.1016/j.jclepro.2020.123952. Accessed 13 Nov 2023

Miller V, Yusuf S, Chow CK, Dehghan M, Corsi DJ, Lock K, Popkin B, Rangarajan S, Khatib R, Lear SA, Mony P, Kaur M, Mohan V, Vijayakumar K, Gupta R, Kruger A, Tsolekile L, Mohammadifard N, Rahman O, Rosengren A, Avezum A, Orlandini A, Ismail N, Lopez-Jaramillo P, Yusufali A, Karsidag K, Iqbal R, Chifamba J, Oakley SM, Ariffin F, Zatonska K, Poirier P, Wei L, Jian B, Hui C, Xu L, Xiulin B, Teo K, Mente

A (2016) Availability, affordability, and consumption of fruits and vegetables in 18 countries across income levels: findings from the Prospective Urban Rural Epidemiology (PURE) study. Lancet Global Health 4(10):e695–e703. https://doi.org/10.1016/S2214-109X(16)30186-3. Accessed 28 May 2024

Mottet A, de Haan C, Falcucci A, Tempio G, Opio C, Gerber P (2017) Livestock: on our plates or eating at our table? A new analysis of the feed/food debate. Glob Food Secur 14:1–8. https://doi.org/10.1016/j.gfs.2017.01.001. Accessed 4 Feb 2022

Muth MK, Birney C, Cuéllar A, Finn SM, Freeman M, Galloway JN, Gee I, Gephart J, Jones K, Low L, Meyer E, Read Q, Smith T, Weitz K, Zoubek S (2019) A systems approach to assessing environmental and economic effects of food loss and waste interventions in the United States. Sci Total Environ 685:1240–1254. https://doi.org/10.1016/j.scitotenv.2019.06.230. Accessed 13 Jul 2022

Neff RA, Spiker M, Rice C, Schklair A, Greenberg S, Leib EB (2019) Misunderstood food date labels and reported food discards: a survey of U.S. consumer attitudes and behaviors. Waste Manag 86:123–132. https://doi.org/10.1016/j.wasman.2019.01.023. Accessed 15 Jul 2022

Nelson G, Bogard J, Lividini K, Arsenault J, Riley M, Sulser TB, Mason-D'Croz D, Power B, Gustafson D, Herrero M, Wiebe K, Cooper K, Remans R, Rosegrant M (2018) Income growth and climate change effects on global nutrition security to mid-century. Nat Sustain 1:773–781. https://doi.org/10.1038/s41893-018-0192-z. Accessed 6 Nov 2023

Obanubi M, Okoruwa A, Morris S, Nordhagen S, Djimeu Wouabe E, Yalch T (2021) Business opportunities to reduce post-harvest loss of nutritious food. GAIN working paper no. 16. Global Alliance for Improved Nutrition, Geneva. https://doi.org/10.36072/wp.16. Accessed 20 May 2024

Odoli CO, Owiti H, Kobingi N, Obiero M, Ogari Z, Mugo J, Nyamweya C, Aura CM (2019) Post-harvest interventions in small-scale fisheries: a boon or bane to food and nutritional security in Kenya? Food Secur 11:855–868. https://doi.org/10.1007/s12571-019-00950-x. Accessed 1 Sep 2023

Otto S, Strenger M, Maier-Nöth A, Schmid M (2021) Food packaging and sustainability—consumer perception vs. correlated scientific facts: a review. J Clean Prod 298:126733. https://doi.org/10.1016/j.jclepro.2021.126733. Accessed 29 May 2024

Parlasca MC, Qaim M (2022) Meat consumption and sustainability. Annu Rev Resour Econ 14:17–41. https://doi.org/10.1146/annurev-resource-111820-032340. Accessed 11 Dec 2023

Pingali P (2015) Agricultural policy and nutrition outcomes—getting beyond the preoccupation with staple grains. Food Secur 7:583–591. https://doi.org/10.1007/s12571-015-0461-x. Accessed 25 Aug 2021

Pinotti L, Luciano A, Ottoboni M, Manoni M, Ferrari L, Marchis D, Tretola M (2021) Recycling food leftovers in feed as opportunity to increase the sustainability of livestock production. J Clean Prod 294:126290. https://doi.org/10.1016/j.jclepro.2021.126290. Accessed 20 Nov 2023

Qi D, Penn J, Li R, Roe BE (2022) Winning ugly: profit maximizing marketing strategies for ugly foods. J Retail Consum Serv 64:102834. https://doi.org/10.1016/j.jretconser.2021.102834. Accessed 19 Dec 2024

Raak N, Symmank C, Zahn S, Aschemann-Witzel J, Rohm H (2017) Processing- and product-related causes for food waste and implications for the food supply chain. Waste Manag 61:461–472. https://doi.org/10.1016/j.wasman.2016.12.027. Accessed 9 Mar 2023

Schmidhuber J, Sur P, Fay K, Huntley B, Salama J, Lee A, Cornaby L, Horino M, Murray C, Afshin A (2018) The Global Nutrient Database: availability of macronutrients and micronutrients in 195 countries from 1980 to 2013. Lancet Planet Health 2(8):E353–E368. https://doi.org/10.1016/S2542-5196(18)30170-0. Accessed 6 Nov 2023

Sheahan M, Barrett CB (2017) Review: food loss and waste in sub-Saharan Africa. Food Policy 70:1–12. https://doi.org/10.1016/j.foodpol.2017.03.012. Accessed 4 Sep 2021

Sheldon RA, Norton M (2020) Green chemistry and the plastic pollution challenge: towards a circular economy. Green Chem 22:6310–6322. https://doi.org/10.1039/D0GC02630A. Accessed 19 Dec 2024

Siegel KR, Ali MK, Srinivasiah A, Nugent RA, Narayan KMV (2014) Do we produce enough fruits and vegetables to meet global health need? PLoS One 9:e104059. https://doi.org/10.1371/journal.pone.0104059. Accessed 2 Sep 2021

Soethoudt JM, Pedrotti M, Bos-Brouwer HEJ, Castelein RB (2021) Adoption of food loss and waste-reducing interventions in low- and middle-income countries. Report No. 2196. Wageningen Food & Biobased Research, Wageningen. https://doi.org/10.18174/554051. Accessed 27 Sep 2021

Soma T, Kozhikode R, Krishnan R (2021) Tilling food under: barriers and opportunities to address the loss of edible food at the farm-level in British Columbia, Canada. Resour Conserv Recycl 170:105571. https://doi.org/10.1016/j.resconrec.2021.105571. Accessed 8 Feb 2024

Soysal M, Bloemhof-Ruwaard JM, van der Vorst JGAJ (2014) Modelling food logistics networks with emission considerations: the case of an international beef supply chain. Int J Prod Econ 152:57–70. https://doi.org/10.1016/j.ijpe.2013.12.012. Accessed 30 May 2024

Sridhar A, Kapoor A, Senthil Kumar P, Ponnuchamy M, Balasubramanian S, Prabhakar S (2021) Conversion of food waste to energy: a focus on sustainability and life cycle assessment. Fuel 302:121069. https://doi.org/10.1016/j.fuel.2021.121069. Accessed 2 Dec 2024

Subramanian R (2016) India processing tomato segment: current status, trends and opportunies for engagement. World Vegetable Center, no. 16-806. https://avrdc.org/download/publications/technical-reports/reports/

GIZ_India-Processed-Tomato-Study_16Sept2016.pdf. Accessed 8 Sep 2021

Swensson LFJ, Tartanac F (2020) Public food procurement for sustainable diets and food systems: the role of the regulatory framework. Glob Food Secur 25:100366. https://doi.org/10.1016/j.gfs.2020.100366. Accessed 29 Nov 2024

UN (United Nations) (2015) A/RES/70/1 transforming our world: the 2030 agenda for sustainable development. Resolution adopted by the General Assembly on 25 September 2015. https://www.un.org/en/development/desa/population/migration/generalassembly/docs/globalcompact/A_RES_70_1_E.pdf. Accessed 4 Mar 2024

US EPA (United States Environmental Protection Agency) (2023) Wasted food scale. https://www.epa.gov/sustainable-management-food/wasted-food-scale. Accessed 1 May 2024

Vernooij DM, Oostewechel RJA, Verschoor JA, Axmann HB, Pereira da Silva FIDG, Castelein RB (2022) Understanding the mechanisms to strengthen food systems through value chain-driven action. Report no. 2334. Wageningen Food & Biobased Research, Wageningen. https://doi.org/10.18174/577463. Accessed 22 Dec 2024

Vorley B, Ferris S, Seville D, Lundy M (2009) Linking worlds: new business models for sustainable trading relations between smallholders and formalized markets. Gates Open Research. https://gatesopenresearch.org/documents/3-1375. Accessed 31 May 2024

Wang Y, Yuan Z, Tang Y (2021) Enhancing food security and environmental sustainability: a critical review of food loss and waste management. Resour Environ Sustain 4:100023. https://doi.org/10.1016/j.resenv.2021.100023. Accessed 1 Sep 2021

Wang R, Liu G, Zhou L, Yang Z, Tang Z, Lu S, Zhao M, Sun H, Ma C, Cheng G (2023) Quantifying food loss along the animal products supply chain in China with large-scale field-survey based primary data. Resour Conserv Recycl 188:106685. https://doi.org/10.1016/j.resconrec.2022.106685. Accessed 5 Jan 2023

White K, Lin L, Dahl DW, Ritchie RJB (2016) When do consumers avoid imperfections? Superficial packaging damage as a contamination cue. J Mark Res 53(1):110–123. https://doi.org/10.1509/jmr.12.0388. Accessed 31 Mar 2024

Wilson NLW, Rickard BJ, Saputo R, Ho S-T (2017) Food waste: the role of date labels, package size, and product category. Food Qual Prefer 55:35–44. https://doi.org/10.1016/j.foodqual.2016.08.004. Accessed 14 Jul 2022

Xu Y, Jeong E, Jang S, Shao X (2021) Would you bring home ugly produce? Motivators and demotivators for ugly food consumption. J Retail Consum Serv 59:102376. https://doi.org/10.1016/j.jretconser.2020.102376. Accessed 8 Mar 2024

Index

J. Boiteau, P. Pingali, *Wasted Potential*, Sustainable Development Goals Series,
https://doi.org/10.1007/978-3-031-97411-3

MIX
Papier aus verantwortungsvollen Quellen
Paper from responsible sources
FSC® C105338

If you have any concerns about our products,
you can contact us on
ProductSafety@springernature.com

In case Publisher is established outside the EU,
the EU authorized representative is:
Springer Nature Customer Service Center GmbH
Europaplatz 3, 69115 Heidelberg, Germany

Printed by Libri Plureos GmbH
in Hamburg, Germany